Springer Monographs in Mathematics

More information about this series at http://www.springer.com/series/3733

Gareth A. Jones • Jürgen Wolfart

Dessins d'Enfants on Riemann Surfaces

Springer

Gareth A. Jones
School of Mathematics
University of Southampton
Southampton, United Kingdom

Jürgen Wolfart
Johann Wolfgang Goethe-Universität
Frankfurt am Main, Germany

ISSN 1439-7382 ISSN 2196-9922 (electronic)
Springer Monographs in Mathematics
ISBN 978-3-319-24709-0 ISBN 978-3-319-24711-3 (eBook)
DOI 10.1007/978-3-319-24711-3

Library of Congress Control Number: 2016931853

Mathematics Subject Classification: 14H57, 11G32 (primary); 05C10, 05C25, 14H45, 14H25, 14H55, 20F65, 30F10, 57M15, 57M60 (secondary)

Springer Cham Heidelberg New York Dordrecht London

Printed on acid-free paper

Springer International Publishing AG Switzerland is part of Springer Science+Business Media (www.springer.com)

To Ingrid and Mary

Preface

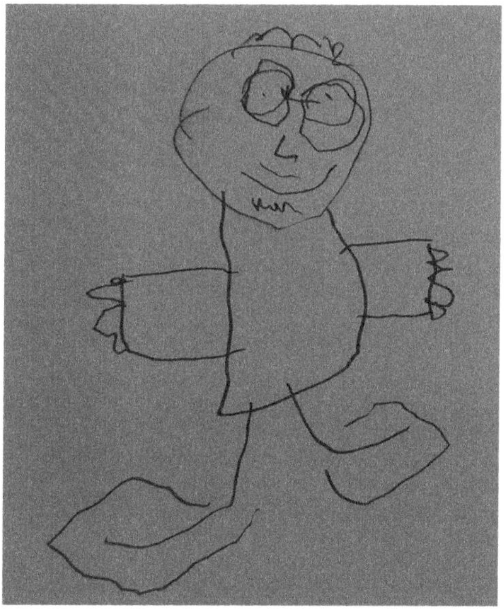

Dessin d'Annie

The term *dessin d'enfant* was introduced by Alexander Grothendieck, who died as we were completing this manuscript. It appears in his *Esquisse d'un Programme*, a set of notes written and circulated in 1984 but not published until 1997. Graphs embedded in surfaces, or, more precisely, in oriented compact 2-manifolds, can indeed look as simple as children's drawings, especially if they are drawn on the Riemann sphere. However, this does not explain why Grothendieck—and the authors of this book—were so attracted by these simple objects of geometric topology. The reason why dessins have received so much attention from the mathematical community during the last 25 years is probably the fact that they open up a rich world of unexpected links between apparently rather distant mathematical ideas.

Southampton, UK Gareth A. Jones
Frankfurt am Main, Germany Jürgen Wolfart
January 2016

Acknowledgements

The authors gratefully acknowledge all they have learned during the last 20 years at very useful meetings and from many other mathematicians. Among the meetings, there were the two Luminy conferences organised in 1993 and 1995 by Leila Schneps and Pierre Lochak, workshops in Oberwolfach 2000 and 2001, ICMS Edinburgh 2008, Castro Urdiales 2010 and Madrid 2012 and a series of mostly smaller meetings in the universities of Southampton and Frankfurt. We are grateful to all these institutions for their financial support, but most of all to the Deutsche Forschungsgemeinschaft, London Mathematical Society and EPSRC.

Many people have contributed to the content of this book through their discussions and their constant interest. In the first instance, we must mention David Singerman, but also Ernesto Girondo, Gabino González-Diez, Manfred Streit, the late Claude Itzykson and in more recent times David Torres-Teigell. Special hints for some details have been given by Friedrich Berg, Marston Conder, Philipp Habegger, Rubén Hidalgo, Bernhard Koeck, Roman Nedela, Martin Škoviera and Alexander Zvonkin.

The authors have given graduate courses on material contained in this book in Southampton, Frankfurt, Lancaster and finally in METU Ankara (2011, on invitation by Ayberk Zeytin and Hursit Onsiper), but the strongest roots of this book were Summer School courses in Jyväskylä 2006 (organised by Tapani Kuusalo) and Würzburg 2009 (organised by Jörn Steuding and Peter Müller). Almost one quarter of this volume is based on the perfect notes of our Jyväskylä courses written by Tuomas Puurtinen (directly typed in LaTeX from the blackboard during our lectures!). Some of the diagrams in Part I come from his original notes, while for some others we thank David Torres-Teigell and Martin Fluch. We are also grateful to Annie Jones (age 3), who contributed the dessin which accompanies the Preface.

Finally we thank David Singerman and Mary Tyrer-Jones for their careful reading of the manuscript and for suggesting numerous improvements.

Contents

Part I Basic Material

1 Historical and Introductory Background ... 3
1.1 Introduction ... 3
1.1.1 History: Topics of the Book ... 3
1.1.2 Prerequisities: Suggestions for the Reader ... 5
1.2 Compact Riemann Surfaces and Algebraic Curves ... 6
1.2.1 Examples and Some Basic Facts ... 6
1.2.2 Algebraic Curves ... 13
1.2.3 An Alternative Approach to Riemann Surfaces of Genus 1 ... 15
1.2.4 A Moduli Problem for Tori ... 17
1.2.5 Sketch of a Proof of Theorem 1.1 ... 22
1.3 Appendix: Existence of Enough Meromorphic Functions ... 24
1.4 Belyĭ Functions and Their Dessins ... 26
1.4.1 Belyĭ's Theorem ... 26
1.4.2 Existence of Belyĭ Functions: Simple Examples ... 27
1.4.3 Dessins ... 29
1.4.4 Belyĭ Algorithms ... 33
References ... 37

2 Graph Embeddings ... 41
2.1 Graphs, Maps, and Hypermaps ... 41
2.1.1 Bipartite Maps ... 41
2.1.2 Algebraic Bipartite Maps ... 45
2.1.3 Dessins as Hypermaps ... 48
2.1.4 Morphisms of Hypermaps ... 51
2.1.5 Maps and Hypermaps ... 51
2.1.6 An Instructive Example ... 53
2.2 Appendix: The Finite Simple Groups ... 54
2.2.1 The Cyclic Groups of Prime Order ... 54
2.2.2 The Alternating Groups ... 55

2.2.3 The Simple Groups of Lie Type 55
2.2.4 The Sporadic Simple Groups 56
References 57

3 Dessins and Triangle Groups 59
3.1 Uniformisation and Fuchsian Groups 59
3.1.1 Uniformisation 60
3.1.2 Triangle Groups 64
3.1.3 More General Facts About Fuchsian Groups 68
3.1.4 Triangle Groups and Belyĭ Functions 71
3.1.5 Inclusions Between Fuchsian Groups 72
3.1.6 Klein's Quartic Curve 76
3.2 Appendix: Group Presentations 78
3.3 From Dessins to Holomorphic Structures 80
3.3.1 Coverings 80
3.3.2 Triangle Groups and Bipartite Maps 81
3.3.3 Holomorphic Structures 83
3.3.4 Non-cocompact Triangle Groups 85
References 86

4 Galois Actions 89
4.1 Galois Theory 89
4.1.1 Basic Galois Theory 89
4.1.2 The Absolute Galois Group 91
4.1.3 Coverings and Galois Groups 94
4.2 Moduli Fields and Fields of Definition 96
4.2.1 Basic Facts and Definitions 96
4.2.2 Galois Action: An Example and Some Invariants 101
4.2.3 Non-invariants of Galois Action 103
4.2.4 Faithful Galois Action on Families of Dessins 104
4.3 Appendix: Another Proof of Theorem 4.9 107
References 110

Part II Regular Dessins

5 Quasiplatonic Surfaces, and Automorphisms 115
5.1 Quasiplatonic Surfaces: Construction and Counting 115
5.1.1 Definitions and Properties 115
5.1.2 Hurwitz Groups and Surfaces 117
5.1.3 Kernels and Epimorphisms 121
5.1.4 Direct Counting 123
5.1.5 Counting by Character Theory 125
5.1.6 Counting by Möbius Inversion 128
5.2 Low Genera 131
5.2.1 Genus 2 131
5.2.2 Genus 3 134
5.2.3 Genus 4 136

5.3 Infinite Families 139
5.4 Fields of Definition Again 142
5.5 Appendix: Linear and Projective Groups 145
References 149

6 Regular Maps 153
6.1 Regularity 153
6.2 Classification by Genus 156
6.3 Classification by Group 159
References 161

7 Regular Embeddings of Complete Graphs 163
7.1 Examples of Regular Embeddings 163
7.1.1 Examples of Genus 0 and 1 164
7.1.2 Cayley Maps 166
7.2 The Biggs Maps 167
7.2.1 Construction of the Biggs Maps 167
7.2.2 Frobenius Groups 169
7.2.3 Classifying the Regular Embeddings 170
7.2.4 Regular Embeddings and Cyclotomic Polynomials 175
References 177

8 Wilson Operations 179
8.1 A Class of Map Operations 179
8.2 Wilson Operations and Galois Actions 182
8.3 The Group of Operations on Dessins 185
References 191

9 Further Examples 193
9.1 Galois Orbits 193
9.1.1 Generalised Paley Maps 194
9.1.2 Complete Bipartite Maps 197
9.2 Regular Dessins and Equations 203
9.2.1 $K_{p,q}$-Dessins, Classification 203
9.2.2 $K_{p,q}$-Dessins, Abelian Case 203
9.2.3 $K_{p,q}$-Dessins, Semidirect Product Case 205
9.2.4 Equations 207
References 209

Part III Applications

10 Arithmetic Aspects 213
10.1 abc for Function Fields 213
10.1.1 Digression: Motivation from Number Theory 213
10.1.2 The Polynomial Case 215
10.1.3 abc on Riemann Surfaces 216
10.1.4 Belyĭ Functions: The Worst Case 217

10.2 Genus 1 Dessins and Complex Multiplication ... 218
10.2.1 Unramified Self-covers of Elliptic Curves ... 219
10.2.2 Complex Multiplication by Roots of Unity ... 220
10.2.3 Complex Multiplication, General Case ... 221
References ... 224

11 Beauville Surfaces ... 225
11.1 Basic Properties and First Examples ... 225
11.1.1 Basic Definitions ... 225
11.1.2 Beauville's Original Example ... 227
11.1.3 Enumeration of Beauville Surfaces ... 230
11.2 Beauville Structures for Specific Families of Groups ... 233
11.2.1 Beauville Surfaces and Simple Groups ... 233
11.2.2 Beauville Structures for Symmetric Groups ... 235
11.2.3 Beauville Structures for Alternating Groups ... 237
11.2.4 Beauville Structures for $L_2(q)$ ... 238
11.3 Further Properties of Beauville Surfaces ... 239
11.3.1 Invariants of a Beauville Surface ... 239
11.3.2 Real Beauville Surfaces ... 243
11.3.3 The Action of the Absolute Galois Group on Beauville Surfaces ... 245
References ... 247

Hints for Selected Exercises ... 251

Index ... 257

Part I
Basic Material

The first part of this book is an introduction to the basic ideas of the theory of dessins d'enfants. We give three definitions of a dessin. The simplest is as a bipartite graph embedded in a compact oriented surface; this can be redefined as a pair of permutations (of the edges of the graph) generating a transitive group, called the monodromy group. This group is a quotient of a triangle group, a group of isometries of the hyperbolic plane (or occasionally the complex plane or Riemann sphere), and the complex structure on this surface imposes a complex structure on the underlying surface of the dessin, making it a compact Riemann surface equipped with a Belyĭ function (a meromorphic function with no critical values outside $\{0, 1, \infty\}$). This gives us a third definition of a dessin (though the first we will introduce), as the pre-image of the unit interval under a Belyĭ function. In order to prove that these definitions are mutually equivalent we use ideas from function theory, group theory, hyperbolic geometry and combinatorics, outlining a number of classical concepts required or giving references to standard sources.

Compact Riemann surfaces are equivalent to complex algebraic curves, defined by polynomial equations, an idea which we illustrate in some detail in the case of elliptic curves (Riemann surfaces of genus 1). The most fundamental result about dessins is Belyĭ's Theorem, that the algebraic curves obtained as above from dessins are those for which the coefficients of the defining polynomials can be chosen to be algebraic numbers. This remarkable result leads us into the Galois theory of algebraic number fields, and in particular the action of the absolute Galois group, the automorphism group of the field of all algebraic numbers, on dessins and on their underlying curves.

In order to make individual chapters more self-contained, we have included some sections, called Appendices, which give important background information on the existence of suitable meromorphic functions, on the finite simple groups, and on group presentations; these are all important topics, used here and later in the book, but readers who are familiar with them can safely omit these sections.

Chapter 1
Historical and Introductory Background

Abstract This chapter begins with a brief historical introduction to the theory of dessins d'enfants, from the early discovery of the platonic solids, through nineteenth-century work on Riemann surfaces, algebraic curves and holomorphic functions, and twentieth-century research on regular maps, to the fundamental and far-reaching ideas circulated by Grothendieck in the 1980s, and subsequent efforts to implement his programme. After this we summarise the background knowledge we will assume, together with suggestions for further reading.

The second section gives a brief introduction to compact Riemann surfaces, including the Riemann-Hurwitz formula for the genus of a surface, and the equivalence of the categories of *compact Riemann surfaces* and of *smooth complex projective algebraic curves*. Elliptic curves (Riemann surfaces of genus 1) are treated in detail, as simple examples of subtler phenomena encountered later. The third section contains technical results on the existence of meromorphic functions with specific properties.

In the final section we define Belyĭ functions and prove one direction of Belyĭ's theorem, that such functions characterise algebraic curves defined over number fields, by using an algorithm which constructs a Belyĭ function on such a curve. We give a first definition of dessins d'enfants as the pre-images of the unit real interval [0, 1] under Belyĭ functions, and we discuss several simple examples of dessins.

Keywords Algebraic curve • Belyĭ function • Belyĭ's theorem • Dessin d'enfant • Elliptic curve • Meromorphic function • Riemann-Hurwitz formula • Riemann surface

1.1 Introduction

1.1.1 History: Topics of the Book

The oldest of the topics in this book is the theory of maps. The regular maps on the sphere, those with the greatest degree of symmetry, are named after Plato, but they were certainly known in times much earlier than his. For us they are the prototypes of regular dessins, a finite number of which exist in every genus greater than 1. Their classification in higher genera began with work of Brahana [6] in 1927 for genus 2.

Gareth A. Jones, J. Wolfart, *Dessins d'Enfants on Riemann Surfaces*, Springer Monographs in Mathematics, DOI 10.1007/978-3-319-24711-3_1

After several handmade generalisations to genera $g \leq 6$ by Threlfall [45], Sherk [42] and Garbe [14], this classification is nowadays the object of powerful algorithms of computational group theory (Conder [7]), currently covering all genera up to 301. Research on these higher genera maps would have been impossible without the understanding of hyperbolic tessellations and of Fuchsian groups developed in the late nineteenth century by Fricke and Klein [13] and Poincaré [35].

This line of research leads us to the link between maps and Riemann surfaces. Grothendieck [18] observed that dessins can be defined by purely topological means and that they induce on the underlying surface a unique conformal structure; he attributed the proof to Malgoire and Voisin [47], and in the meantime there have been further proofs of this important fact, for example one by Voevodsky and Shabat [46]. However, the first proof goes back to a paper by Singerman from pre-dessin times, see [23, 43].

Already from Riemann's work [38] one might have guessed that—in modern language—the category of compact Riemann surfaces and the category of smooth projective algebraic curves are equivalent. It took a long time to make this equivalence precise through work of Poincaré and Koebe [28–30, 36, 37] on the uniformisation of Riemann surfaces; for the historical background and details the reader may consult Scholz's book [41]. However, apart from some very exceptional examples with many symmetries, such as Klein's quartic [27], the Fricke-Macbeath curve [12] or the Bolza curves [5], this equivalence was far from explicit: until 1979, there was no function-theoretic criterion giving a necessary and sufficient condition for a compact Riemann surface X of genus $g \geq 1$ to be defined (as an algebraic curve) over a number field, that is, given in suitable coordinates by polynomial equations with coefficients in the field $\overline{\mathbb{Q}}$ of algebraic numbers.

This criterion was provided by Belyĭ's theorem [3]: *X can be defined over a number field if and only if there is a non-constant meromorphic function β on X ramified over at most three points.* Nowadays such functions β are called *Belyĭ functions*, and Grothendieck's "Esquisse d'un programme" [18] showed that Belyĭ functions can be characterized in a simple way by maps on their surfaces X. Later on, it turned out that the slight generalisation to *hypermaps*, introduced by Cori [8] in 1975 with motivation from computer graphics, was an even better adapted tool to treat Belyĭ functions and dessins.

Belyĭ's own work [3] was done in the framework of inverse Galois theory, that is, the question of whether and how it is possible to construct Galois extensions of number fields (or function fields) with a given Galois group, or more generally, to get as much information as possible about the absolute Galois group (the automorphism group of the field $\overline{\mathbb{Q}}$). It has long been known that this group acts faithfully on dessins; an important recent development has been the proof by González-Diez and Jaikin-Zapirain [17] that it acts faithfully on regular dessins, so that in a sense one can see the entire Galois theory of algebraic number fields through these simple and highly symmetric combinatorial objects.

Grothendieck broadened this viewpoint by linking dessins to questions about moduli spaces and motives. This so-called Grothendieck-Teichmüller theory is beyond the scope of this book. For the reader who wants to learn more about this branch of dessin theory we refer to the volumes edited by Schneps and Lochak [39, 40] and to the surveys by Guillot [19] and Oesterlé [34]. Concerning Galois theory, we restrict our coverage to elementary matters such as defining Galois actions on dessins, their invariants and their interpretation as map and hypermap operations, with the emphasis on concrete examples rather than abstraction and generalisation.

Another important link between dessins and the rest of the mathematical world is given by explicit uniformisation. The classical uniformisation theorem says that each Riemann surface X—for simplicity, say compact and of genus greater than 1—can be written as a quotient $\Gamma\backslash\mathbb{H}$ of the upper half plane $\mathbb{H}$ by some Fuchsian group Γ. However in general it is impossible to determine generators of Γ from the equations for X or to determine the explicit equations for X from group theoretic properties of Γ. With dessins, we are now in a much more favourable situation: there is a Belyĭ function on X if and only if we can choose Γ as a subgroup of a certain triangle group [1], so we have a kind of explicit uniformisation theory for curves defined over number fields. However this does not mean that it is always easy to determine explicit curve equations or coefficients of Belyĭ functions from dessins. In particular, we leave aside all hard questions concerning these computational aspects.

We also leave aside the possible connections with Physics (see the short account in [1]), but in Chap. 10 we briefly indicate links with the abc-theorem for function fields [32, 53], and with complex multiplication [2, 49, 51]. At the moment, apart from algebraic curves, maps and Galois theory, the most important application of (regular) dessins seems to be the construction and the properties of Beauville surfaces, the subject of the last chapter.

1.1.2 Prerequisities: Suggestions for the Reader

The reader of this volume should have a sufficient basic knowledge of complex functions (including Möbius transformations, the monodromy theorem and Schwarz's reflection principle) and group theory. It would also be useful to have some familiarity with covering spaces and the basics about hyperbolic geometry in the Poincaré model. Several less common concepts and results from function theory and group theory are presented in the appendices to the respective sections. Most topics needed about Riemann surfaces and Galois theory are briefly explained in Chaps. 1 and 4, but for the inexperienced reader these explanations may be rather short. All other chapters of Part I contain important results about Belyĭ functions and dessins developed during the last 40 years and essential for the understanding of everything in the later parts of the book.

Much of the material presented in Part I and many examples can also be found in the excellent and much more detailed introduction [15] by Girondo and González-Diez. Other sources of information about these basic questions are the survey articles [19, 23, 25, 50] or the dessin sections of the books by Lando and Zvonkin [32] or Degtyarev [9].

Part II deals with regular dessins (roughly speaking, those with the largest possible symmetry groups) and their underlying 'quasiplatonic' surfaces. Chapters 5 and 8 contain the most important basic results; the other chapters describe how regular dessins can be constructed and classified. There we discuss examples of families of regular dessins and quasiplatonic surfaces for which the Galois action is completely understood.

Part III contains two chapters, one about the abc theorem and complex multiplication, which can be read without the results of Part II. The last chapter about Beauville surfaces depends on Chap. 5 from Part II.

1.2 Compact Riemann Surfaces and Algebraic Curves

This section does not replace a book about Riemann surfaces. Here we simply collect some important facts, arguments, and examples serving as a guideline for what follows. For a more detailed account we refer to the many books about the topic, such as [10, 11, 24, 26].

Riemann surfaces are Hausdorff spaces which have a countable base for their topology. They are manifolds whose chart maps (also called local coordinates) take their values in the complex plane $\mathbb{C}$, and are defined with biholomorphic transition functions where their domains overlap. Here we will restrict our attention to connected Riemann surfaces.

1.2.1 Examples and Some Basic Facts

Example 1.1 Our first example is the *Riemann sphere*, or *complex projective line*,

$$\hat{\mathbb{C}} = \mathbb{P}^1(\mathbb{C}) = \mathbb{C} \cup \{\infty\}.$$

We take two open subsets, for example $U_1 = \mathbb{C}$ and $U_2 = (\mathbb{C}\setminus\{0\})\cup\{\infty\}$, with chart maps $U_1 \to \mathbb{C},\ z \mapsto z$ and $U_2 \to \mathbb{C},\ z \mapsto 1/z$ (where, by convention, we interpret $1/\infty$ as 0). Then $z \mapsto 1/z$ is a biholomorphic transition function between local coordinates on the intersection $U_1 \cap U_2 = \mathbb{C} \setminus \{0\}$. We therefore have a Riemann surface. It can be identified, by stereographic projection, with the unit sphere in euclidean 3-space, so it is compact.

Example 1.2 The *Fermat curve* of degree $n > 1$ (as an affine curve) is defined to be

$$F_n^{\mathrm{aff}} := \{\, (x, y) \in \mathbb{C}^2 \mid x^n + y^n = 1 \,\}.$$

We can take as chart maps $(x, y) \mapsto y$, which is a homeomorphism on suitable neighbourhoods of all points except those where $x = 0$ and $y^n = 1$, and $(x, y) \mapsto x$ which behaves similarly except where $y = 0$ and $x^n = 1$. As transition functions we take holomorphic branches of $x = \sqrt[n]{1 - y^n}$ and $y = \sqrt[n]{1 - x^n}$. Unlike $\hat{\mathbb{C}}$, this Riemann surface is not compact: for instance, the continuous real-valued function $(x, y) \mapsto |x|$ is unbounded on F_n^{aff}.

Example 1.3 More generally we can take any *smooth* affine algebraic curve

$$X^{\text{aff}} := \{ (x, y) \in \mathbb{C}^2 \mid f(x, y) = 0 \}$$

where f is a polynomial such that at each point $p \in X^{\text{aff}}$ either

$$\frac{\partial f}{\partial x}(p) \neq 0 \quad \text{or} \quad \frac{\partial f}{\partial y}(p) \neq 0.$$

The implicit function theorem implies that locally around p all solutions of $f(x, y) = 0$ are of the form $(h(y), y)$ or $(x, g(x))$ respectively where h and g are holomorphic. Then the projections onto the coordinates y or x can be used as chart maps.

Example 1.4 As a special case of this, an affine *hyperelliptic curve* is given by an equation

$$y^2 = (x - a_1) \dots (x - a_n)$$

with pairwise distinct $a_1, \dots, a_n \in \mathbb{C}$. Taking $f(x, y) = y^2 - \prod_j (x - a_j)$ we have

$$\frac{\partial f}{\partial y}(p) = 2y = 0$$

only at the points $p = p_j = (a_j, 0)$, so away from these points we can use x as a local coordinate. At the points $p = p_j$ we have

$$\frac{\partial f}{\partial x}(p) \neq 0,$$

so near them we can use y as a local coordinate instead.

The chart maps allow us to define *holomorphic* and *meromorphic functions* on Riemann surfaces and *holomorphic mappings* between Riemann surfaces by tracing back all these properties locally to the usual definitions in domains of the complex plane. Thus holomorphic and meromorphic functions on Riemann surfaces inherit the usual properties from holomorphic and meromorphic functions in the plane.

Exercise 1.1 Prove that there are no non-constant holomorphic functions on compact Riemann surfaces.

Exercise 1.2 Prove that the meromorphic functions on a Riemann surface X, with the usual addition and multiplication of meromorphic functions, form a field $\mathbb{C}(X)$.

Exercise 1.3 Prove that the field of meromorphic functions on the Riemann sphere $\hat{\mathbb{C}}$ is isomorphic to the rational function field $\mathbb{C}(z)$.

Proposition 1.1 *Let $f : X \to Y$ be a non-constant holomorphic mapping between connected Riemann surfaces X and Y, let $p \in X$, and let $p' = f(p)$. Then there exist chart maps $z : U(p) \to V \subset \mathbb{C}$ and $w : U'(p') \to V' \subset \mathbb{C}$ with $z(p) = 0$ and $w(p') = 0$, and an integer $n \in \mathbb{N}$ such that the diagram*

$$\begin{array}{ccc} U & \xrightarrow{f} & U' \\ z \downarrow & & \downarrow w \\ \mathbb{C} \ni z & \longmapsto & w = z^n \in \mathbb{C} \end{array}$$

is commutative. This integer n, which is independent of the choice of the charts, is called the 'multiplicity' $\mathrm{mult}_p f$ *of f at p.*

If $n = 1$ then f is locally biholomorphic (*unramified at p*); otherwise it is *ramified* with order $n > 1$. Here are some consequences.

1. If $f : X \to \hat{\mathbb{C}}$ is meromorphic and non-constant, then its zeros and poles form a discrete subset of X.
2. The ramification points of f form a discrete subset of X .
3. The identity theorem, the maximum principle, and the open mapping theorem, familiar from complex function theory for domains in the complex plane, are also valid on Riemann surfaces.
4. If X is a compact Riemann surface then a non-constant meromorphic function $f : X \to \hat{\mathbb{C}}$ has only a finite number of zeros or poles, and also only a finite number of ramification points. A holomorphic function $f : X \to \mathbb{C}$ must be constant (see Exercise 1.1).
5. If X is a compact Riemann surface then any non-constant holomorphic function $f : X \to Y$ is surjective, and Y is also compact.
6. Under the same hypotheses the *degree*

$$\deg f := \sum_{p \in f^{-1}(y)} \mathrm{mult}_p f$$

of f is independent of the choice of $y \in Y$.

Before continuing with basic facts we give some more examples.

Example 1.5 Fermat curves again: the projective version of the Fermat curve of degree n is

$$F_n := \{ [x, y, z] \in \mathbb{P}^2(\mathbb{C}) \mid x^n + y^n = z^n \}.$$

(It is sometimes more useful, because of the greater symmetry, to define F_n by the equation

$$x^n + y^n + z^n = 0,$$

obtained by replacing z with ζz where $\zeta^n = -1$.) By taking $z = 1$, $y = 1$ and $x = 1$ respectively, we see that F_n is covered by three copies of the affine curve F_n^{aff}, which omit the points of F_n where z, y and x are zero. For the first affine curve we can take chart maps of the form $[x, y, z] \mapsto x/z$ or y/z, and similarly for the other two. This is a typical example of a *smooth projective algebraic curve*. Of course, because $\mathbb{P}^2(\mathbb{C})$ is compact, its closed subset F_n is also compact, whereas affine Fermat curves are not.

The great advantage of using projective algebraic curves is that they are compact. There are a number of very useful theorems about Riemann surfaces which have compactness among their hypotheses; these include various numerical results such as the Riemann-Hurwitz formula, which we shall state shortly. However, there is a disadvantage in passing from an affine model of a Riemann surface to a projective model: we need an extra coordinate (generally three, rather than two, when we use plane models), and each point no longer determines its coordinates uniquely, but rather their ratios, which can be less convenient (in defining chart maps, for instance). A good compromise is to represent a projective algebraic curve as the union of two or more affine curves, and then to work with whichever model is most convenient.

Example 1.6 Let us try to compactify the hyperelliptic curve X^{aff}, given by

$$y^2 = \prod_{j=1}^{n}(x - a_j),$$

which we considered in Example 1.4. The corresponding projective curve X^{proj} is given by the equation

$$y^2 z^{n-2} = \prod_{j=1}^{n}(x - a_j z).$$

As in the case of the Fermat curves, this is compact. If $z \neq 0$ then since these are projective coordinates we can take $z = 1$, giving the original affine curve $X^{\text{aff}} \subset X^{\text{proj}}$; on the other hand, if $z = 0$ then $x^n = 0$ (provided $n \geq 3$) and so $x = 0$, giving a single point $p_\infty = [0, 1, 0]$ as the complement of X^{aff} in X^{proj}. As we saw in Example 1.4, we can use either x or y as a local coordinate on X^{aff}, as we are away from or near a point $p_j = [a_j, 0, 1]$.

Similarly, if $y \neq 0$ we can take $y = 1$, giving an affine curve $Y^{\text{aff}} \subset X^{\text{proj}}$ with equation

$$z^{n-2} = \prod_{j=1}^{n} (x - a_j z) ;$$

its complement consists of the n points p_j, so X^{proj} is the union of these two affine curves.

In order to define local coordinates near p_∞, which is in Y^{aff} but not in X^{aff}, we would like to apply the implicit function theorem to the polynomial

$$h(x,z) = z^{n-2} - \prod_{j=1}^{n} (x - a_j z) .$$

Unfortunately, logarithmic differentiation shows that if $n > 3$ then $\partial h/\partial x = \partial h/\partial z = 0$ at p_∞, so the theorem does not apply. Instead, let us go back to X^{aff}, and write its defining equation as

$$y^2 = q(x) := \prod_{j=1}^{n} (x - a_j) .$$

We may assume that each $a_j \neq 0$, by replacing x with $x - a$ for some constant a if necessary. Let us define new variables s and t by

$$t := \frac{1}{x} \quad \text{and} \quad s := \frac{y}{x^{g+1}} ,$$

where $g := \lfloor (n-1)/2 \rfloor$, so that

$$\deg q = n = \begin{cases} 2g+1 & (n \text{ odd}), \\ 2g+2 & (n \text{ even}). \end{cases} \tag{1.1}$$

The equation $y^2 = q(x)$ is then equivalent to $s^2 = k(t)$ at all points with $x \neq 0$, where

$$k(t) := t^{2g+2} q\left(\frac{1}{t}\right) = \frac{q(x)}{x^{2g+2}}$$

is a polynomial of degree $2g+2$ in $\mathbb{C}[t]$ with simple zeros at the points $t = a_j^{-1}$, and also at $t = 0$ if n is odd. Note that

$$x = \infty \Leftrightarrow t = 0 \Rightarrow s = \sqrt{k(0)} = \begin{cases} 0 & (n \text{ odd}), \\ \pm 1 & (n \text{ even}). \end{cases}$$

We can now apply the implicit function theorem to the polynomial $f^*(s,t) = s^2 - k(t)$ since, as in Example 1.4, its two partial derivatives do not simultaneously vanish. Specifically, if n is even then we can use t as a local coordinate since $\partial f^*/\partial s = 2s \neq 0$, while if n is odd, so that $\partial f^*/\partial s = 2s = 0$, we can use s since $\partial f^*/\partial t \neq 0$.

This calculation illustrates an important general point about the Riemann surface X associated with the hyperelliptic curve $y^2 = q(x)$. The projection $(x,y) \mapsto x$ realises it as a 2-sheeted covering of $\hat{\mathbb{C}}$, branched over the roots a_j of q, and also over ∞ if n is odd: in the latter case there is a single point $(s,t) = (0,0)$ where $x = \infty$, whereas if n is even there are two points $(s,t) = (\pm 1, 0)$. This can be explained by writing y as a 2-valued function

$$y = \sqrt{(x-a_1)\dots(x-a_n)}\,,$$

so that each point $x \neq a_j, \infty$ in $\hat{\mathbb{C}}$ is covered by two points $(x, \pm y)$. Now let $x = re^{i\theta}$ for fixed $r > |a_1|, \dots, |a_n|$, and let θ increase by 2π, so that x follows a circular path enclosing all the roots of q, or equivalently, enclosing the point $\infty \in \hat{\mathbb{C}}$; each factor $\sqrt{(x-a_j)}$ is multiplied by $e^{i\pi} = -1$, so y changes sign, that is, (x,y) passes from one sheet to the other, if and only if n is odd. (A similar argument explains the branching at roots a_j of q for all n: if x follows a small closed path enclosing a_j but no other roots, then just one factor $\sqrt{(x-a_j)}$ is multiplied by -1, while the rest are unchanged.)

The projective model X^{proj} of X discussed earlier obscures this distinction between odd and even values of n: in either case, it has a single point $p_\infty = [0,1,0]$ at infinity. When n is odd this is where the covering is branched, but when n is even it represents a singularity in the model, where two sheets of the Riemann surface X intersect.

We can use the 2-sheeted covering $X \to \hat{\mathbb{C}}$ to construct a topological model of X by taking two copies of $\hat{\mathbb{C}}$, one for each branch of $\sqrt{q(x)}$, and joining them across disjoint cuts between $g+1$ pairs of branch-points, namely the roots of q if $n = 2g+2$ is even, together with ∞ if $n = 2g+1$ is odd. A topological model of X^{proj} is then formed by identifying the two points $(s,t) = (\pm 1, 0)$ of X over ∞ if n is even.

Every Riemann surface is orientable! This is because the transition functions are biholomorphic, and therefore preserve the orientation.

Riemann surfaces can also be triangulated. In fact, this is true more generally for all topological surfaces. This may seem intuitively obvious, but in fact the proof, by Radó in 1925, is not straightforward. (The corresponding result for three-dimensional manifolds is also true, but it is false in dimension 4, Freedman's E_8 manifold providing a counterexample.) However, there is a simpler proof for compact Riemann surfaces X if one accepts the existence of a non-constant (and hence surjective) meromorphic function $f : X \to \hat{\mathbb{C}}$ (see Exercise 1.12): construct a triangulation $\mathscr{T}$ of $\hat{\mathbb{C}}$ such that the vertices include the *critical* values, that is, the images of all the ramification points of f, and each face is sufficiently small that

f is injective on each connected component of its inverse image; then $f^{-1}(\mathscr{T})$ is a triangulation of X.

Any triangulation of a compact surface X has finite numbers V, E and F of vertices, edges and faces. The *Euler characteristic* of X is defined to be

$$\chi(X) := V - E + F;$$

this can be shown to be independent of the choice of a triangulation. For example $\chi(\hat{\mathbb{C}}) = 2$, since one can triangulate a sphere with three vertices, three edges and two faces. Similarly, $\chi(X) = 0$ if X is a torus (easy exercise!).

For compact orientable surfaces X (including compact Riemann surfaces), the *genus* $g(X)$ is a more commonly-used invariant than the Euler characteristic; this is defined by

$$2 - 2g(X) = \chi(X)\,.$$

In topology, one shows that such a surface X is homeomorphic to a sphere with $g(X)$ handles attached. Thus $g(X)$ is a non-negative integer, so that $\chi(X)$ is an even integer, with $\chi(X) \leq 2$.

The reader may find more information about all these topological aspects in textbooks on surface topology or in [44].

Proposition 1.2 (Riemann-Hurwitz Formula) *If $f : X \to Y$ is a non-constant holomorphic mapping of compact Riemann surfaces, then*

$$2g(X) - 2 = (\deg f)(2g(Y) - 2) + \sum_{p\in X}(\mathrm{mult}_p f - 1).$$

(Note that the sum on the right-hand side is finite, since $\mathrm{mult}_p f = 1$ for all but finitely many points $p \in X$.)

Outline Proof We can choose a triangulation $\mathscr{T}$ of Y such that the vertices include all the finitely many points of Y over which f is ramified. Since there is no ramification away from the vertices, the pre-image of $\mathscr{T}$ is a triangulation $\mathscr{S}$ of X. If $\mathscr{T}$ has v vertices, e edges and f faces, then $\mathscr{S}$ has de edges and df faces, where $d = \deg f$. If there were no ramification then $\mathscr{S}$ would also have dv vertices, but in fact we 'lose' $\mathrm{mult}_p f - 1$ vertices at each ramification point $p \in X$, so $\mathscr{S}$ has $dv - \sum_p(\mathrm{mult}_p f - 1)$ vertices. Thus

$$\chi(X) = d\chi(Y) - \sum_{p\in X}(\mathrm{mult}_p f - 1),$$

giving the required formula. □

Here we give some applications of this important result.

1. We have $g(Y) \leq g(X)$, with equality if and only if either f is an isomorphism (unramified), or $g(X) = g(Y) = 1$ and f is unramified. If $g(X) > g(Y) = 0$ or 1, then f is ramified.
2. The Fermat curve F_n has genus $(n-1)(n-2)/2$. This can be seen by considering the function $f : F_n \to \hat{\mathbb{C}},\ [x, y, z] \mapsto x/z$. On the affine part of the curve given by $z = 1$ and $x^n + y^n = 1$, we have $f : (x, y) \mapsto x$. This shows that $\deg f = n$, since for general x there are n solutions $y \in \mathbb{C}$ of the equation $x^n + y^n = 1$. The exceptions are those points x with $x^n = 1$, where f has only one pre-image, giving us n points $p = (x, 0)$ with $\mathrm{mult}_p f = n$. The points of F_n on the line at infinity $z = 0$ are those of the form $[\zeta, 1, 0]$ where $\zeta^n = -1$, giving n simple (therefore unramified) poles of f. The Riemann-Hurwitz formula now implies that
$$2g(F_n) - 2 = n(-2) + n(n-1) = n^2 - 3n,$$
so
$$g(F_n) = \frac{(n-1)(n-2)}{2}.$$
3. More generally, if $f : X \to \hat{\mathbb{C}}$ is a non-constant meromorphic function on a compact Riemann surface X, then
$$g(X) = 1 - \deg f + \frac{1}{2}\sum_{p \in X}(\mathrm{mult}_p f - 1).$$

Exercise 1.4 Use the Riemann-Hurwitz formula to show that the genus of the hyperelliptic Riemann surface X considered in Example 1.6 is just the integer g given by Eq. (1.1).

1.2.2 Algebraic Curves

For our purposes, the following is a crucial result:

Theorem 1.1 *There is an equivalence between the two categories of compact Riemann surfaces and of smooth complex projective algebraic curves.*

We can represent this schematically as follows:

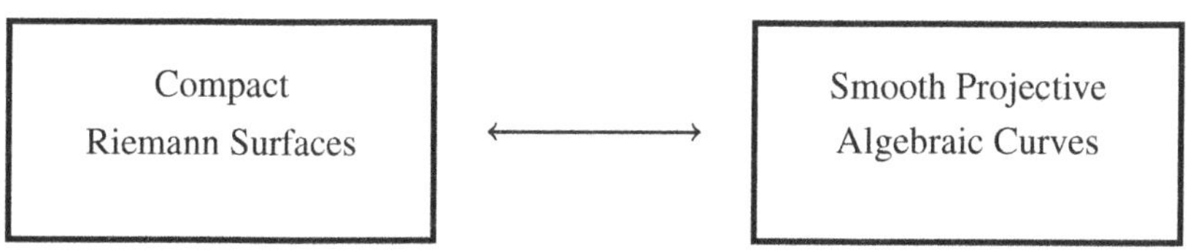

Later, in Sect. 1.2.5, we will make this equivalence more explicit. An equally important result for us is Belyĭ's theorem, which we will also discuss in detail later. This shows that within these two categories, the compact Riemann surfaces which are obtained from maps (that is, embedded graphs) correspond to the smooth projective algebraic curves which are defined over algebraic number fields (more precisely: those defined in suitable coordinates by equations in which the coefficients are algebraic numbers). The automorphisms of these algebraic number fields, and their effect on the corresponding Riemann surfaces and maps, lead us naturally into Galois theory:

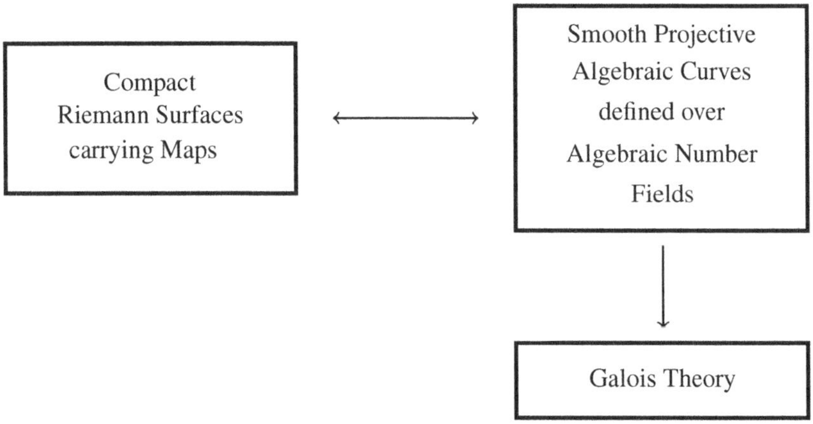

In due course we have to explain what *maps* are, and how Galois theory comes into the story, but first we illustrate the main equivalence of categories in the special case of Riemann surfaces of genus 1 .

An *elliptic curve* is an algebraic curve E of the form

$$y^2 = p(x) ,$$

where p is a cubic polynomial in $\mathbb{C}[x]$ with distinct roots. A cubic polynomial with roots e_1, e_2, e_3 has discriminant

$$\Delta = 16(e_1 - e_2)^2(e_2 - e_3)^2(e_3 - e_1)^2 ,$$

so that p has distinct roots if and only if $\Delta \neq 0$. By applying an affine substitution $x \mapsto ax + b$ with $a \neq 0$ we can put the equation for E in *Weierstrass normal form*

$$y^2 = 4x^3 - c_2x - c_3, \quad (c_2, c_3 \in \mathbb{C}).$$

Then it is an easy exercise to show that $\Delta = c_2^3 - 27c_3^2$. Alternatively, by applying affine substitutions to x and y, we get the *Legendre normal form*

$$y^2 = x(x-1)(x-\lambda), \quad (\lambda \in \mathbb{C}\backslash\{0,1\}).$$

Exercise 1.5 Find Δ and these normal forms for the elliptic curve

$$y^2 = x^3 - 9x^2 + 23x - 15.$$

In the original equation $y^2 = p(x)$, if we multiply y by a suitable constant and then cancel leading coefficients, we can assume that $p(x)$ is a monic polynomial

$$(x-e_1)(x-e_2)(x-e_3).$$

This shows that E is a special case of the hyperelliptic curves $y^2 = q(x)$ considered in Examples 1.4 and 1.6, with $n = \deg q = 3$. The arguments discussed there show that E is a 2-sheeted covering $(x,y) \mapsto x$ of $\hat{\mathbb{C}}$, branched over the roots e_1, e_2, e_3 of p, and also over ∞ since $\deg p$ is odd: if $x = e_j$ for $j = 1, 2$ or 3 then only $y = 0$ is possible, while if $x = \infty$ then only $y = \infty$ is possible.

One can construct the Riemann surface of E by taking two copies of $\hat{\mathbb{C}}$, one for each branch of $\sqrt{p(x)}$, and joining them across two disjoint cuts between e_1 and e_2, and between e_3 and ∞. The result is a torus, of genus 1, as shown by Exercise 1.4.

1.2.3 An Alternative Approach to Riemann Surfaces of Genus 1

For this and the next section, much more detailed information can be found in [24], especially Chaps. 3, 4 and 6. Let ω_1 and ω_2 be elements of $\mathbb{C}$ which are linearly independent over $\mathbb{R}$. The additive subgroup

$$\Lambda = \Lambda(\omega_1, \omega_2) = \{\, m\omega_1 + n\omega_2 \mid m, n \in \mathbb{Z}\,\}$$

which they generate is called a *lattice*, and we call ω_1 and ω_2 a *basis* for Λ. This subgroup Λ is discrete, meaning that every $\omega \in \Lambda$ has an open neighbourhood in $\mathbb{C}$ containing no other elements of Λ.

Given a lattice Λ, there is an equivalence relation on $\mathbb{C}$ given by $z_1 \equiv z_2 \mod \Lambda$ if and only if $z_1 - z_2 \in \Lambda$: the equivalence classes are the cosets $z + \Lambda$ of Λ in $\mathbb{C}$, and the quotient space is denoted by $\mathbb{C}/\Lambda$. Since the group $\mathbb{C}$ is abelian, Λ is a normal subgroup and hence $\mathbb{C}/\Lambda$ is a group, with its structure inherited from $\mathbb{C}$.

The parallelogram $P = \{\, x\omega_1 + y\omega_2 \mid x, y \in [0,1]\,\}$ is a fundamental region for Λ, that is, each $z \in \mathbb{C}$ is equivalent to an element of P, and if two elements of P are equivalent, then they lie on the boundary ∂P of P. Since each coset of Λ has a representative in P, we can form $\mathbb{C}/\Lambda$ by identifying equivalent points $z_1, z_2 \in \partial P$, specifically by identifying pairs of opposite sides $x = 0, 1$ and $y = 0, 1$ of P to form a torus. The holomorphic structure on $\mathbb{C}$ yields a holomorphic structure on $\mathbb{C}/\Lambda$, making it a Riemann surface. This surface is compact, since it is a continuous image of P, which is compact by the Heine-Borel theorem. It is an easy exercise to find a triangulation of $\mathbb{C}/\Lambda$ (draw one in P !) which shows that this surface has genus 1.

It is also possible to obtain this Riemann surface $\mathbb{C}/\Lambda$ from an elliptic curve. To establish the link between these two approaches, we need to use *elliptic functions*. These are meromorphic functions on $\mathbb{C}$ which are doubly periodic with respect to some lattice Λ. To say that f is *doubly periodic* with respect to Λ means that

$$f(z + \omega) = f(z) \quad \text{for all } z \in \mathbb{C} \text{ and all } \omega \in \Lambda.$$

(This is an extension of the concept of a simply periodic function, such as the functions $\sin(z), \cos(z)$ or $\exp(z)$ which are periodic with respect to the subgroups $2\pi\mathbb{Z}$ or $2\pi i\mathbb{Z}$ of $\mathbb{C}$.)

For a given lattice Λ, the elliptic functions with respect to Λ form a field $F(\Lambda)$ under the obvious operations, such as $(f+g)(z) = f(z)+g(z)$. We can think of these functions as the meromorphic functions on $\mathbb{C}/\Lambda$ by defining $f(z+\Lambda) = f(z)$ (well-defined by double periodicity). Since $\mathbb{C}/\Lambda$ is compact, the theory of meromorphic functions works nicely for this Riemann surface.

The constant functions are obvious examples of elliptic function, but we also need some non-constant examples. The *Weierstrass function* is defined to be

$$\wp(z) = \wp_\Lambda(z) = \frac{1}{z^2} + \sum\nolimits_{\omega}' \left(\frac{1}{(z-\omega)^2} - \frac{1}{\omega^2} \right),$$

where $\sum'$ denotes the sum over all $\omega \neq 0$ in Λ. This series is uniformly convergent on compact subsets of $\mathbb{C}\backslash\Lambda$, so $\wp$ is meromorphic on $\mathbb{C}$, with poles of order 2 at the lattice points. To show that $\wp$ is double periodic, first consider

$$\wp'(z) = -2 \sum_{\omega} \frac{1}{(z-\omega)^3}.$$

This is meromorphic, with poles of order 3 at the lattice-points.

Exercise 1.6 Show that $\wp'$ is doubly periodic with respect to Λ. Deduce that $\wp$ is also doubly periodic.

Thus $\wp, \wp' \in F(\Lambda)$ so the field $\mathbb{C}(\wp, \wp')$ of rational functions of $\wp$ and $\wp'$ is contained in $F(\Lambda)$. In fact, $F(\Lambda) = \mathbb{C}(\wp, \wp')$, so that every elliptic function can be expressed as a rational function of $\wp$ and $\wp'$. The functions $\wp$ and $\wp'$ are not algebraically independent: by comparison of the Laurent series on both sides, it is not hard to prove that they satisfy a differential equation

$$(\wp')^2 = 4\wp^3 - g_2\wp - g_3$$

where $g_2 = g_2(\Lambda) = 60G_4$ and $g_3 = g_3(\Lambda) = 140G_6$, and where $G_k = G_k(\Lambda) = \sum'_\omega \omega^{-k}$ for $k = 4, 6$ (these are called the *Eisenstein series* for Λ). Now compare this equation with the Weierstrass normal form

$$y^2 = 4x^3 - c_2x - c_3$$

for an elliptic curve E. Given any $c_2, c_3 \in \mathbb{C}$ such that the discriminant $\Delta := c_2^3 - 27c_3^2$ is non-zero, one can show that there is a lattice Λ in $\mathbb{C}$ with $g_2(\Lambda) = c_2$ and $g_3(\Lambda) = c_3$. We can then write $x = \wp(z)$ and $y = \wp'(z)$, where $\wp = \wp_\Lambda$ for this lattice Λ. This allows us to identify each point $(x, y) \in E$ with the corresponding point $z + \Lambda \in \mathbb{C}/\Lambda$, so we identify the affine elliptic curve E with $\mathbb{C}/\Lambda$ minus the point $0 + \Lambda$, where $\wp$ and $\wp'$ have poles. Embedding E in its projective closure $E^{\text{proj}} = E \cup \{p_\infty\}$, we can extend this identification to the entire torus $\mathbb{C}/\Lambda$ by identifying the point $p_\infty = [0, 1, 0]$ at infinity in E^{proj}, with $0 + \Lambda$. (Compare this with parametrising the circle $x^2 + y^2 = 1$ by putting $x = \sin(z)$ and $y = \cos(z)$, where $z \in \mathbb{R}/2\pi\mathbb{Z}$.) This identification is in fact a biholomorphic map, so we get the following special case of Theorem 1.1:

Theorem 1.2 *Every torus is isomorphic to a smooth projective algebraic curve whose affine part can be described by a Weierstrass equation.* □

In Chap. 3, when we study uniformisation, we will see that every compact Riemann surface of genus 1 is a torus, so this theorem applies to all such surfaces.

1.2.4 A Moduli Problem for Tori

We will now consider the problem of identifying and parametrising the isomorphism classes of tori. First we need to know the automorphisms of $\mathbb{C}$ (as a Riemann surface):

Exercise 1.7 Show that every automorphism a of the Riemann surface $\mathbb{C}$ is of the form

$$z \mapsto \mu z + c \quad \text{for constants} \quad \mu \neq 0 \quad \text{and} \quad c \in \mathbb{C}.$$

Proposition 1.3 *Suppose that Λ and Λ' are lattices in $\mathbb{C}$. Then the Riemann surfaces $\mathbb{C}/\Lambda$ and $\mathbb{C}/\Lambda'$ are isomorphic (as Riemann surfaces) if and only if Λ and Λ' are similar lattices, in the sense that $\Lambda' = \mu\Lambda$ for some $\mu \in \mathbb{C}\backslash\{0\}$.*

Proof If there is an automorphism $a : \mathbb{C} \to \mathbb{C}$, $z \mapsto \mu z$ with $\mu(\Lambda) = \Lambda'$, then it clearly induces an isomorphism of the respective quotient surfaces. On the other hand, any isomorphism

$$\alpha : \mathbb{C}/\Lambda \to \mathbb{C}/\Lambda'$$

lifts to an automorphism $a : \mathbb{C} \to \mathbb{C}$. Then $a(z) = \mu z + c$ by Exercise 1.7. We may assume that $c = 0$ since translations in $\mathbb{C}$ induce automorphisms of all tori; because α has to be well defined, a has to satisfy $a(\Lambda) = \Lambda'$, and μ has to satisfy $\mu\Lambda = \Lambda'$. □

If Λ has basis ω_1, ω_2, then elements ω_1', ω_2' of Λ form a basis for Λ if and only if $\omega_2' = a\omega_2 + b\omega_1$ and $\omega_1' = c\omega_2 + d\omega_1$ with $a, b, c, d \in \mathbb{Z}$ and $ad - bc = \pm 1$. The 2×2 integer matrices with $ad - bc = \pm 1$ form the *general linear group* $\mathrm{GL}_2(\mathbb{Z})$ under multiplication, and those with $ad - bc = 1$ form the *special linear group* $\mathrm{SL}_2(\mathbb{Z})$, a normal subgroup of index 2 in $\mathrm{GL}_2(\mathbb{Z})$.

The *modulus* $\tau := \omega_2/\omega_1$ of a basis is invariant under the similarity transformation of multiplying Λ by μ. Changing basis by a matrix

$$\begin{pmatrix} a & b \\ c & d \end{pmatrix} \in \mathrm{GL}_2(\mathbb{Z})$$

transforms τ to

$$\tau' = \frac{\omega_2'}{\omega_1'} = \frac{a\omega_2 + b\omega_1}{c\omega_2 + d\omega_1} = \frac{a\tau + b}{c\tau + d}.$$

These transformations $\tau \mapsto \tau'$ form the *projective general linear group*

$$\mathrm{PGL}_2(\mathbb{Z}) \cong \mathrm{GL}_2(\mathbb{Z})/\{\pm I\};$$

the kernel of this action of $\mathrm{GL}_2(\mathbb{Z})$ is the normal subgroup $\{\pm I\}$, so we factor it out.

Transposing ω_1 and ω_2 if necessary, we can assume that $\operatorname{Im}\tau > 0$, that is, that τ lies in the hyperbolic upper half plane $\mathbb{H} = \{\tau \in \mathbb{C} \mid \operatorname{Im}\tau > 0\}$. This allows us to restrict our attention to transformations $\tau \mapsto (a\tau + b)/(c\tau + d)$ with $ad - bc = 1$, those which map $\mathbb{H}$ to itself. These form the *modular group*, or *projective special linear group*

$$\Gamma = \mathrm{PSL}_2(\mathbb{Z}) \cong \mathrm{SL}_2(\mathbb{Z})/\{\pm I\},$$

a normal subgroup of index 2 in $\mathrm{PGL}_2(\mathbb{Z})$. Then two tori $\mathbb{C}/\Lambda$ and $\mathbb{C}/\Lambda'$, having bases with moduli τ and τ' in $\mathbb{H}$, are isomorphic if and only if τ and τ' are equivalent under the action of Γ on $\mathbb{H}$. In other words, the isomorphism classes of tori $\mathbb{C}/\Lambda$ correspond to the orbits of Γ on $\mathbb{H}$. This makes it important for us to understand how Γ acts on $\mathbb{H}$.

The region $F \subset \mathbb{H}$ defined by $|\tau| \geq 1$, $|\operatorname{Re}\tau| \leq \frac{1}{2}$ is a *fundamental region* for Γ. This means that every orbit of Γ contains a point in F, and if two points in F are in the same orbit, they lie on the boundary ∂F of F. The element $X : \tau \mapsto -1/\tau$ of order 2 fixes $\tau = i$, the value of τ corresponding to the square lattice $\Lambda = \mathbb{Z}[i]$ with basis $1, i$; it transposes the two halves of the side $|\tau| = 1$ of F. The element $Z : \tau \mapsto \tau + 1$ of infinite order has no fixed points in $\mathbb{H}$, and sends the side $\operatorname{Re}\tau =$

$-\frac{1}{2}$ of F to the side $\mathrm{Re}\,\tau = \frac{1}{2}$. The element $Y : \tau \mapsto -(\tau+1)/\tau$ of order 3 fixes $\tau = \omega := e^{2\pi i/3}$, corresponding to the hexagonal lattice $\Lambda = \mathbb{Z}[\omega]$ with basis $1, \omega$.

One can show (see [24, §6.8], for instance) that Γ has a presentation

$$\Gamma = \langle X, Y \mid X^2 = Y^3 = 1 \rangle \cong C_2 * C_3$$

as the free product of the cyclic groups C_2 and C_3 generated by X and Y. (See Appendix 3.2 for presentations and free products.)

For each integer $n \geq 2$, reduction mod n is a ring homomorphism from $\mathbb{Z}$ to $\mathbb{Z}/n\mathbb{Z}$. Applying this to matrix entries induces a group homomorphism $\theta_n : \mathrm{SL}_2(\mathbb{Z}) \to \mathrm{SL}_2(\mathbb{Z}/n\mathbb{Z})$, which in turn induces a group homomorphism

$$\phi_n : \Gamma = \mathrm{PSL}_2(\mathbb{Z}) \to \mathrm{PSL}_2(\mathbb{Z}/n\mathbb{Z}) := \mathrm{SL}_2(\mathbb{Z}/n\mathbb{Z})/\{\pm I\}.$$

Exercise 1.8 Show that θ_n and ϕ_n are epimorphisms.

We define the *principal congruence subgroup* of level n in Γ to be

$$\Gamma(n) := \ker \phi_n,$$

a normal subgroup of finite index in Γ, with $\Gamma/\Gamma(n) \cong \mathrm{PSL}_2(\mathbb{Z}/n\mathbb{Z})$. For example, one can show that $\Gamma(2)$ is a free group of rank 2, generated by the elements $\tau \mapsto \tau/(-2\tau+1)$ (fixing 0) and $\tau \mapsto (-\tau+2)/(-2\tau+3)$ (fixing 1).

Exercise 1.9 Show that Γ acts transitively on the rational projective line $\hat{\mathbb{Q}} := \mathbb{P}^1(\mathbb{Q}) = \mathbb{Q} \cup \{\infty\}$, and that $\Gamma(2)$ has three orbits on $\hat{\mathbb{Q}}$. Deduce that $\Gamma/\Gamma(2) \cong S_3$ (the symmetric group of degree 3).

Since the isomorphism classes of tori correspond to the orbits of Γ on $\mathbb{H}$, we would like to have a 'nice' function on $\mathbb{H}$, taking a single value on each orbit of Γ, and different values on different orbits. We can regard g_2, g_3 and $\Delta = g_2^3 - 27g_3^2$ as functions of $\tau \in \mathbb{H}$ by evaluating them for the lattice $\Lambda = \Lambda(1, \tau)$ with basis elements $\omega_1 = 1$ and $\omega_2 = \tau$, and with modulus $\omega_2/\omega_1 = \tau$. Unfortunately, there is a problem with this approach: if we replace Λ with a similar lattice $\Lambda' = \mu\Lambda$, corresponding to an isomorphic torus, then g_2 and g_3 are multiplied by μ^{-4} and μ^{-6}, and Δ by μ^{-12}. However, if we define

$$J(\tau) := \frac{g_2(\tau)^3}{\Delta(\tau)} = \frac{g_2(\tau)^3}{g_2(\tau)^3 - 27g_3(\tau)^2}$$

then these powers of μ cancel, so $J(\tau)$ depends only on the similarity class of Λ. Moreover, g_2, g_3 and hence J depend only on the lattice Λ, and not on our choice of a basis for Λ, so J is invariant under the action of Γ on $\mathbb{H}$, that is,

$$J(T(\tau)) = J(\tau)$$

for all $\tau \in \mathbb{H}$ and all $T \in \Gamma$. This function J is called the *elliptic modular function* (but note that it is not an elliptic function!—the name is derived from its connection with elliptic curves). By their definition as constant multiples of Eisenstein series, the functions g_2 and g_3 are holomorphic on $\mathbb{H}$, and hence so is J since $\Delta \neq 0$. Using this one can show that J maps $\mathbb{H}$ onto $\mathbb{C}$, so it induces a bijection between the orbits of Γ on $\mathbb{H}$ and the complex numbers, and hence between the isomorphism classes of tori and the complex numbers.

Exercise 1.10 Evaluate $J(\tau)$ at $\tau = i$ and at $\tau = \omega = e^{2\pi i/3}$, and find the corresponding elliptic curves.

Remark 1.1 A second way to construct J is provided by the Riemann mapping theorem: there is a biholomorphic mapping of the open left half of the fundamental region F, defined by $|\tau| > 1$, $-\frac{1}{2} < \mathrm{Re}\,\tau < 0$, onto the upper half plane, continuously extendable to the boundary of this hyperbolic triangle. This extension is uniquely determined if we fix three values on the boundary, for example by putting $\lim_{y\to\infty} J(iy) = \infty$ and choosing values at $\tau = i$ and ω, see Exercise 1.10. By successive applications of Schwarz's reflection principle on the boundaries of hyperbolic triangles, J can be extended to a holomorphic function $\mathbb{H} \to \mathbb{C}$; at the corners of the triangles one can use Riemann's theorem about removable singularities. The construction shows moreover that J is everywhere locally biholomorphic, except at the orbits of Γ containing i and ω, where J has multiplicity 2 and 3 respectively. This method of construction can be made very explicit by means of hypergeometric functions as inverse mappings of so-called Schwarz triangle functions, see [52].

Exercise 1.11 Show that for all $J \neq 0, 1$, the Weierstrass normal form

$$y^2 = 4x^3 - \frac{27J}{J-1}(x-1)$$

defines an elliptic curve with invariant J.

There is a third approach to the construction of the elliptic modular function J, starting with elliptic curves E in Legendre form

$$y^2 = x(x-1)(x-\lambda)$$

where $\lambda \in \mathbb{C}\setminus\{0,1\}$. Here we regard λ as a function of the modulus τ corresponding to E. The difficulty here is that the Legendre form for an elliptic curve $y^2 = p(x)$ is not quite unique. This is because there are six ways of sending two of the three roots e_j of $p(x)$ to 0 and 1 by an affine transformation, with the third going to λ.

For instance, if we replace x with $1-x$ (transposing the roots 0 and 1) the right-hand side of the Legendre equation becomes

$$(1-x)(-x)(1-x-\lambda) = -x(x-1)(x-(1-\lambda)).$$

If we also replace y with iy the left-hand side becomes $-y^2$, so we have an isomorphic elliptic curve with Legendre form

$$y^2 = x(x-1)(x-(1-\lambda)).$$

Thus λ is replaced with $1-\lambda$. Another substitution (find it!) replaces λ with $1/\lambda$. These two substitutions generate a group isomorphic to S_3 (corresponding to permuting the three roots e_1, e_2 and e_3 of $p(x)$), and the six permutations in S_3 give rise to six values

$$\lambda,\ 1-\lambda,\ \frac{1}{\lambda},\ \frac{1}{1-\lambda},\ \frac{\lambda}{\lambda-1},\ \frac{\lambda-1}{\lambda}.$$

We can resolve this ambiguity, and define λ uniquely as a function of τ, as follows. For any lattice $\Lambda = \Lambda(\omega_1, \omega_2)$ we have $\wp'(z) = 0$ at $z = \frac{1}{2}\omega_1, \frac{1}{2}\omega_2$ and $\frac{1}{2}(\omega_1 + \omega_2)$ (why?), so the differential equation

$$(\wp')^2 = p(\wp)$$

implies that the roots e_1, e_2 and e_3 of $p(x)$ are at the points $x = \wp(\frac{1}{2}\omega_1)$, $\wp(\frac{1}{2}\omega_2)$ and $\wp(\frac{1}{2}(\omega_1 + \omega_2))$. In our case we take $\omega_1 = 1$ and $\omega_2 = \tau$, and we can number the roots as $e_1 = \wp(\frac{1}{2})$, $e_2 = \wp(\frac{1}{2}\tau)$ and $e_3 = \wp(\frac{1}{2}(1+\tau))$. An affine transformation $x \mapsto ax + b$ of $\mathbb{C}$ sending e_2 and e_3 to 0 and 1 respectively sends e_1 to

$$\lambda = \frac{e_1 - e_2}{e_3 - e_2},$$

and this depends only on τ.

As a function of τ, λ is holomorphic on $\mathbb{H}$, and is invariant under $\Gamma(2)$ (a normal subgroup of index 6 in Γ), but not under Γ. By applying elements T from the six cosets of $\Gamma(2)$ in Γ to τ, we obtain the six possible values for λ (some of these coincide for certain values of τ). The functions λ and J are related by the identity

$$J(\tau) = \frac{4(1-\lambda(\tau)+\lambda(\tau)^2)^3}{27\lambda(\tau)^2(1-\lambda(\tau))^2},$$

which shows that in general there are six values of λ corresponding to each value of J. The function

$$\beta(x) = \frac{4(1-x+x^2)^3}{27x^2(1-x)^2} \tag{1.2}$$

giving J in terms of λ is a special case of a class of functions we will study in the next chapters under the name *Belyĭ functions*. This example has triple zeros at

$e^{\pm 2\pi i/6}$, double zeros of $\beta - 1$ at $-1, \frac{1}{2}$ and 2, and double poles of β at 0, 1 and ∞; there are no other ramification points.

1.2.5 *Sketch of a Proof of Theorem 1.1*

We now return to compact Riemann surfaces X of arbitrary genus, and give a brief account of the ideas needed to prove that they can be considered as smooth complex projective algebraic curves. The basic idea is to find a 'good' pair of meromorphic functions f and g on X, to show that f and g are algebraically dependent, that is that $a(f, g)$ is identically zero for some non-zero polynomial $a(z, w) \in \mathbb{C}[z, w]$, and then to show that X is isomorphic to the algebraic curve $a(z, w) = 0$, or more precisely, to a nonsingular projective model of this affine curve.

1. If X is any compact Riemann surface then there is a non-constant meromorphic function $f : X \to \hat{\mathbb{C}}$: for this see Sect. 1.3, in particular Exercises 1.12 and 1.13. (This apparently obvious result is not so straightforward: in higher dimensions there are analogues of Riemann surfaces which have no non-constant global meromorphic functions.) This realises X as an n-sheeted branched covering of $\hat{\mathbb{C}}$, where $n = \deg f$.
2. Let g be any other meromorphic function on X, and let F be the finite subset of $\hat{\mathbb{C}}$ consisting of ∞, the branch-points of f, and the images under f of the poles of g. Then for each $q \in \hat{\mathbb{C}} \setminus F$ there are n distinct points $p_1, \ldots, p_n \in X$ such that $f(p_i) = q$.
3. Since $g(p_1), \ldots, g(p_n) \in \mathbb{C}$ we can define the elementary symmetric functions

$$S_1 = \sum_i g(p_i),$$

$$S_2 = \sum_{i<j} g(p_i)g(p_j),$$

$$\vdots$$

$$S_n = \prod_i g(p_i).$$

   By their construction, these are single-valued analytic functions of q on $\hat{\mathbb{C}} \setminus F$.
4. Around each point $q_0 \in F$ we can use $z = q - q_0$ as a local coordinate (or $z = 1/q$ if $q_0 = \infty$), so that each S_r is represented near q_0 as a Laurent series in $z^{1/k}$ for some k. Since S_r is single-valued, only integer powers of z can appear in this series, and since g is meromorphic only finitely many negative powers can appear. Thus each S_r is meromorphic on the whole of $\hat{\mathbb{C}}$, so—by Exercise 1.3—it is a rational function of q.

5. Composing each $S_r : \hat{\mathbb{C}} \to \hat{\mathbb{C}}$ with the projection $f : X \to \hat{\mathbb{C}}$ we obtain a meromorphic function $s_r : X \to \hat{\mathbb{C}},\ p \mapsto S_r(f(p))$, which is a rational function of f, that is, $s_r = S_r(f) \in \mathbb{C}(f)$.
6. By the well-known relationship between the coefficients of a polynomial and the symmetric functions of its roots, $(-1)^r S_r$ is the coefficient of t^{n-r} in the polynomial

$$A(t) := \prod_{i=1}^{n} (t - g(p_i)),$$

so that

$$A(t) = t^n - S_1 t^{n-1} + \cdots + (-1)^n S_n.$$

For any $p \in X \setminus f^{-1}(F)$ we have $p = p_i$ for some $p_i \in f^{-1}(q), q = f(p)$, so $A(g(p)) = 0$ since the factor $g(p) - g(p_i)$ vanishes. It follows that g satisfies an algebraic equation

$$a(g) := g^n - s_1 g^{n-1} + \cdots + (-1)^n s_n = 0, \tag{1.3}$$

with coefficients $s_r \in \mathbb{C}(f)$, on the complement of a finite set in X, and hence it satisfies this equation identically on X.
7. We need to show that the polynomial

$$a(t) = t^n - s_1 t^{n-1} + \cdots + (-1)^n s_n \in \mathbb{C}(f)[t]$$

is irreducible in $\mathbb{C}(f)[t]$, so suppose that it factorises as $a(t) = b(t)c(t)$ where $b(t), c(t) \in \mathbb{C}(f)[t]$. Meromorphic functions can be chosen to separate pairs of points (that is, to have distinct values at any given pair of points), and hence, by a simple argument—see Exercise 1.13—to separate any finite set of points. We can therefore choose the above function g so that it takes distinct values at the points $p_1, \ldots, p_n \in f^{-1}(q_0)$ for some $q_0 \in \hat{\mathbb{C}} \setminus F$. Near p_1 we can write g as a power series $s(z) = \sum a_n z^n$ in the local coordinate $z = q - q_0$. This series must satisfy $b(s(z)) = 0$ or $c(s(z)) = 0$, and without loss we can assume it is the former. By analytic continuation along paths in $X \setminus f^{-1}(F)$ from p_1 to p_i $(i > 1)$, we find that $b(g) = 0$ at each p_i. Thus b has n distinct roots $g(p_i)$, so $\deg b \geq n = \deg a$. Hence $\deg b = n$ and $\deg c = 0$, so a is irreducible.
8. We can use the values $z = f(p)$ and $w = g(p)$ of the functions f and g at points $p \in X$ as coordinates for p, satisfying the algebraic equation

$$a(w) = w^n - s_1 w^{n-1} + \cdots + (-1)^n s_n = 0,$$

where each s_r is a rational function $S_r(z) \in \mathbb{C}(z)$ of z. Multiplying through by the least common multiple of the denominators of these rational functions

$S_1(z), \dots, S_n(z)$ gives a polynomial $P(z, w) \in \mathbb{C}[z, w]$ with

$$P(z, w) = 0$$

at all points $p \in X$. This equation gives an affine model of X in $\mathbb{C}^2$, and by replacing the complex cartesian coordinates z and w with three homogeneous coordinates in the usual way, we obtain a projective model of X in $\mathbb{P}^2(\mathbb{C})$.

9. In general, this projective model may be singular, if f and g are both ramified at some point $p \in X$ (we have already met this problem in connection with hyperelliptic curves, see Example 1.6). Algebraic geometers have developed extensive machinery for resolving singularities, but here more elementary arguments are available: the Riemann-Roch theorem or the more elementary Exercise 1.12 in Sect. 1.3 allows one to replace g with a finite set of meromorphic functions $g_j : X \to \hat{\mathbb{C}}$ such that at least one of them is unramified at each ramification point of f. The resulting set of algebraic equations (1.3) yields a nonsingular projective model of X in $\mathbb{P}^M(\mathbb{C})$ for some integer M.
10. This outline proof establishes the required equivalence between compact Riemann surfaces and smooth complex projective algebraic curves. To complete the proof of Theorem 1.1 we must also deal with the morphisms in these categories. Specifically, we need to prove that any holomorphic map $f : X \to Y$ between compact Riemann surfaces gives a rational map on the corresponding algebraic curves. This is easy if we consider X and Y as algebraic curves in $\mathbb{P}^M(\mathbb{C})$ and $\mathbb{P}^N(\mathbb{C})$ respectively. The graph of f, that is, the set

$$G_f = \{(p, f(p)) \in X \times Y \mid p \in X\},$$

is an algebraic curve in $\mathbb{P}^M(\mathbb{C}) \times \mathbb{P}^N(\mathbb{C})$ isomorphic to X via the first projection $\pi_1 : G_f \to X$. Composing π_1^{-1} with the second projection $\pi_2 : G_f \to Y$ gives f as a rational map $X \to Y$.
11. This completes the outline proof of Theorem 1.1, but in fact we have proved rather more, for instance that every meromorphic function g on X is a root of a polynomial of degree at most $n = \deg f$ with coefficients in $\mathbb{C}(f)$. Indeed, the field $\mathbb{C}(X)$ of meromorphic functions on X can be identified with the quotient of the polynomial ring $\mathbb{C}(f)[w] \cong \mathbb{C}(z)[w]$ by the ideal $(a(w))$, which is maximal by the irreducibility of $a(w)$.

1.3 Appendix: Existence of Enough Meromorphic Functions

In addition to elementary facts of field algebra, for the proofs of Theorem 1.1 and later of Theorem 4.5 (which relates coverings of Riemann surfaces to extensions of their meromorphic function fields) we need the existence of nontrivial meromorphic functions on compact Riemann surfaces, and moreover the existence of meromorphic functions taking pairwise distinct values at finite sets of points.

Both results can be derived from the Riemann-Roch theorem (see almost any textbooks about Riemann surfaces, such as [10, 11, 26]), or from the construction of enough automorphic functions for the covering group, as in the recent book by González-Diez and Girondo [15]—but then one relies on the Main Theorem of Uniformisation. A slightly simpler approach, but also based on heavy machinery involving real analysis and Hilbert space theory, comes from the construction of harmonic differentials in connection with Dirichlet's problem. We quote two basic results, Theorems II.5.1.a and II.5.3 from [10]:

Proposition 1.4 *Let X be a compact Riemann surface, P a point in X, and z a local coordinate (that is, a chart) on X vanishing at P. Then for each integer $n \geq 1$ there is a meromorphic differential ω on X, holomorphic on $X \setminus \{P\}$, which can be written near P as*

$$\omega = f(z)dz \quad \textit{where} \quad f(z) = \frac{1}{z^{n+1}} + h(z)$$

for some holomorphic function $h(z)$.

Proposition 1.5 *Let X be a compact Riemann surface, $P_1, \ldots, P_k$ pairwise distinct points on X for some $k > 1$, and $c_1, \ldots, c_k$ arbitrary non-zero complex numbers with $\sum_{j=1}^{k} c_j = 0$. Then there is a meromorphic differential ω on X, holomorphic on $X \setminus \{P_1, \ldots, P_k\}$, with simple poles at all points P_j and with residues c_j at P_j $(j = 1, \ldots, k)$.*

In other words: in terms of a local coordinate z on X vanishing at each P_j, the differential can be written locally around P_j as

$$\omega = f(z)dz\,, \quad f(z) = \frac{c_j}{z} + \text{ a holomorphic function.}$$

Remember that quotients of meromorphic differentials give meromorphic functions on X, and now prove:

Exercise 1.12 Let X be a compact Riemann surface. For every $P \in X$ there is a meromorphic function on X having a simple zero at P.

Exercise 1.13 Let X be a compact Riemann surface, and $P_1, \ldots, P_k$ pairwise distinct points on X. There is a meromorphic function f on X taking pairwise distinct finite values at $P_1, \ldots, P_k$.

1.4 Belyĭ Functions and Their Dessins

1.4.1 Belyĭ's Theorem

It is not at all easy to make explicit the correspondence between Riemann surfaces and algebraic curves. There is particular interest in this problem in the case of curves defined over the field of algebraic numbers. This is partly because these are the complex numbers which are easiest to deal with, requiring only algebraic techniques (as opposed to the transcendental numbers, which require analytic methods). Much less obviously, we shall see that there is a strong connection between such curves and maps drawn on the corresponding Riemann surfaces, offering the prospect of an insight into the Galois theory of algebraic number fields.

We say that a smooth algebraic curve X is *defined over a subfield* K of $\mathbb{C}$ if it is isomorphic to the set of zeros, in some affine or projective space over $\mathbb{C}$, of a finite set of polynomials with coefficients in K; we then call K a *field of definition* of X. For instance, the Fermat curves are all defined over $\mathbb{Q}$. An *algebraic number field* (or simply a *number field*) is a subfield $K \subset \mathbb{C}$ which is a finite extension of $\mathbb{Q}$. The elements of such a field K are all algebraic over $\mathbb{Q}$, so K is contained in the field $\overline{\mathbb{Q}}$ of all algebraic numbers, that is, the algebraic closure of $\mathbb{Q}$ in $\mathbb{C}$. It follows that if an algebraic curve X is defined over a number field, then it is defined over $\overline{\mathbb{Q}}$. The converse is also true, since if X is defined over $\overline{\mathbb{Q}}$ then the finitely many coefficients of the defining polynomials all lie in some finite extension of $\mathbb{Q}$, that is, in a number field. Thus we have proved the following:

Lemma 1.1 *An algebraic curve is defined over a number field if and only if it is defined over* $\overline{\mathbb{Q}}$.

In 1979 Belyĭ [3] gave a function theoretic criterion for an algebraic curve to be defined over a number field. It is one of the main subjects of this book, and it can be stated as follows:

Theorem 1.3 *Let X be a compact Riemann surface, that is, a smooth projective algebraic curve in $\mathbb{P}^N(\mathbb{C})$ for some N. Then X can be defined over $\overline{\mathbb{Q}}$ if and only if there exists a non-constant meromorphic function $\beta : X \to \hat{\mathbb{C}}$ ramified over at most three points.*

Such a function β is called a *Belyĭ function*; if a Belyĭ function exists on a compact Riemann surface X, this surface is called a *Belyĭ surface* or a *Belyĭ curve*, and (X, β) is called a *Belyĭ pair*. Thus, according to Theorem 1.3, Belyĭ curves are those isomorphic to projective algebraic curves defined over number fields. The group $\mathrm{Aut}\,\hat{\mathbb{C}} = \mathrm{PGL}_2(\mathbb{C})$ of automorphisms of $\hat{\mathbb{C}}$ acts triply transitively on $\hat{\mathbb{C}}$, so by composing β with a suitable automorphism we can (and generally will) assume that its critical values are contained in $\{0, 1, \infty\}$. In the rest of this chapter we will consider some examples of Belyĭ functions, we will show how they can be illustrated combinatorially using dessins, and we will describe two algorithms for constructing them.

These algorithms provide proofs of one half of Belyĭ's Theorem, namely that the existence of a Belyĭ function is a necessary condition for an algebraic curve to be defined over $\overline{\mathbb{Q}}$. The converse, known as the 'obvious' part of Belyĭ's Theorem since he sketched it in just a few lines in [3], is in fact far from obvious, and a full proof was not published until about 25 years later: see [16, 22, 31, 33, 48] for details. The main ideas of two such proofs will be presented later in Sects. 4.2.1 and 4.3.

1.4.2 Existence of Belyĭ Functions: Simple Examples

We start by giving a series of examples of Belyĭ functions.

Example 1.7 Take $X := \hat{\mathbb{C}} = \mathbb{P}^1(\mathbb{C})$ and $\beta : \hat{\mathbb{C}} \to \hat{\mathbb{C}}$, $z \mapsto z$ (unramified).

Example 1.8 Also on the Riemann sphere take $\beta : \hat{\mathbb{C}} \to \hat{\mathbb{C}}$, $z \mapsto z^n$ for some integer $n > 1$. This function is now ramified over $z = 0$ and $z = \infty$.

Example 1.9 On the Riemann sphere again, recall that the Chebyshev polynomials $T_n(z)$ can be defined recursively by

$$\begin{aligned} T_0(z) &= 1, \\ T_1(z) &= z, \\ T_2(z) &= 2z^2 - 1, \\ &\vdots \\ T_{n+1}(z) &= 2zT_n(z) - T_{n-1}(z) \ , \end{aligned}$$

or equivalently by the property that $\cos(n\vartheta) = T_n(\cos(\vartheta))$. This shows that T_n is a polynomial of degree n, and that it maps the interval $[-1, 1]$ to itself. It also shows that if $n \geq 1$ then T_n has simple zeros at the points $\cos \frac{2k-1}{2n}\pi$ where $k = 1, \ldots, n$, has double values ± 1 between them, and has simple values ± 1 at the points ± 1. It follows that the function T_n^2 has n double zeros, $n-1$ double values 1 between them, and simple values 1 at ± 1 (see Fig. 1.1 for the case $n = 5$).

Now the ramification points for T_n^2 in $\mathbb{C}$ are the zeros of its derivative, so there are $\deg(T_n^2)' = 2n - 1$ of these, counting multiplicities. We have already accounted

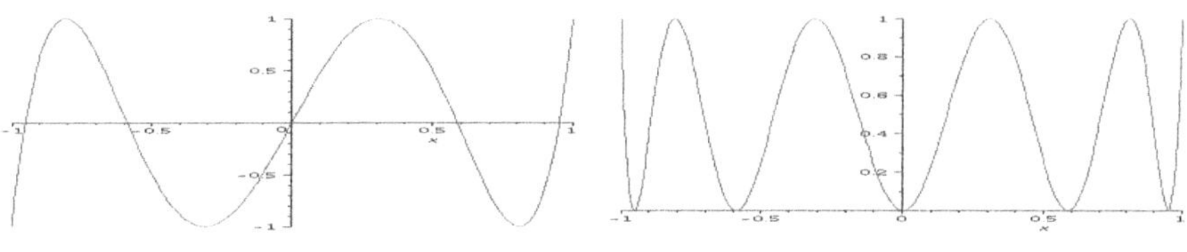

Fig. 1.1 The Chebyshev polynomial T_5 and its square T_5^2 in the interval $[-1, 1] \subset \mathbb{R}$

for $n + (n-1) = 2n-1$ points where T_n^2 has double values, so these are the only ramification points. Thus the only critical values of T_n^2 in $\mathbb{C}$ are in $\{0, 1\}$, so this is a Belyĭ function on $X = \hat{\mathbb{C}}$ for each $n \geq 1$. (In fact, as for all non-linear polynomials, ∞ is a critical value if $n \geq 1$.)

Example 1.10 Let X be the Fermat curve F_n, given by $x^n + y^n = z^n$ as in Example 1.5. We will show that

$$\beta : F_n \to \hat{\mathbb{C}},\ [x, y, z] \mapsto \frac{x^n}{z^n}$$

is an example of a Belyĭ function on F_n. On the affine part $z = 1$ of F_n we have $\beta : (x, y) \mapsto x^n$, so $\deg \beta = n^2$. Defining

$$\zeta_n := e^{2\pi i/n}$$

we have $|\beta^{-1}(x^n)| < n^2$ at

- affine points with $x^n = 0$, so that $(x, y) = (0, \zeta_n^k)$ for some k, and $\beta = 0$;
- affine points with $x^n = 1$, so that $(x, y) = (\zeta_n^k, 0)$ for some k, and $\beta = 1$;
- points with $z = 0$, so that $[x, y, z] = [1, y, 0]$ with $y^n = -1$, and $\beta = \infty$.

This shows that β is a Belyĭ function. Note that in each of these three cases, there are n critical points and the ramification order is n at each of them.

The following exercises give several ways of constructing Belyĭ functions.

Exercise 1.14 Suppose that m and n are positive integers. Find the critical points of the function

$$\beta = \beta_{m,n} : \hat{\mathbb{C}} \to \hat{\mathbb{C}},\ z \mapsto \frac{(m+n)^{m+n}}{m^m n^n} z^m (1-z)^n,$$

and show that it is a Belyĭ function.

Remark 1.2 Polynomials $\beta(z) \in \mathbb{C}[z]$ which induce Belyĭ functions $\beta : \hat{\mathbb{C}} \to \hat{\mathbb{C}}$, that is, which have no critical values in $\mathbb{C} \setminus \{0, 1\}$, are known as *Shabat polynomials*. They include the squares of the Chebyshev polynomials (see Example 1.9) and the polynomials $\beta_{m,n}$ considered here.

Exercise 1.15 Show that if $\beta : X \to \hat{\mathbb{C}}$ is a Belyĭ function, then so are the functions

$$\frac{1}{\beta},\quad 1-\beta,\quad 1-\frac{1}{\beta},\quad \frac{1}{1-\beta} \quad \text{and} \quad \frac{\beta}{\beta-1}.$$

If β is replaced with another of these *renormalisations*, what effect does this have on the critical points above 0, 1 and ∞?

Exercise 1.16 Show that if $\alpha : X \to Y$ is a holomorphic mapping between compact Riemann surfaces, and $\beta : Y \to \hat{\mathbb{C}}$ is a Belyĭ function such that the critical values of α are all contained in $\beta^{-1}(\{0, 1, \infty\})$, then the composition $\beta \circ \alpha : X \to \hat{\mathbb{C}}$ is also a Belyĭ function.

1.4.3 Dessins

If $\beta : X \to \hat{\mathbb{C}}$ is a Belyĭ function on a compact Riemann surface X, then it is useful to illustrate β by means of a map on X, that is, a graph embedded in X, dividing the surface into a finite number of simply connected faces. There are several ways of doing this, and not only do they give a good way of visualising Belyĭ functions, they also play a major role in both the general theory and the specific examples arising from Belyĭ's Theorem. These maps uniquely determine both X, as a Riemann surface or equivalently an algebraic curve, and also the Belyĭ function β; indeed, one can interpret Belyĭ's Theorem as stating that X is defined over an algebraic number field if and only if its complex structure is obtained from a map. On the subject of this remarkable characterisation, Grothendieck wrote in [18]:

> Il y a une identité profonde entre la combinatoire des cartes finies d'une part, et la géométrie des courbes algébriques définies sur des corps de nombres, de l'autre. Ce résultat profond, joint à l'interprétation algébrico–géométrique des cartes finies, ouvre la porte sur un monde nouveau, inexploré — et à portée de main de tous, qui passent sans le voir.
>
> [There is a profound identity between the combinatorics of finite maps on the one hand, and the geometry of algebraic curves defined over number fields on the other. This deep result, together with the algebraic–geometric interpretation of maps, opens the door onto a new, unexplored world — within reach of all, who pass without seeing it.]

Perhaps the most natural way of illustrating a Belyĭ function $\beta : X \to \hat{\mathbb{C}}$ is to use triangulations. We start with the triangulation $\mathscr{T}_1$ of $\hat{\mathbb{C}}$ which has three vertices at $0, 1$ and ∞, three edges along $\mathbb{R}$, and two faces (the upper and lower half-planes). We then lift this via β to a triangulation $\mathscr{T} = \beta^{-1}(\mathscr{T}_1)$ of X. The vertices of $\mathscr{T}$ are partitioned into three sets, lying above $0, 1$ and ∞, and it is useful to assign different colours to these vertices, say white, black and red respectively. There is no ramification over interior points of the two faces of $\mathscr{T}_1$, so each of these faces lifts to $\deg \beta$ triangular faces of $\mathscr{T}$; similarly, each of the three edges of $\mathscr{T}_1$ lifts, without ramification, to $\deg \beta$ edges of $\mathscr{T}$. However there is, in general, ramification at the vertices: if β has multiplicity n at a vertex v of $\mathscr{T}$, so that it is locally n-to-1 near v, then since each vertex of $\mathscr{T}_1$ has valency 2 it follows that v has valency $2n$.

If β is a Belyĭ function, then so are $\frac{1}{\beta}$, $1 - \beta$, $1 - \frac{1}{\beta}$, $\frac{1}{1-\beta}$ and $\frac{\beta}{\beta-1}$ (see Exercise 1.15). Using these instead of β permutes the critical values 0, 1 and ∞; this leaves the triangulation $\mathscr{T}$ invariant since the inverse image of the real projective line $\hat{\mathbb{R}} := \mathbb{P}^1(\mathbb{R}) = \mathbb{R} \cup \{\infty\}$ is unchanged, and it simply permutes the colours of the vertices.

Exercise 1.17 Describe the triangulation $\mathcal{T}$ of $\hat{\mathbb{C}}$ corresponding to the Belyĭ function $\beta : \hat{\mathbb{C}} \to \hat{\mathbb{C}}$ given by $z \mapsto z^n$ for $n \in \mathbb{N}$ (see Example 1.8).

Exercise 1.18 Show that if X is the Fermat curve F_n, and β is given by $[x, y, z] \mapsto x^n/z^n$ as in Example 1.10, then the vertices and edges of the corresponding triangulation $\mathcal{T}$ form a complete tripartite graph $K_{n,n,n}$, that is, there are n vertices of each of the three colours, and each pair of vertices of different colours are joined by a single edge. Draw $\mathcal{T}$ in the case $n = 2$.

One disadvantage of using a triangulation to illustrate a Belyĭ function is that it has many edges ($3 \deg \beta$, in fact). A more economical method is to use a bipartite map. (A graph or a map is said to be *bipartite* if the vertices can be partitioned into two sets, say white and black, so that each edge joins vertices of different colours.) We start with the bipartite map $\mathcal{B}_1 \subset \mathcal{T}_1$ on $\hat{\mathbb{C}}$ with white and black vertices at 0 and 1, joined by an edge along the unit interval, and a single face. This lifts, via β, to a bipartite map $\mathcal{B} = \beta^{-1}(\mathcal{B}_1)$ on X: the embedded graph consists of the white and black vertices of $\mathcal{T}$, representing the zeros of β and $\beta - 1$, and the $\deg \beta$ edges between them, consisting of the points where β takes values in the open interval $(0, 1)$. Since the vertices of $\mathcal{B}_1$ both have valency 1, each vertex v of $\mathcal{B}$ has valency equal to the multiplicity of β at v. Similarly each face of $\mathcal{B}$ is topologically a $2n$-gon, formed from $2n$ triangular faces of $\mathcal{T}$ with a common red vertex $v \in \beta^{-1}(\infty)$ (called the *face centre*) where β has pole order n, that is, multiplicity n at v.

This map $\mathcal{B}$ can be obtained from $\mathcal{T}$ simply by deleting all its red vertices and their incident edges. Conversely, starting with $\mathcal{B}$, one can reconstruct a version of $\mathcal{T}$ which is at least topologically correct, by placing one red vertex in each face of $\mathcal{B}$, and joining it by non-intersecting edges to successive white and black vertices in cyclic order around the boundary of the face. Thus these two representations of β are essentially equivalent. Whereas $\mathcal{B}$ has the advantage of greater simplicity, $\mathcal{T}$ has the advantage of giving equal status to each of the three critical values 0, 1 and ∞.

Definition 1 We will call this bipartite map $\mathcal{B} = \beta^{-1}(\mathcal{B}_1)$ a *dessin*, or more precisely the *Belyĭ dessin* associated with the Belyĭ function β.

Here 'dessin' is the French word for 'drawing': Grothendieck used the phrase 'dessins d'enfants', meaning 'children's drawings' to describe both the general theory and the surface embeddings of graphs which arise in it. (He used this phrase to recognise the way in which, as we shall see later, very simple examples of maps can represent very complicated mathematical structures.) In fact, those working in this area often use the term 'dessin' to describe any combinatorial structure on a surface, such as a triangulation or bipartite map, which illustrates or defines a Belyĭ function. In Chap. 2 we will give two more general definitions of a dessin, which include that given here as a particular case, but which are, as we shall prove, equivalent to it.

Example 1.11 The bipartite map $\mathcal{B}$ on $\hat{\mathbb{C}}$ associated with the Belyĭ function $\beta = T_n^2$ in Example 1.9 is, up to homeomorphism (that is, without giving the correct

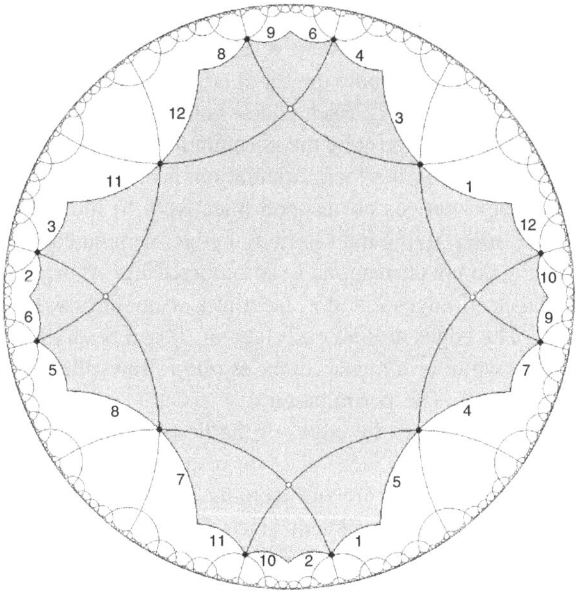

Fig. 1.2 Dessin on the Fermat quartic F_4 of genus $g = 3$

distances between the vertices), as follows:

-1 •——○——···——•——○——···——○——• 1

This is a path of $2n + 1$ alternating black and white vertices, representing the zeros of $\beta - 1$ and β, along the interval $[-1, 1] \subset \hat{\mathbb{C}}$, with black vertices at the end-points ± 1.

Example 1.12 It follows from Exercise 1.18 that if X is the Fermat curve F_n, with $\beta : [x, y, z] \mapsto x^n/z^n$ as in Example 1.10, then $\mathscr{B}$ is an embedding of the complete bipartite graph $K_{n,n}$, with n white and n black vertices, each white and black pair joined by a single edge. The faces of $\mathscr{B}$ are all $2n$-gons.

Exercise 1.19 Describe the bipartite maps $\mathscr{B}$ corresponding to the Belyĭ functions $\beta : \hat{\mathbb{C}} \to \hat{\mathbb{C}}$ given by $z \mapsto z^n$ for $n \in \mathbb{N}$ (see Example 1.8).

Exercise 1.20 Describe the bipartite maps $\mathscr{B}$ corresponding to the Belyĭ functions $\beta_{m,n}$ in Exercise 1.14.

Exercise 1.21 Draw the bipartite map $\mathscr{B}$ corresponding to the Fermat curve F_3 in Example 1.12.

To visualise the bipartite map $\mathscr{B}$ on F_n for $n > 3$ it is better to use hyperbolic geometry, that is, to draw it on the fundamental region of a uniformising group, see Chap. 3. In Fig. 1.2 we illustrate the case $n = 4$, using the unit disc model of the hyperbolic plane for greater symmetry and ease of drawing. The shaded area is the fundamental region, and the numbers indicate the required identifications of its sides; these are not edges of the dessin.

There is an even simpler representation of Belyĭ functions, which is purely algebraic and is therefore very suitable for calculations, either by hand or by computer. Let E denote the set of edges of the bipartite map $\mathscr{B}$ representing a Belyĭ function $\beta : X \to \hat{\mathbb{C}}$. Each edge e is incident with one white vertex u and one black vertex v. Let x and y be the permutations of E which send each edge e to the next edge, following the local orientation, around u and v respectively. Thus the white and black vertices correspond bijectively to the cycles of x and y on E, with the cyclic order giving the rotation of edges around each vertex. (Note that, in general, x and y do not correspond to automorphisms of maps or their embedded graphs: for instance, if edges e and e' meet at a white vertex, then their images under y need not.) The edges around each face of $\mathscr{B}$ can be divided into two types, as they start with a white or a black vertex as one follows the orientation around the boundary of the face. The permutation $z = (xy)^{-1} = y^{-1}x^{-1}$ of E (that is, y^{-1} followed by x^{-1}) permutes the edges of the first type around each face, again following the orientation, so the faces of $\mathscr{B}$ correspond bijectively to the cycles of z on E. These three permutations are sufficient to determine $\mathscr{B}$ uniquely, up to homeomorphism; in fact x and y are sufficient, since they determine z uniquely.

The above construction, representing $\mathscr{B}$ by means of a pair of permutations x and y, can in fact be applied to *any* bipartite map on a compact oriented surface, not just one obtained from a Belyĭ function. We will return to this point in Sect. 2.1.1, where more general bipartite maps are considered.

Example 1.13 If we number the edges of the bipartite map $\mathscr{B}$ in Example 1.11 from left to right as $1, 2, \ldots, 2n-2$, then

$$x = (1,2)(3,4)\ldots(2n-3, 2n-2),$$
$$y = (1)(2,3)\ldots(2n-4, 2n-3)(2n-2),$$
$$z = (1,2,4,6,\ldots,2n-2,2n-3,2n-5,\ldots,3).$$

By Exercise 1.14, with $m = n = 1$, the function $z \mapsto 4z(1-z)$ is a Belyĭ function $\hat{\mathbb{C}} \to \hat{\mathbb{C}}$; it sends $\infty \mapsto \infty$, $0 \mapsto 0$, $1 \mapsto 0$ and $\frac{1}{2} \mapsto 1$. It then follows from Exercise 1.16 that if $\beta : X \to \hat{\mathbb{C}}$ is a Belyĭ function then so is the function $\beta^* = 4\beta(1-\beta) : X \to \hat{\mathbb{C}}$. If $\mathscr{B}$ is the bipartite map on X associated with β, then the bipartite map $\mathscr{B}^*$ associated with β^* is formed from $\mathscr{B}$ by colouring all its vertices white (since β^* sends them to 0), and placing black vertices at the points in $\beta^{-1}(\frac{1}{2})$, that is, one on each edge of $\mathscr{B}$ (since β^* sends these points to 1). The faces of $\mathscr{B}^*$ are the same as those of $\mathscr{B}$, though each $2n$-gonal face of $\mathscr{B}$ is now regarded as a $4n$-gonal face of $\mathscr{B}^*$. Since all the zeros of $\beta^* - 1$ are of order 2 (such a Belyĭ function is called a *clean* Belyĭ function), these new black vertices all have valency 2. We can therefore ignore them and regard $\mathscr{B}^*$ as a simpler one-colour map $\mathscr{M}$ on X, obtained from $\mathscr{B}$ by ignoring its vertex-colouring. This leads us to the more general theory of *maps*, to be discussed in the next chapter.

The maps $\mathcal{M}$ obtained by the above method may or may not be bipartite. However, every map $\mathcal{M}$, bipartite or not, on a compact oriented surface, can be converted into a bipartite map $\mathcal{B}$ on the same surface, by colouring its vertices white and then placing a black vertex in every edge. The edges of $\mathcal{B}$ then correspond to the pairs consisting of an edge and an incident vertex of $\mathcal{M}$, or equivalently to the directed edges of $\mathcal{M}$ (see Sect. 2.1.5). This method of representing maps by permutations is equivalent to that introduced by Hamilton [20, 21] as early as 1856, though it remained undeveloped until it was rediscovered and generalised several times by other mathematicians, many years later.

It is remarkable that such complicated structures as Riemann surfaces and algebraic curves defined over number fields, together with their Belyĭ functions, can be described by such apparently simple objects as maps, bipartite maps and triangulations. Even more remarkably, as we will show later in Chap. 3, the converse is also true: each of these simple objects determines a unique conformal structure on the underlying topological surface X, together with a Belyĭ function β, making X into a compact Riemann surface which is defined, as an algebraic curve, over $\overline{\mathbb{Q}}$. Motivated by the often unsophisticated nature of these combinatorial objects, Grothendieck called them *dessins d'enfants* (children's drawings), or simply *dessins* (see Sect. 2.1.6 for a good example).

1.4.4 Belyĭ Algorithms

Our aim in this section is to prove part of Theorem 1.3 by describing algorithms which produce a Belyĭ function for any compact Riemann surface defined over $\overline{\mathbb{Q}}$. In order to motivate this, we first discuss another example, taken from [50], which will play a prominent role in our study of Galois actions (see Example 4.7).

Example 1.14 We will construct a Belyĭ function on the elliptic curve X given in Legendre form

$$y^2 = x(x-1)(x-\lambda),$$

where we take

$$\lambda = \alpha := \frac{1}{\sqrt[3]{2}} \in \mathbb{R}.$$

(Clearly, this curve is defined over $\overline{\mathbb{Q}}$.) We need a meromorphic function on X, so we start with the projection

$$X \to \hat{\mathbb{C}}\,,\ (x,y) \mapsto x\,,$$

which is ramified at the critical points (∞,∞), $(0,0)$, $(1,0)$ and $(\alpha,0)$, with critical values $\infty, 0, 1$ and α. In order to find a Belyĭ function, we need to eliminate the critical value α, so we compose this projection with the function

$$x \mapsto x^3,$$

which sends the algebraic number α to the rational number $\frac{1}{2}$, while fixing the other three critical values. To deal with the resulting critical value $\frac{1}{2}$, we now apply the function

$$\beta_{1,1} : z \mapsto 4z(1-z)$$

in Exercise 1.14, which sends $\frac{1}{2}$ to 1, and sends $0, 1$ and ∞ to $0, 0$ and ∞. The composition

$$\beta : (x, y) \mapsto x \mapsto x^3 \mapsto 4x^3(1-x^3).$$

of these three steps therefore maps the critical values of the first step into $\{0, 1, \infty\}$ in the following way:

$$\infty \mapsto \infty\,, \quad 0 \mapsto 0\,, \quad 1 \mapsto 0\,, \quad \alpha \mapsto 1.$$

Using polynomials like this to deal with algebraic critical values may induce new ramifications. Here $x \mapsto x^3$ is ramified at $x = 0$ and $x = \infty$, so the composition $(x, y) \mapsto x^3$ is ramified over $0, 1, \infty$ and $\frac{1}{2}$. However, the last step is ramified only at ∞ and $\frac{1}{2}$, and sends these points to ∞ and 1, so the composition β is a Belyĭ function $X \to \hat{\mathbb{C}}$, sending

$$\begin{aligned}
(\infty, \infty) &\mapsto \infty && \text{(pole of order 12)}\\
(0, 0) &\mapsto 0 && \text{(multiplicity 6)}\\
(1, 0) &\mapsto 0 && \text{(multiplicity 2)}\\
(\zeta_3^k, \pm y'_k) &\mapsto 0 && \text{(multiplicity 1)}\\
(\alpha, 0) &\mapsto 1 && \text{(multiplicity 4)}\\
(\zeta_3^k\alpha, \pm y_k) &\mapsto 1 && \text{(multiplicity 2).}
\end{aligned}$$

(Here we take $k = 1, 2$, so that $\zeta_3^k\alpha$ represents the non-real cube roots of $\frac{1}{2}$, that is, the algebraic conjugates of α, and $\pm y_k$ represents the values of y corresponding to these values of x. Similarly $\pm y'_k$ gives the values of y corresponding to the simple zeros of β at the four points with $x = \zeta_3^k$ for $k = 1, 2$.) The pre-image of the unit interval is (up to homeomorphism) as in Fig. 1.3, with opposite sides of the rectangle identified to form a torus. The centre of the face represents the unique pole (∞, ∞) of β.

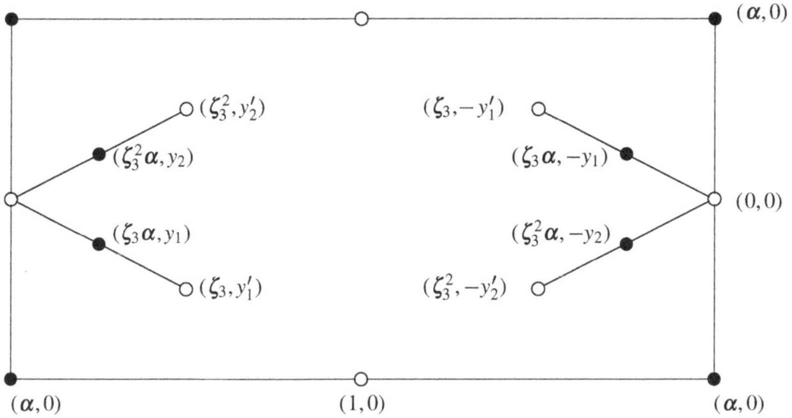

Fig. 1.3 Belyĭ dessin on the elliptic curve $y^2 = x(x-1)(x-\alpha)$, $\alpha = 1/\sqrt[3]{2}$

The following two exercises are very instructive for developing an understanding of how dessins can arise from Belyĭ functions, and of how Galois conjugation can act on them (we will return to this example in Sect. 4.2.2):

Exercise 1.22 Construct the dessin in Example 1.14 for yourself, by using β to lift the unit interval from $\hat{\mathbb{C}}$ back to X via the three successive steps described above.

Exercise 1.23 Find Belyĭ functions and dessins for the elliptic curves

$$y^2 = x(x-1)(x - \zeta_3^{\pm 1}\alpha).$$

To prove at least one direction of Theorem 1.3, suppose that X is given as an algebraic curve defined over a number field. In this case, Belyĭ gave an algorithm to construct a Belyĭ function $\beta : X \to \hat{\mathbb{C}}$ by systematically applying steps which generalise those used in Example 1.14. First take some non-constant meromorphic function $X \to \hat{\mathbb{C}}$ defined over $\overline{\mathbb{Q}}$, for example a coordinate projection, ramified over finitely many points $\alpha_1, \ldots, \alpha_n \in \overline{\mathbb{Q}}$ and possibly over ∞; compose it with a polynomial $p : \hat{\mathbb{C}} \to \hat{\mathbb{C}}$ of minimal degree defined over $\mathbb{Q}$ sending $\alpha_1, \ldots, \alpha_n$ to $\mathbb{Q}$; if new ramifications arise, repeat the procedure and use the following facts which are easy to prove.

Lemma 1.2 *Let X be a Riemann surface, $f : X \to \hat{\mathbb{C}}$ a non-constant meromorphic function, and W its set of critical values. Let $g : \hat{\mathbb{C}} \to \hat{\mathbb{C}}$ be a non-constant rational function with critical values in a set V. Then the critical values of $g \circ f$ are contained in $g(W) \cup V$.*

Lemma 1.3 *Let $p \in \mathbb{Q}[x]$ be a minimal polynomial of a finite set $W \subset \overline{\mathbb{Q}}$ (that is, of least degree, vanishing on W), and let V be the set of critical values of p. Then each minimal polynomial for V has degree less than* deg p.

Exercise 1.24 Prove these two lemmas.

So, after a finite number of rather obvious steps we will have a meromorphic function $f : X \to \hat{\mathbb{C}}$ with all its critical values in $\hat{\mathbb{Q}}$; using an automorphism of $\hat{\mathbb{C}}$ in $\mathrm{PGL}_2(\mathbb{Q})$ if necessary we may suppose that these include $0, 1$ and ∞. Surprisingly, one can continue composing with a finite number of rational functions $\hat{\mathbb{C}} \to \hat{\mathbb{C}}$ so that all these points are sent to $\{0, 1, \infty\}$ and no other critical values are created. For example if f has a rational critical value r such that $0 < r < 1$, we can choose $m, n \in \mathbb{N}$ so that $r = m/(m+n)$, and compose f with the function

$$\beta_{m,n} : z \mapsto \frac{(m+n)^{m+n}}{m^m n^n} z^m (1-z)^n$$

which sends

$$0 \mapsto 0\,, \quad 1 \mapsto 0\,, \quad \infty \mapsto \infty\,, \quad r \mapsto 1.$$

At this step, the only ramifications occur at $0, 1, r$ and ∞ (see Exercise 1.14), so by Lemma 1.2 the critical value r is eliminated, without any new ones being created. By iterating finitely many steps like this (see the following exercise), we eventually obtain a meromorphic function $X \to \hat{\mathbb{C}}$ with all its critical values in $\{0, 1, \infty\}$, that is, a Belyĭ function.

Exercise 1.25 In the above algorithm, how can one deal with rational critical values $r \notin (0, 1)$?

It is easy to see that in typical implementations of this algorithm, the degree of the Belyĭ function grows astronomically with the number of rational points r to be treated. In the meantime more effective procedures have been found to perform this last part of the construction. Here we follow Belyĭ's second proof of his theorem [4], which relies on the following:

Lemma 1.4 *Let $r_1, \ldots, r_n \in \mathbb{Q}$ be pairwise different and put*

$$y_i := \Big(\prod_{j \neq i} (r_i - r_j)\Big)^{-1}.$$

Then we have

$$\sum_i \frac{y_i}{x - r_i} = \frac{1}{\prod_i (x - r_i)}.$$

Proof Consider the polynomials $p(x) = \prod_i (x - r_i)$ and

$$q(x) = p(x) \sum_i \frac{y_i}{x - r_i} = \sum_i y_i \prod_{j \neq i} (x - r_j) \in \mathbb{Q}[x].$$

Then $\deg q \leq n-1$ and $q(r_1) = \ldots = q(r_n) = 1$, so $q(x) = 1$ for all x, and the result immediately follows. □

To begin the Belyĭ algorithm we use Lemmas 1.2 and 1.3 as before to construct a meromorphic function $f : X \to \hat{\mathbb{C}}$ with at most ∞ and finitely many rational numbers $r_1, \ldots, r_n$ as its critical values. Let N be the least common denominator of the resulting non-zero rational numbers y_i appearing in Lemma 1.4. Then $a_i := Ny_i \in \mathbb{Z}$, and the rational function

$$g(x) := \prod_i (x - r_i)^{a_i} \in \mathbb{Q}(x)$$

satisfies

$$\frac{g'(x)}{g(x)} = \sum_i \frac{a_i}{x - r_i} = \sum_i \frac{Ny_i}{x - r_i} = \frac{N}{\prod_i (x - r_i)}$$

by Lemma 1.4. Therefore g has ∞ and at most the rational numbers r_i as critical points. Since it sends these to 0 or ∞, it follows from Lemma 1.2 that $g \circ f$ has at most 0, 1 and ∞ as critical values.

Exercise 1.26 Find a Belyĭ function and a dessin for the curve $y^2 = x^n - 1$, where $n > 3$.

References

1. Beazley Cohen, P., Itzykson, C., Wolfart, J.: Fuchsian triangle groups and Grothendieck dessins: variations on a theme of Belyi. Commun. Math. Phys. **163**, 605–627 (1994)
2. Beazley Cohen, P., Wolfart, J.: Dessins de Grothendieck et variétés de Shimura. C. R. Acad. Sci. Paris Sér. I Math. **315**, 1025–1028 (1992)
3. Belyĭ, G.V.: On Galois extensions of a maximal cyclotomic field. Izv. Akad. Nauk SSSR Ser. Mat. **43**, 267–276, 479 (1979)
4. Belyĭ, G.V.: A new proof of the three-point theorem. Sb. Math. **193**, 329–332 (2002)
5. Bolza, O.: On binary sextics with linear transformations onto themselves. Am. J. Math. **10**, 47–70 (1888)
6. Brahana, H.R.: Regular maps and their groups. Am. J. Math. **49**, 268–284 (1927)
7. Conder, M.D.E.: Regular maps and hypermaps of Euler characteristic −1 to −200. J. Comb. Theory Ser. B **99**, 455–459 (2009). Associated lists of computational data available at http://www.math.auckland.ac.nz/~conder/hypermaps.html
8. Cori, R.: Un code pour les graphes planaires et ses applications. Astérisque, vol. 27. Société Mathématique de France, Paris (1975)
9. Degtyarev, A.: Topology of Algebraic Curves. An Approach via Dessins d'Enfants. De Gruyter, Berlin (2012)
10. Farkas, H.M., Kra, I.: Riemann Surfaces. Springer, Berlin (1991)
11. Forster, O.: Lectures on Riemann Surfaces. Springer, Berlin (1991)
12. Fricke, R.: Ueber eine einfache Gruppe von 504 Operationen. Math. Ann. **52**, 321–339 (1899)

13. Fricke, R., Klein, F.: Vorlesungen über die Theorie der automorphen Funktionen 1, 2. Teubner, Leipzig (1897/1912)
14. Garbe, D.: Über die regulären Zerlegungen geschlossener orientierbarer Flächen. J. Reine Angew. Math. **237**, 39–55 (1969)
15. Girondo, E., González-Diez, G.: Introduction to Compact Riemann Surfaces and Dessins d'Enfants. London Mathematical Society Student Texts, vol. 79. Cambridge University Press, Cambridge (2012)
16. González-Diez, G.: Variations on Belyi's theorem. Q. J. Math. **57**, 339–354 (2006)
17. González-Diez, G., Jaikin-Zapirain, A.: The absolute Galois group acts faithfully on regular dessins and on Beauville surfaces. Proc. London Math. Soc. (3) **111**(4), 775–796 (2015)
18. Grothendieck, A.: Esquisse d'un Programme. In: Schneps, L., Lochak, P. (eds.) Geometric Galois Actions 1. Around Grothendieck's Esquisse d'un Programme. London Mathematical Society Lecture Note Series, vol. 242, pp. 5–48. Cambridge University Press, Cambridge (1997)
19. Guillot, P.: An elementary approach to dessins d'enfants and the Grothendieck-Teichmüller group. Enseign. Math. **60**(3–4), 293–375 (2014)
20. Hamilton, W.R.: Letter to John T. Graves on the icosian. In: Halberstam, H., Ingram, R.E. (eds.) The Mathematical Papers of Sir William Rowan Hamilton, vol. III. Algebra, pp. 612–625. Cambridge University Press, Cambridge (1967)
21. Hamilton, W.R.: Account of the icosian calculus. Proc. R. Ir. Acad. **6**, 415–416 (1858); In: Halberstam, H., Ingram, R.E. (eds.) The Mathematical Papers of Sir William Rowan Hamilton, vol. III. Algebra, p. 609. Cambridge University Press, Cambridge (1967)
22. Hammer, H., Herrlich, F.: A remark on the moduli field of a curve. Arch. Math. (Basel) **81**, 5–10 (2003)
23. Jones, G.A., Singerman, D.: Theory of maps on orientable surfaces. Proc. Lond. Math. Soc. (3) **37**, 273–307 (1978)
24. Jones, G.A., Singerman, D.: Complex Functions. An algebraic and geometric viewpoint. Cambridge University Press, Cambridge (1986)
25. Jones, G.A., Singerman, D.: Belyi functions, hypermaps and Galois groups. Bull. Lond. Math. Soc. **28**, 561–590 (1996)
26. Jost, J.: Compact Riemann Surfaces. An Introduction to Contemporary Mathematics. Springer, Berlin (1997)
27. Klein, F.: Über die Transformationen siebenter Ordnung der elliptischen Functionen. Math. Ann. **14**, 428–497 (1878/1879)
28. Klein, F.: Zum Kontinuitätsbeweise des Fundamentaltheorems. In: Gesammelte Mathematische Abhandlungen, Band 3, pp. 731–741. Springer, Berlin (1923)
29. Koebe, P.: Über die Uniformisierung beliebiger analytischer Kurven. Göttinger Nachr. 191–210, 633–669 (1907)
30. Koebe, P.: Über die Uniformisierung der algebraischen Kurven IV. Math. Ann. **75**, 42–129 (1914)
31. Koeck, B.: Belyi's theorem revisited. Beitr. Algebra Geom. **45**, 253–275 (2004)
32. Lando, S.K., Zvonkin, A.K.: Graphs on Surfaces and Their Applications. Springer, Berlin (2004)
33. Malle, G., Matzat, B.H.: Inverse Galois Theory. Springer, Berlin (1999)
34. Oesterlé, J.: Dessins d'enfants. Astérisque (Sém. Bourbaki 2001/02, Exp. 907) **290**, 285–305 (2003)
35. Poincaré, H.: Théorie des groupes Fuchsiens. Acta Math. **1**, 1–62 (1882)
36. Poincaré, H.: Les fonctions Fuchsiennes. Acta Math. **1**, 193–294 (1882)
37. Poincaré, H.: Sur l'uniformisation des fonctions analytiques. Acta Math. **31**, 1–63 (1907)
38. Riemann, B.: Grundlagen für eine allgemeine Theorie der Funktionen einer veränderlichen Größe. Dissertation Göttingen (1851). In: Mathematische Werke und wissenschaftlicher Nachlaß, Teubner, Leipzig (1876)
39. Schneps, L. (ed.): The Grothendieck Theory of Dessins d'Enfants. London Mathematical Society Lecture Note Series, vol. 200. Cambridge University Press, Cambridge (1994)

40. Schneps, L., Lochak, P. (eds.): Geometric Galois Actions 1, 2. London Mathematical Society Lecture Note Series, vols. 242, 243. Cambridge University Press, Cambridge (1997)
41. Scholz, E.: Geschichte des Mannigfaltigkeitsbegriffs von Riemann bis Poincaré. Birkhäuser, Boston (1984)
42. Sherk, F.A., The regular maps on a surface of genus 3. Can. J. Math. **11**, 452–480 (1959)
43. Singerman, D.: Automorphisms of maps, permutation groups and Riemann surfaces. Bull. Lond. Math. Soc. **8**, 65–68 (1976)
44. Stillwell, J.: Geometry of Surfaces. Universitext. Springer, New York (1992)
45. Threlfall, W.: Gruppenbilder. Abh. Sächs. Akad. Wiss. Math. Phys. Kl. **41**, 1–59 (1932)
46. Voevodsky, V.A., Shabat, G.: Equilateral triangulations of Riemann surfaces and curves over algebraic number fields. Sov. Math. Dokl. **39**, 38–41 (1989)
47. Voisin, C., Malgoire, J.: Cartes Cellulaires. Cahiers Mathématiques, vol. 12. Université de Montpellier, Montpellier (1977)
48. Wolfart, J.: The 'Obvious' part of Belyi's theorem and Riemann surfaces with many automorphisms. In: Schneps, L., Lochak, P. (eds.) Geometric Galois Actions 1. Around Grothendieck's Esquisse d'un Programme. London Mathematical Society Lecture Note Series, vol. 242, pp. 97–112. Cambridge University Press, Cambridge (1997)
49. Wolfart, J.: Regular dessins, endomorphisms of Jacobians, and transcendence. In: Wüstholz, G. (ed.) A Panorama of Number Theory or the View from Baker's Garden, pp. 107–120. Cambridge University Press, Cambridge (2002)
50. Wolfart, J.: ABC for polynomials, dessins d'enfants, and uniformization – a survey. In: Schwarz, W., Steuding, J. (eds.) Elementare und Analytische Zahlentheorie (Tagungsband). Proceedings ELAZ-Conference, 24–28 May 2004, pp. 313–345. Steiner, Stuttgart (2006)
51. Wolfart, J.: Triangle groups and Jacobians of CM type. http://www.uni-frankfurt.de/50936228/Publikationen (2011). Accessed 12 Dec 2014
52. Yoshida, M.: Fuchsian Differential Equations. Vieweg, Braunschweig (1987)
53. Zannier, U.: On Davenport's bound for the degree of $f^3 - g^2$ and Riemann's existence theorem. Acta Arith. **71**, 107–137 (1995)

Chapter 2
Graph Embeddings

Abstract In this chapter we introduce maps and hypermaps, firstly as topological structures on surfaces, and then as equivalent algebraic objects, described by means of their monodromy groups. We give two further topological and group theoretic definitions of dessins, equivalent to that given in Chap. 1 in terms of Belyĭ functions; these definitions involve bipartite graphs embedded in surfaces, and 2-generator permutation groups. Morphisms, automorphisms and quotients of maps and dessins are defined, together with their regularity properties. Various other possibilities for the graphical representation of dessins are briefly discussed. The chapter includes an instructive example, in which a very simple dessin gives rise to a group of order 95040, the Mathieu group M_{12}. The chapter closes with a summary of the finite simple groups, which appear first here and then again in later chapters of this book.

Keywords Automorphism group • Bipartite map • Dessin d'enfant • Hypermap • Map • Mathieu group • Monodromy group • Regular dessin • Simple group

2.1 Graphs, Maps, and Hypermaps

2.1.1 Bipartite Maps

A graph $\mathscr{G}$ can be regarded as a pair (V, E) consisting of a set V of vertices and a set E of edges. We will suppose the graph to be connected and finite (later on this last condition will be relaxed), and we allow loops and multiple edges:

For our purposes, a *map* is an embedding $\mathscr{M} : \mathscr{G} \hookrightarrow X$ of a graph $\mathscr{G}$, where X is a surface which is connected, compact, without boundary, and oriented—in our diagrams, the chosen orientation is anticlockwise. (There are generalisations of this theory, in which all of these condition on X can be relaxed, but we will not explore them here.) The faces—connected components of $X \setminus \mathscr{G}$—must be simply-

Gareth A. Jones, J. Wolfart, *Dessins d'Enfants on Riemann Surfaces*, Springer Monographs in Mathematics, DOI 10.1007/978-3-319-24711-3_2

connected, that is homeomorphic to an open disc. Examples are given by the platonic solids on the sphere $X = S^2$.

Motivated by the need to describe Belyĭ functions, as in Sect. 1.4.3, we will generally assume that $\mathcal{G}$ is bipartite, that is, one can colour the vertices black and white so that each edge joins a black vertex to a white vertex •——○ (possible if and only if each circuit in $\mathcal{G}$ has even length).

Definition 2 We call such a map, consisting of a connected, finite, bipartite graph embedded in a connected, compact, oriented surface without boundary, a *bipartite map*, or a *dessin* (= *dessin d'enfant*), and we will denote it by $\mathcal{B}$.

Note that this definition includes that given in Sect. 1.4.3 for a Belyĭ dessin, arising from a Belyĭ function.

Example 2.1 The function

$$\beta(z) = \frac{4(1 - z + z^2)^3}{27z^2(1 - z)^2}$$

relating the functions J and λ at the end of Sect. 1.2.4 is an example of a Belyĭ function (this is not an accident!). The dessin $\mathcal{B}_1$ corresponding to this function looks as in Fig. 2.1 (up to homeomorphism, as always):

Example 2.2 It is easy to see that $\mathcal{B}_1$ is invariant under the transformation $a : z \mapsto 1 - z$, which is a half-turn of $\mathbb{C}$, fixing the black vertex at $\frac{1}{2}$ and the face-centre at ∞. (This corresponds to the fact that J is invariant under the substitution $\lambda \mapsto 1 - \lambda$ considered in Sect. 1.2.4). The quotient $\mathcal{B}_2$ of $\mathcal{B}_1$ by the cyclic group $\langle a \rangle$ looks as in Fig. 2.2.

Exercise 2.1 Find a Möbius transformation b of order 3 which leaves $\mathcal{B}_1$ invariant. What does $\mathcal{B}_1 / \langle b \rangle$ look like? What group do a and b generate?

Example 2.3 Identifying opposite sides of the outer hexagon in Fig. 2.3a gives a bipartite map $\mathcal{B}_3$ on a torus. Each black and white pair are joined by a single edge, so $\mathcal{G}$ is the complete bipartite graph $K_{3,3}$, with three black and three white vertices. This is in fact a dessin on the Fermat curve F_3, representing the Belyĭ function $[x, y, z] \mapsto x^3/z^3$ (see Example 1.12).

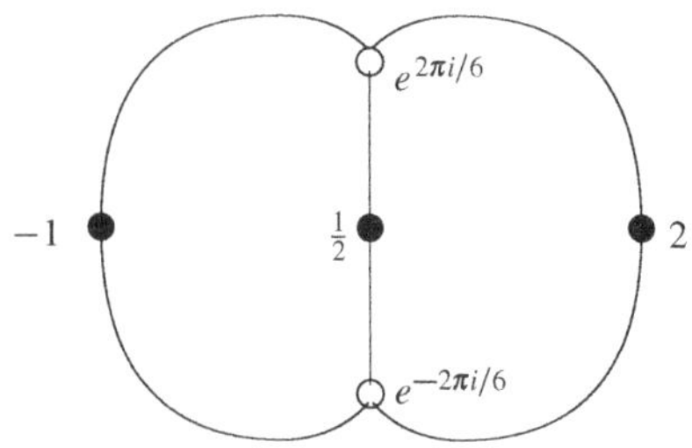

Fig. 2.1 Pre-image of $[0, 1]$ under β in Example 2.1

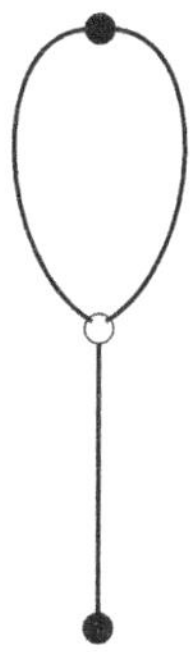

Fig. 2.2 Quotient of the Fig. 2.1 dessin

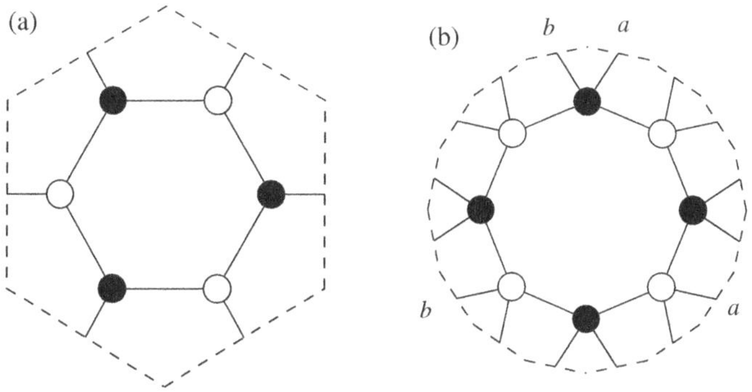

Fig. 2.3 Embeddings of $K_{3,3}$ and $K_{4,4}$ in the Fermat curves F_3 for (**a**) and F_4 for (**b**)

Similarly the map $\mathscr{B}_4$ in Fig. 2.3b shows an embedding of $K_{4,4}$ on the Fermat curve F_4, corresponding to the Belyĭ function $[x, y, z] \mapsto x^4/z^4$. In this case pairs of sides of the outer 16-gon are identified to give a surface of genus 3; the letters indicate two such pairs, and the other identifications can be obtained by rotation through multiples of $\pi/2$.

Whereas Fig. 2.3a is geometrically correct, since the flat printed page (or computer screen) accurately models the euclidean geometry of the complex plane and its quotient torus, Fig. 2.3b is only combinatorially and topologically correct: the equivalent Fig. 1.2 correctly represents the hyperbolic geometry of $\mathscr{B}_4$ on a quotient of the unit disc. Nevertheless, as we will see later in this book, it is remarkable that the very basic combinatorial information conveyed by a simple picture such as Fig. 2.3b is sufficient to determine the Riemann surface, the algebraic curve, and the Belyĭ function.

To describe a bipartite map $\mathscr{B}$ algebraically, we use the orientation of X (as outlined in Sect. 1.4.3) to define two permutations x and y of the set E of edges of $\mathscr{B}$. For each $e \in E$, we define ex and ey to be the next edges around the unique white and black vertices incident with e, following the orientation of X. Warning: these permutations are not generally automorphisms. We have bijections

White vertices	$\longleftrightarrow$	cycles of x on E
Black vertices	$\longleftrightarrow$	cycles of y on E
Faces	$\longleftrightarrow$	cycles of xy on E

The last bijection should be handled with care: in general, xy provides a cyclic permutation of only half of the boundary edges of a face; the other edges should be considered as boundary edges of their respective neighbouring face.

The orders l, m and n of x, y and xy are the least common multiples of their cycle-lengths. We call the triple (l, m, n) the *type* of $\mathcal{B}$. For instance, $\mathcal{B}_1$ and $\mathcal{B}_2$ have type $(3, 2, 2)$, while $\mathcal{B}_3$ has type $(3, 3, 3)$. The *monodromy group* of $\mathcal{B}$ is the subgroup $G = \langle x, y \rangle$ generated by x and y in the symmetric group $\mathrm{Sym}(E)$ of all permutations of E. If $\mathcal{B}$ is the dessin corresponding to a Belyĭ function β (in Sect. 3.3 we will see that this is always the case), G is in fact the monodromy group of the covering β : observe that the edges of $\mathcal{B}$ represent the different branches of the inverse function β^{-1}, that is, the different sheets of the covering, and that the generators x and y correspond to the permutations of the sheets arising from analytic continuation of β^{-1} along positively oriented loops around 0 and 1, respectively.

As $\mathcal{G}$ is connected, G acts transitively on E, so this action of G is equivalent to its action on the cosets Hg ($g \in G$) of the stabiliser $H = G_e$ of an edge $e \in E$. We say that G acts *regularly* if $G_e = 1$, in which case this action is equivalent to G acting on itself by right multiplication.

Example 2.4 The trivial bipartite map, defined as the Belyĭ dessin for Example 1.7, has a single white vertex at 0, a single black vertex at 1, and a single edge along the unit interval $[0, 1]$, so that there is a single face $\hat{\mathbb{C}} \setminus [0, 1]$. In the corresponding algebraic bipartite map, E consists of a single element, x and y are the identity permutation, and G is the trivial group, acting regularly on E.

Example 2.5 In the bipartite map $\mathcal{B}_1$ in Example 2.1 we have $x^3 = y^2 = (xy)^2 = 1$; these relations define the dihedral group D_3 of order 6, so G is a quotient of D_3. Since G is transitive on the six edges, the stabiliser of an edge e has index $|G : G_e| = 6$, so $G \cong D_3$ with $G_e = 1$. Thus G acts regularly on E.

In $\mathcal{B}_2$ we again have $G \cong D_3$, but now $|G_e| = 2$, so the action of G on E is not regular.

In $\mathcal{B}_3$ we have $x^3 = y^3 = 1$ and $xy = yx$, $x \neq y$, both $\neq 1$, so that $G \cong C_3 \times C_3$, acting regularly on E.

2.1.2 Algebraic Bipartite Maps

An *algebraic bipartite map* is a quadruple (G, x, y, E) where $G = \langle x, y\rangle$ is a permutation group acting transitively on a set E. As explained in the preceding section, every bipartite map gives rise to an algebraic bipartite map. Conversely, we can reconstruct a bipartite map $\mathscr{B}$ from an algebraic bipartite map (G, x, y, E) as follows: define

$$\text{edges} := \text{elements of } E$$
$$\text{white and black vertices} := \text{cycles of } x \text{ and } y$$
$$\text{faces} := \text{cycles of } xy,$$

and define an edge to be incident with a vertex or a face if the corresponding element of E is contained in the corresponding cycle of x, y or xy. In general, G (or equivalently E) could be finite or infinite, but since we are dealing here with compact surfaces, our main interest is in the case where they are finite. In this situation we have the following definition:

Definition 3 A *dessin* is an algebraic bipartite map (G, x, y, E) in which G and E are finite.

At this point we have not described the techniques needed to show that this definition is equivalent to Definitions 1 and 2, given in Sects. 1.4.3 and 2.1.1, so we will postpone this until Chap. 3.

We will call G the *monodromy group* of $\mathscr{B}$. (It is also sometimes called the *cartographic group*.) We will see in Chap. 3 that every algebraic bipartite map comes from an essentially unique Belyĭ function, and is the monodromy group of the branched covering induced by that function, so the word *monodromy* is justified in this context. Independently of Belyĭ functions one may introduce *coverings* of algebraic bipartite maps, see Sect. 3.3.1; in this terminology, G is the monodromy group of $\mathscr{B}$ regarded as a branched covering of the trivial bipartite map.

Exercise 2.2

(a) Taking $x = (1, 2, \ldots, N)$ and $y = (1, 2)$ in the symmetric group S_N, find $\mathscr{B}$ and G.
(b) With the same x, but $y = (1, N)(2, N-1)\ldots$, find $\mathscr{B}$ and G.
(c) With the same x, but $y = (1, 2)(3, 4)\ldots$ and $N = 2m$ even, find $\mathscr{B}$ and G.

We define an *automorphism* of $\mathscr{B}$ to be a permutation of the set E of edges of $\mathscr{B}$ which preserves the cyclic order of edges around each vertex, that is, which commutes with x and y, or equivalently, commutes with G. The automorphisms of $\mathscr{B}$ form a group

$$\operatorname{Aut}\mathscr{B} = \{\, c \in \operatorname{Sym}(E) \mid cg = gc \text{ for all } g \in G \,\},$$

the centraliser $C = C(G)$ of G in the symmetric group $\mathrm{Sym}(E)$. This group permutes the cycles of x and those of y, preserving incidence, so it induces a group of automorphisms of the embedded bipartite graph $\mathscr{G}$, preserving the colours of the vertices. Similarly, the cycles of xy are permuted, again preserving incidence with those of x and y, so these automorphisms permute the faces of the map, preserving their incidence with edges and vertices. The automorphisms of $\mathscr{B}$ therefore induce self-homeomorphisms of X, preserving its orientation since they preserve the cyclic order within the cycles of x, y and xy.

In many simple cases, the automorphisms of a bipartite map can easily be seen from a diagram.

Example 2.6 In Example 2.1, $\mathrm{Aut}\,\mathscr{B}_1$ is a group of six rotations, isomorphic to D_3, whereas in Example 2.2, $\mathscr{B}_2$ has only the identity automorphism. The bipartite map $\mathscr{B}_3$ in Example 2.3 has nine automorphisms, namely the identity, and rotations through $\pm 2\pi/3$ about the centre of the central face in Fig. 2.3a and about the three white vertices; these form a group isomorphic to $C_3 \times C_3$. Similarly, though less obviously, for the map $\mathscr{B}_4$ in Fig. 2.3b the automorphism group is $C_4 \times C_4$.

Exercise 2.3 Verify that these nine transformations really are automorphisms of $\mathscr{B}_3$, and that we have not omitted any others, such as rotations about the black vertices or about other face-centres.

A permutation group is *semiregular* (acts freely) if each stabiliser is trivial.

$$\text{The group is} \left\{ \begin{array}{c} \text{semiregular} \\ \text{transitive} \\ \text{regular} \end{array} \right\} \text{ as } \left\{ \begin{array}{c} \text{at most} \\ \text{at least} \\ \text{exactly} \end{array} \right\} \begin{array}{c} \text{one group element} \\ \text{takes one point} \\ \text{to another.} \end{array}$$

Thus a permutation group is regular if and only if it is both transitive and semiregular.

Theorem 2.1 *Let G be any transitive permutation group on a set E, and let $C = C(G)$ be its centraliser in the symmetric group on E.*

1. *C acts semiregularly on E.*
2. *C acts regularly on E if and only if G does.*
3. *If C and G act regularly on E then $C \cong G$.*

Proof

(1) Let $c \in C$ fix $e \in E$. Any $e' \in E$ has the form $e' = eg$ for some $g \in G$ by transitivity. Then $e'c = egc = ecg = eg = e'$, so $c = 1$.

(2) Let C act regularly on E. Then C is transitive, so its centraliser is semiregular by (1) applied to C; but G commutes with C, so G is semiregular, and being transitive it must be regular.

Conversely, if G acts regularly, then E can be identified with G so that G acts on itself by right-multiplication $\rho_g : e \mapsto eg$; then left-multiplication $\lambda_c : e \mapsto$

$c^{-1}e$ commutes with right-multiplication (since $c^{-1}(eg) = (c^{-1}e)g$), and acts transitively, so C is transitive; since C is semiregular by (1), C is regular.

(3) When C and G act regularly, $\lambda_{g^{-1}} \leftrightarrow \rho_g$ gives the isomorphism $C \cong G$. (Why do we need $\lambda_{g^{-1}}$, and not λ_g, here?) □

If $\mathcal{B}$ is any bipartite map then its monodromy group G acts transitively on the set E of edges, so Theorem 2.1(1) implies that $\mathrm{Aut}\,\mathcal{B}$ acts semiregularly on E. We define $\mathcal{B}$ to be a *regular dessin* if $\mathrm{Aut}\,\mathcal{B}$ acts transitively on E, so that $\mathcal{B}$ is as symmetric as possible. By Theorem 2.1 this is equivalent to $\mathrm{Aut}\,\mathcal{B}$ being regular on E, and also to G being regular on E, and it implies that $\mathrm{Aut}\,\mathcal{B} \cong G$. In the case of a Belyĭ dessin $\mathcal{B}$, induced by a Belyĭ function $\beta : X \to \hat{\mathbb{C}}$ (see Sect. 1.4.3), this is equivalent to β being a regular covering, induced by a group G of covering transformations $X \to X$ acting transitively (and hence regularly) on the sheets of the covering, with quotient $\hat{\mathbb{C}}$.

Example 2.7 In our earlier examples, $\mathcal{B}_1, \mathcal{B}_3$ and $\mathcal{B}_4$ are regular, and they all satisfy $\mathrm{Aut}\,\mathcal{B} \cong G$, whereas $\mathcal{B}_2$ is not regular. Similarly the cube, regarded as a bipartite map, is regular, with G isomorphic to the alternating group A_4 (not S_4, since this group contains elements which transpose vertex-colours).

Theorem 2.2 *Let G be a transitive permutation group, acting on a set E, let G_e be the stabiliser in G of an element $e \in E$, and let C be the centraliser of G in* $\mathrm{Sym}(E)$. *Then $C \cong N_G(G_e)/G_e$, where $N_G(G_e)$ is the normaliser of G_e in G.*

Proof The elements of E correspond to the cosets G_eg of G_e in G. Under the action

$$h : G_eg \mapsto G_ehg\ ,$$

only permutations $h \in G$ can permute these cosets; this action is well defined if and only if $h \in N_G(G_e)$, in which case it commutes with the action of G. Moreover, h induces the trivial action if and only if $h \in G_e$, so $C \cong N_G(G_e)/G_e$. □

Exercise 2.4 Let G be the symmetric group S_N, acting on the set E of ordered k-tuples of distinct elements of $\{1, 2, \ldots, N\}$ for some $k = 1, 2, \ldots, N$. Describe G_e, $N_G(G_e)$ and C in this case, and verify that $C \cong N_G(G_e)/G_e$.

Corollary 2.1 *Let $\mathcal{B}$ be an algebraic bipartite map (G, x, y, E), let e be an edge of $\mathcal{B}$, let G_e be the stabiliser of e in G, and let C be the centraliser of G in* $\mathrm{Sym}(E)$. *Then*

$$\mathrm{Aut}\,\mathcal{B} \cong C \cong N_G(G_e)/G_e$$

where $N_G(G_e)$ is the normaliser of G_e in G.

Proof The first isomorphism is simply the definition of $\mathrm{Aut}\,\mathcal{B}$, and the second is a particular case of the preceding theorem. □

Example 2.8 If $\mathscr{B}$ is regular then $G_e = 1$, so $N_G(G_e) = G$ and we deduce that $\mathrm{Aut}\,\mathscr{B} \cong G$. In fact, the converse is also true, at least for finite maps: if $\mathrm{Aut}\,\mathscr{B} \cong G$ then $|N_G(G_e)/G_e| = |G|$, and for finite groups G this is possible only if $G_e = 1$, so that $\mathscr{B}$ is regular. However, this argument fails if E (and hence G) is infinite, and indeed there are counterexamples in this case [14].

Exercise 2.5 Find G_e, $N_G(G_e)$ and C for each of the bipartite maps $\mathscr{B}$ defined in Exercise 2.2.

If a bipartite map $\mathscr{B}$ is regular then $\mathrm{Aut}\,\mathscr{B}$ acts transitively on the edges, and hence also on the white vertices, so these all have the same valency, namely the order l of x. Similarly, the black vertices have the same valency, the order m of y.

Exercise 2.6 Show that if $\mathscr{B}$ is regular then $\mathrm{Aut}\,\mathscr{B}$ acts transitively on the faces, and these are all $2n$-gons where n is the order of xy. (It may happen—see Fig. 2.2—that both sides of an edge belong to the same face: in that case the edge has to be counted twice.)

It follows that if a bipartite map $\mathscr{B}$ is regular then it has $|G|/l$ white vertices, $|G|/m$ black vertices, and $|G|/n$ faces. Recall that the triple (l, m, n) is called the *type* of $\mathscr{B}$. Since there are $|E| = |G|$ edges, the underlying surface of $\mathscr{B}$ has Euler characteristic

$$\chi = \frac{|G|}{l} + \frac{|G|}{m} - |G| + \frac{|G|}{n} = |G|\left(\frac{1}{l} + \frac{1}{m} + \frac{1}{n} - 1\right),$$

so it has genus

$$g = 1 - \frac{\chi}{2} = 1 + \frac{|G|}{2}\left(1 - \frac{1}{l} - \frac{1}{m} - \frac{1}{n}\right). \tag{2.1}$$

Exercise 2.7 Show that if $\frac{1}{l} + \frac{1}{m} + \frac{1}{n} \neq 1$ then there are (up to isomorphism, which you will need to define!) only finitely many regular bipartite maps of a fixed type (l, m, n) and of a given genus g. What happens if $\frac{1}{l} + \frac{1}{m} + \frac{1}{n} = 1$?

2.1.3 Dessins as Hypermaps

We need to mention briefly some other possible visualisations of dessins than those we use in most parts of this book. Instead of bipartite graphs we may use *hypermaps* consisting of *hypervertices, hyperedges* and *hyperfaces*: see [7] for an excellent survey. Hypermaps differ from the usual concept of maps, with their vertices, edges and faces, in that we no longer require that hyperedges must be incident with at most two hypervertices and at most two hyperfaces. In our usual representation by bipartite graphs we consider white vertices as hypervertices, faces as hyperfaces and black vertices as hyperedges; a hypervertex and a hyperedge are incident if the corresponding white and black vertices are neighbours in the bipartite graph. This representation of a hypermap as a bipartite graph embedded in a surface, first considered by Walsh [22], is called the *Walsh representation.*

As the pre-image of the unit interval $[0, 1]$, the Walsh representation gives a 'picture' of a Belyĭ function β which is both simple and natural; however it has the disadvantage of treating the three critical values $0, 1, \infty$ unequally. These three points can be arbitrarily permuted by *renormalisation*, that is, by replacing β with $1-\beta$, $1/\beta$ and so on (see Exercise 1.15), so the corresponding ramification orders should be visible in the corresponding dessin in similar ways. In the case of the Walsh representation this is only partly true:

- the orders of the zeros of β are the valencies of the white vertices,
- the orders of the zeros of $1-\beta$ are the valencies of the black vertices,
- but the valencies of the faces are twice the orders of the poles (that is, the zeros of $1/\beta$).

(For this last point, note again that an edge is counted twice if both sides belong to the same face, see the convention introduced in Exercise 2.6.) This deficiency of the Walsh representation can be dealt with by adding edges and vertices to form a triangulation, as in Sect. 1.4.3, or by using the James representation, explained later in this section.

An alternative to the Walsh representation is the *Cori representation* of a hypermap, introduced by Cori [6] in genus 0 in connection with the encoding of visual images. One has to blow up the white vertices in the Walsh representation to grey faces B and the black vertices to black faces G, which meet in a common boundary point whenever the corresponding vertices are joined by an edge in the Walsh representation; one may imagine that the trivial bipartite map ○——● is replaced with a lemniscate in $\mathbb{C}$ around 0 and 1 with a singular point $\frac{1}{2}$, dividing $\hat{\mathbb{C}}$ into a grey, a black and a white domain (see the right part of Fig. 2.4). This is the trivial Cori hypermap, and its pre-image under β can be seen as the dessin in its Cori representation. For instance, the left part of Fig. 2.4 shows this representation of the $K_{3,3}$-dessin on the Fermat curve F_3 shown earlier in Fig. 2.3a.

In this Cori representation of a dessin, the boundary lines of the faces form a graph in which all the vertices have valency 4. The orders of the zeros of β and $1-\beta$ are visible as the valencies of the grey and black faces, respectively, while the valencies of the white faces are again twice the orders of the poles of β. The number of vertices—pre-images of the non-critical point $1/2$—is equal to the degree of β. We will come back to this Cori representation in Sect. 7.1.2 since it provides an interesting link with Cayley graphs for automorphism groups of regular dessins.

However, like the Walsh representation, the Cori representation does not show the equivalence between the roles of the three critical points of a Belyĭ function. There are two possible ways of doing this. One way is to use the link between Belyĭ functions and triangle groups to be discussed in the next chapter, and to replace the dessin with a tessellation of the surface by white and black triangles; these are the pre-images under β of the upper and the lower half plane, as explained in Sect. 1.4.3 and shown in Sect. 3.1.6 as the left part of Fig. 3.4.

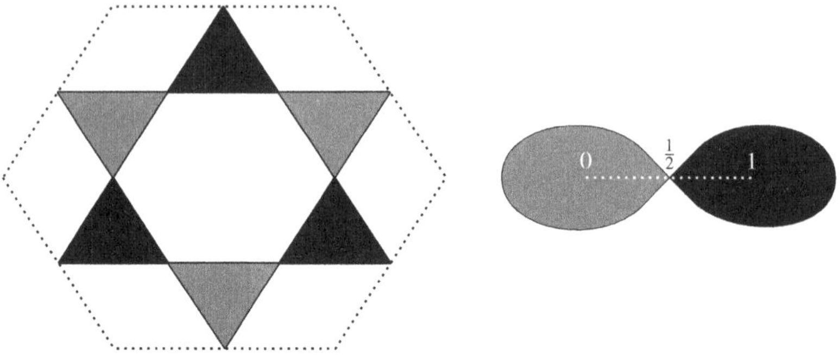

Fig. 2.4 Cori hypermap on the torus F_3 and the trivial Cori hypermap in $\hat{\mathbb{C}}$. Identify the opposite boundary lines of the outer hexagon to obtain the torus; the *dotted lines* do not belong to the hypermap

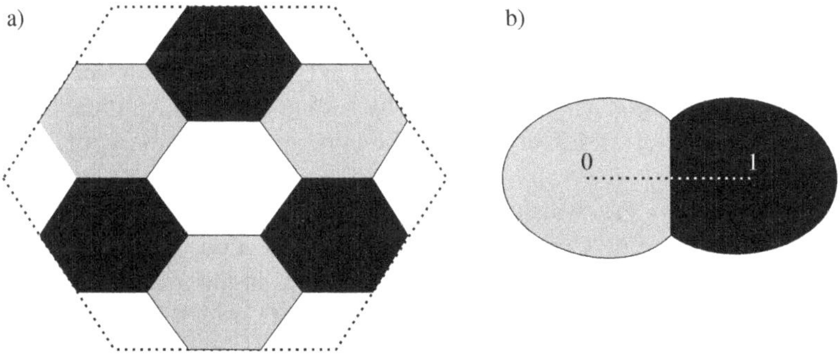

Fig. 2.5 (**a**) The same example as in Figs. 2.3a and 2.4, but in its James representation. (**b**) The trivial hypermap in its James representation

An alternative method is to use another representation introduced in [12] by Lynne James. Like the Cori representation, this *James representation* arises from the Walsh representation by blowing up the vertices, but this time so far that neighbouring black and grey faces touch along a line. This is illustrated in Fig. 2.5 for the Fermat cubic and for the trivial dessin. In the James representation, all vertices of the boundary graph have valency 3, and the orders of the zeros of β, $1 - \beta$ and $1/\beta$ are twice the valencies of the grey, black and white faces, respectively.

Exercise 2.8 What is the connection between the James representation and the triangulations introduced in Sect. 1.4.3?

2.1.4 Morphisms of Hypermaps

Isomorphisms of dessins can be defined as follows. If $\mathcal{B} = (G, x, y, E)$ and $\mathcal{B}' = (G', x', y', E')$ are bipartite maps (equivalent to dessins, as we will see later), then an isomorphism $i : \mathcal{B} \to \mathcal{B}'$ consists of a group-isomorphism $\theta : G \to G'$ sending x to x' and y to y', and a bijection $\phi : E \to E'$ compatible with θ, that is, with the property that $\phi(eg) = \phi(e)\theta(g)$ for all $e \in E$ and all $g \in G$. Thus we have the following commutative diagram, where the horizontal arrows denote the actions of G on E and of G' on E':

$$\begin{array}{ccc} E \times G & \longrightarrow & E \\ \phi \downarrow \downarrow \theta & & \downarrow \phi \\ E' \times G' & \longrightarrow & E' \end{array}$$

Theorem 2.3 *Every dessin $\mathcal{B}$ is isomorphic to the quotient dessin $A\backslash\tilde{\mathcal{B}}$ for some regular dessin $\tilde{\mathcal{B}}$ and some subgroup $A \leq \mathrm{Aut}\,\tilde{\mathcal{B}}$.*

Proof Take G to be the monodromy group of $\mathcal{B}$, and take $\tilde{\mathcal{B}}$ to be the dessin corresponding to the regular representation of G, so $\tilde{\mathcal{B}}$ is regular by Theorem 2.1. Take A to be the group of left multiplications λ_g $(g \in G_e)$ for some $e \in E$; then the orbits of A on E are just the cosets $G_e g$ $(g \in G)$ of G_e in G, so $A\backslash\tilde{\mathcal{B}} \cong \mathcal{B}$. □

We call $\tilde{\mathcal{B}}$ the *canonical regular cover* of $\mathcal{B}$. It has the property that any regular cover of $\mathcal{B}$ also covers $\tilde{\mathcal{B}}$.

Exercise 2.9 Give a reasonable definition for a *cover of a dessin*—topologically and algebraically! (We will return to the subject of these more general morphisms in Sect. 3.3.1.)

Exercise 2.10 Let $\mathcal{B}$ consist of a path of N edges, alternately black and white: •——○——•——○——··· . Find G, C, $\tilde{\mathcal{B}}$ and A for this dessin.

2.1.5 Maps and Hypermaps

What can be said about embeddings of graphs $\mathcal{G}$ which are not necessarily bipartite, for example the graphs of the tetrahedron or octahedron?

One can convert any graph $\mathcal{G}$ into a bipartite graph $\mathcal{G}^{\mathrm{bip}}$ by regarding the vertices of $\mathcal{G}$ as white vertices, and placing a black vertex in each edge of $\mathcal{G}$.

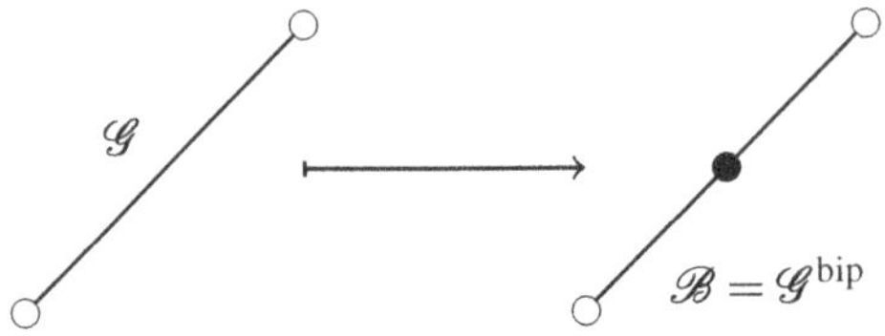

Any embedding of $\mathcal{G}$ in a surface is also an embedding of $\mathcal{G}^{\mathrm{bip}}$, so it gives a bipartite map $\mathcal{B}$. The edges of $\mathcal{B}$ correspond to the directed edges (sometimes called *darts*) of $\mathcal{G}$. (In many papers on the subject, the term *ribbon graph* is used to describe this subdivision of edges.) The rotations x and y of the set E of edges of $\mathcal{B}$ correspond to rotations x and y of the set Ω of darts of $\mathcal{G}$. Thus x rotates darts α around their incident vertices, following the orientation of the surface, and y reverses the direction of each dart, so $y^2 = 1$.

We can define an *algebraic map* (not necessarily bipartite) to be a 4-tuple (G, x, y, Ω) where $G = \langle x, y \rangle$ is a transitive permutation group acting on a set Ω, with $y^2 = 1$. As before, we can identify the vertices, edges and faces with the cycles of x, y and xy on Ω, incidence being given by non-empty intersection. (This idea can be traced back to a long letter written by Hamilton in 1856 [11], in which he described what we now call Hamiltonian cycles in graphs. This appears to be the earliest algebraic representation of maps, though the idea has subsequently been attributed to several other mathematicians. Hamilton introduced it specifically for the icosahedron and the dodecahedron. He represented what we would now call the triangle group $G = \Delta(2, 3, 5)$ as the monodromy group of the icosahedral map—definitely not the automorphism group, though they are in fact isomorphic—as a group of permutations of the ordered pairs of adjacent faces, which are equivalent to directed edges in an oriented map. He mentioned that this idea could also be applied to the other Platonic solids, and intriguingly, in a postscript to his letter and a short published summary [10], he promised further generalisations; however, none appear among his publications.)

The algebraic theory of maps is similar to that for bipartite maps: for instance, the obvious analogues of Theorem 2.2 and its corollaries are true. This theory includes an extension, which we will not explore further in this book, to maps on surfaces which may be non-orientable and may have non-empty boundary, and to automorphisms which may reverse the orientation. In this case the objects permuted are not darts but flags (mutually incident vertex-edge-face triples), with permutations r_i $(i = 0, 1, 2)$ sending each flag to the adjacent flag with the same j-dimensional components for $j \neq i$, or fixing the flag if it lies on the boundary and has no such adjacent flag. These permutations, which satisfy

$$r_0^2 = r_1^2 = r_2^2 = (r_0 r_2)^2 = 1\,,$$

generate a transitive group of permutations of the flags, and the automorphism group of the map is its centraliser in the symmetric group. For details, see [2, 13, 20, 21]. This form of the theory has the advantage of extending naturally to higher-dimensional generalisations of maps, called abstract polytopes [17], where similarly-defined permutations $r_0, \ldots, r_n$ of flags satisfy

$$r_i^2 = 1 \text{ for all } i, \quad (r_i r_j)^2 = 1 \text{ if } |i - j| \geq 2\,.$$

2.1.6 An Instructive Example

The following example illustrates how a simple drawing can encode very rich and complicated mathematical phenomena. *Monsieur Mathieu* $\mathcal{M}$ can be described by the map in Fig. 2.6.

Here $\Omega = \{1, 2, \ldots, 12\}$, with

$$x = (1\,2\,3)(4\,5\,6)(7)(8\,9\,10)(11)(12)$$

and

$$y = (1\,2)(3\,4)(5\,8)(6\,7)(9\,12)(10\,11).$$

Now $G = \langle x, y\rangle$. A calculation with GAP [19] shows that $|G| = 95040$, and that G is isomorphic to the Mathieu group M_{12}, a 5-transitive simple permutation group of degree 12. The stabiliser G_α of a dart $\alpha \in \Omega$ is isomorphic to the Mathieu group

Fig. 2.6 Monsieur Mathieu

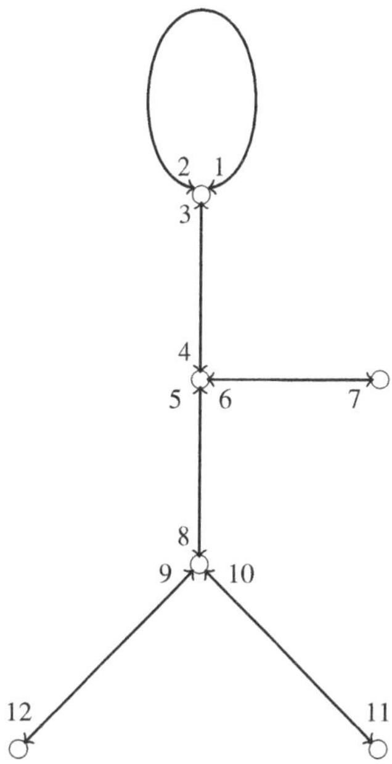

M_{11}, a 4-transitive simple permutation group of degree 11. (See Sect. 2.2 for these and other finite simple groups.)

Clearly the map $\mathcal{M}$ has genus 0, and type $(3, 2, 11)$. The corresponding bipartite map $\mathcal{B}$ has canonical regular cover $\tilde{\mathcal{B}}$ of type $(3, 2, 11)$ and genus $g = 3601$ (see Eq. (2.1) in Sect. 2.1.2), with $\mathrm{Aut}\,\tilde{\mathcal{B}} \cong G \cong M_{12}$.

By Belyĭ's theorem $\tilde{\mathcal{B}}$ corresponds to an algebraic curve defined over an algebraic number field. One can show (see Malle's tables in [16]) that the field of definition is $\mathbb{Q}(\sqrt{-11})$. This has Galois group isomorphic to C_2, generated by complex conjugation. Applying this to the coefficients of the algebraic curve and the Belyĭ function, we get the mirror image $\overline{\mathcal{M}}$ of $\mathcal{M}$. Later we will see more interesting and less obvious actions of Galois groups on families of maps.

2.2 Appendix: The Finite Simple Groups

The Jordan-Hölder Theorem shows that every finite group has a composition series, in which the factors are simple groups. The composition series is not generally unique, but the set of composition factors is unique. This existence and uniqueness result allows many problems about finite groups to be reduced to problems about finite simple groups.

The classification of finite simple groups was announced by Gorenstein in 1983, after over 25 years of intensive effort by many mathematicians. Although the statement of the classification was correct, the full details of the proof were not completed until 2004, with the publication of a 1221-page paper by Aschbacher and Smith. The whole proof is contained in about 100 papers, adding up to over 10,000 pages, although efforts are now under way to simplify the proof.

Wilson [23] gives an excellent description of the finite simple groups. The *ATLAS* [5] contains a much more concise survey, but its main feature consists of very detailed information (conjugacy classes, characters, automorphism groups, and so on) for all the smaller, and many far-from-small, non-abelian finite simple groups. An online version of the *ATLAS*, including detailed information on matrix representations over various fields, is available at http://brauer.maths.qmul.ac.uk/Atlas/v3/.

The finite simple groups can be divided into four families, as follows.

2.2.1 The Cyclic Groups of Prime Order

The only abelian simple groups are the cyclic groups C_p of prime order p. These are by far the easiest of the simple groups to deal with, the only real obstacle being our relative ignorance about the properties and distribution of the prime numbers, for instance whether or not there are infinitely many Fermat or Mersenne primes.

2.2.2 *The Alternating Groups*

Next to the cyclic groups, the alternating groups are the easiest finite simple groups to define. The alternating group A_n, consisting of the even permutations of n objects, is simple for each $n \geq 5$. The properties of A_n are closely related to those of the symmetric group S_n. For instance, the conjugacy classes of S_n correspond, via cycle-structures of permutations, to the partitions of n. By considering centralisers, one easily deduces that the even permutations with a given cycle-structure form a single conjugacy class in A_n, except for those consisting of cycles of distinct odd lengths, which form two conjugacy classes transposed by S_n. Similarly, if $n \neq 6$ then $\mathrm{Aut}\,A_n$ is isomorphic to S_n, acting by conjugation. The exception arises because S_6 has an outer automorphism interchanging the transpositions with the products of three disjoint transpositions, and the 3-cycles with the products of two disjoint 3-cycles, so that $\mathrm{Aut}\,A_6 = \mathrm{Aut}\,S_6$ contains S_6 with index 2. (This can also be seen from the isomorphism $A_6 \cong \mathrm{PSL}_2(9)$, so that $\mathrm{Aut}\,A_6 \cong \mathrm{Aut}\,\mathrm{PSL}_2(9) \cong \mathrm{P\Gamma L}_2(9)$, containing A_6 with index 4.)

The subgroup structures of the alternating groups are rather complicated: by Cayley's Theorem every group of order n is isomorphic to a subgroup of S_n, and hence of A_{n+2} (easy exercise!). Nevertheless, the maximal subgroups of the symmetric and alternating groups have been determined by Liebeck, Praeger, and Saxl [15], though the classification is not easily summarised (see also [8, §8.5]).

2.2.3 *The Simple Groups of Lie Type*

Most of the non-abelian finite simple groups are obtained from Lie algebras. A *Lie algebra* is a vector space L over a field K, with a binary operation $[\ ,\ ] : L \times L \to L$ which is bilinear (linear in each variable), alternating ($[x, x] = 0$ for all $x \in L$) and satisfies the Jacobi identity ($[x, [y, z]] + [y, [z, x]] + [z, [x, y]] = 0$ for all $x, y, z \in L$). The most obvious non-trivial example is the 3-dimensional euclidean space $\mathbb{R}^3$, with $[x, y]$ defined to be the vector product $x \times y$. Other important examples include the algebra $M_n(K)$ of $n \times n$ matrices over K, with $[A, B] = AB - BA$, and the tangent spaces of Lie groups (differentiable manifolds with smooth group operations); see [9] for a detailed study of Lie algebras and Lie groups.

One can define the concept of a *simple* Lie algebra, much as one does for groups and rings. The finite-dimensional simple Lie algebras over $\mathbb{C}$ were classified by Killing and Cartan. They form four infinite families A_l $(l \geq 1)$, B_l $(l \geq 2)$, C_l $(l \geq 3)$ and D_l $(l \geq 4)$, together with five exceptional algebras E_l $(l = 6, 7, 8)$, F_4 and G_2. Here the letter $A, \ldots, G$ indicates the 'shape' of the Lie algebra, in terms of its internal structure, while the subscript l (the *Lie rank*) indicates its 'size', in the sense that for each family the dimension $\dim_{\mathbb{C}} L$ is a simple monotonic function of l. See [9, 18] for details.

By replacing the field $K = \mathbb{C}$ with a finite field $\mathbb{F}_q$, one obtains finite analogues of these simple Lie algebras. With a few small exceptions, their automorphism groups each have a single non-abelian finite simple group among their composition factors; this is denoted by $A_l(q), \ldots, G_2(q)$. For example, the complex Lie algebra A_l is the subalgebra of $M_{l+1}(\mathbb{C})$ consisting of the $(l+1) \times (l+1)$ complex matrices with trace 0; it gives rise to the family of finite groups $A_l(q) = L_{l+1}(q) = \mathrm{PSL}_{l+1}(q)$, which are simple except when $l = 1$ and $q = 2$ or 3. Similarly, the Lie algebras B_l, C_l and D_l give rise to various projective orthogonal and projective symplectic simple groups, known as the classical or Chevalley groups, while those obtained from E_l, F_4 and G_2 are known as the exceptional groups of Lie type. Each of these is generated in a uniform way by l subgroups isomorphic to $A_1(q) = \mathrm{PSL}_2(q)$, with the relationships between these subgroups encoded in a graph on l vertices called a *Dynkin diagram*. See [4, 23] for details.

In addition there are the 'twisted' groups of Lie type. These are subgroups of some of the preceding groups fixed by certain automorphisms. For instance, if $l \geq 2$ then the projective special unitary group ${}^2A_l(q^2) = \mathrm{U}_{l+1}(q) = \mathrm{PSU}_{l+1}(q)$ is the subgroup of $A_l(q^2) = \mathrm{PSL}_{l+1}(q^2)$ corresponding to the group of matrices $M \in \mathrm{SL}_{l+1}(q^2)$ satisfying $M\overline{M}^t = I$, that is, fixed by the automorphism $M \mapsto (\overline{M}^t)^{-1}$, where $\overline{M}$ is obtained by applying the automorphism $x \mapsto x^q$ of order 2 of $\mathbb{F}_{q^2}$ to the entries of M. Other important examples include the Suzuki groups ${}^2B_2(q) = Sz(q)$ and the Ree groups ${}^2G_2(q) = R(q)$, subgroups of $B_2(q) = \mathrm{Sp}_4(q)$ and of $G_2(q)$ where $q = 2^e$ or 3^e respectively for odd $e \geq 3$. Again, see [4, 23] for details.

2.2.4 *The Sporadic Simple Groups*

There are 26 other finite simple groups. These are called the *sporadic groups* since they are not members of infinite families, but instead arise as individual groups or as members of small families of similar groups. Many of them are obtained from the automorphism groups of combinatorial or geometric structures such as graphs, block designs or lattices. See [1, 23] for good surveys.

The first to be discovered were the Mathieu groups M_{11} and M_{12} in 1861, followed by M_{22}, M_{23} and M_{24} in 1873. These are most easily constructed as the automorphism groups of certain block designs called Steiner systems with n points for $n = 11, 12, 22, 23$ and 24 (see [3] or [8, Chap. 6], for example), though they also arise in several other contexts such as error-correcting codes.

The remaining 21 sporadic simple groups appeared between 1966 and 1976, when Janko discovered the first and last of the four groups which bear his name. The largest of the sporadic simple groups is the Fischer-Griess Monster M, of order

$$
\begin{aligned}
& 2^{46} \cdot 3^{20} \cdot 5^9 \cdot 7^6 \cdot 11^2 \cdot 13^3 \cdot 17 \cdot 19 \cdot 23 \cdot 29 \cdot 31 \cdot 41 \cdot 47 \cdot 59 \cdot 71 \\
= \ & 808017424794512875886459904961710757005754368000000000 \\
\approx \ & 8 \cdot 10^{53}.
\end{aligned}
$$

There are mysterious connections, called 'monstrous moonshine', between this group and other areas of mathematics. For instance, the degrees of the irreducible representations of M are closely connected with the coefficients in the Fourier series for the modular function $j(\tau) = 1728J(\tau)$, and the primes p dividing $|M|$ are precisely those for which the compactified surface $\overline{N(\Gamma_0(p))\backslash\mathbb{H}}$ has genus 0. (Here $\Gamma_0(p)$ is the subgroup of the modular group $\Gamma = \mathrm{PSL}_2(\mathbb{Z})$ consisting of the elements $z \mapsto (az+b)/(cz+d)$ with $c \equiv 0 \bmod p$, and $N(\Gamma_0(p))$ denotes its normaliser in $\mathrm{PSL}_2(\mathbb{R})$.) In recent years, considerable progress has been made towards explaining these connections, but there is still much to be discovered about this group.

References

1. Aschbacher, M.: Sporadic Groups. Cambridge University Press, Cambridge (1994)
2. Bryant, R.P., Singerman, D.: Foundations of the theory of maps on surfaces with boundary. Q. J. Math. (2) **36**, 17–41 (1985)
3. Cameron, P.J., van Lint, J.H.: Designs, Graphs, Codes and Their Links. London Mathematical Society Student Texts, vol. 22. Cambridge University Press, Cambridge (1991)
4. Carter, R.W.: Simple Groups of Lie Type. Wiley, London/New York/Sydney (1972)
5. Conway, J.H., Curtis, R.T., Norton, S.P., Parker, R.A., Wilson, R.A.: ATLAS of Finite Groups. Clarendon Press, Oxford (1985)
6. Cori, R.: Un code pour les graphes planaires et ses applications. In: Astérisque, vol. 27. Société Mathématique de France, Paris (1975)
7. Cori, R., Machì, A.: Maps, hypermaps and their automorphisms: a survey, I, II, III. Expo. Math. **10**, 403–427, 429–447, 449–467 (1992)
8. Dixon, J.D., Mortimer, B.: Finite Permutation Groups. Graduate Texts in Mathematics, vol. 163. Springer, Berlin/Heidelberg/New York (1996)
9. Fulton, W., Harris, J.: Representation Theory. Springer, Berlin/Heidelberg/New York (1991)
10. Hamilton, W.R.: Account of the icosian calculus. Proc. R. Ir. Acad. **6**, 415–416 (1858). In: Halberstam, H., Ingram, R.E. (eds.) The Mathematical Papers of Sir William Rowan Hamilton, Vol. III Algebra, p. 609. Cambridge University Press, Cambridge (1967)
11. Hamilton, W.R.: Letter to John T. Graves on the icosian. In: Halberstam, H., Ingram, R.E. (eds.) The Mathematical Papers of Sir William Rowan Hamilton, Vol. III Algebra, pp. 612–625. Cambridge University Press, Cambridge (1967)
12. James, L.D.: Operations on hypermaps and outer automorphisms. Eur. J. Comb. **9**, 551–560 (1988)
13. Jones, G.A.: Graph embeddings, groups and Riemann surfaces. In: Algebraic Methods in Graph Theory I, II, Szeged 1978. Colloqium of Mathematical Society János Bolyai, vol. 25. North-Holland, Amsterdam (1981)
14. Jones, G.A., Jones, J.M., Wolfart, J.: On the regularity of maps. J. Comb. Theory Ser. B **98**, 631–636 (2008)
15. Liebeck, M., Praeger, C., Saxl, J.: A classification of the maximal subgroups of the finite alternating and symmetric groups. J. Algebra **111**, 365–383 (1987)
16. Malle, G.: Fields of definition of some three point ramified field extensions. In: Schneps, L. (ed.) The Grothendieck Theory of Dessins d'Enfants. London Mathematical Society Lecture Note Series, vol. 200, pp. 147–168. Cambridge University Press, Cambridge (1994)
17. McMullen, P., Schulte, E.: Abstract Regular Polytopes. Cambridge University Press, Cambridge (2002)
18. Serre, J.-P.: Complex Semisimple Lie Algebras. Springer, Berlin/Heidelberg/New York (1987)

19. The GAP Group: GAP – Groups, Algorithms, and Programming. Version 4.7.6 (2014). http://www.gap-system.org. Accessed 20 January 2015
20. Tutte, W.T.: What is a map? In: New Directions in the Theory of Graphs, Ann Arbor, 1971, pp. 309–325. Academic, New York (1973)
21. Voisin, C., Malgoire, J.: Cartes cellulaires. In: Cahiers Mathématiques, vol. 12. Université de Montpellier, Montpellier (1977)
22. Walsh, T.R.S.: Hypermaps versus bipartite maps. J. Comb. Theory Ser. B **18**, 155–163 (1975)
23. Wilson, R.A.: The Finite Simple Groups. Springer, London (2009)

Chapter 3
Dessins and Triangle Groups

Abstract The first section of this chapter gives a short account of uniformisation theory for Riemann surfaces, in which the main result is the extended Riemann mapping theorem. We describe the classification of cocompact Fuchsian groups, and the local and global properties of quotients of the hyperbolic plane by such groups. We discuss the inclusion relations between these groups, classified by Singerman, giving particular attention to triangle groups. These groups turn out to be very important, not only for a better understanding of maps and hypermaps, but also because certain automorphic functions related to them act as a source for all Belyĭ functions. These ideas are illustrated with Klein's quartic curve, a classic example of a Riemann surface of genus 3.

In the second section we use triangle groups to prove Grothendieck's important observation, that even a purely topological definition of a dessin as a graph embedded in a compact oriented surface leads to a unique conformal structure on this surface, so that the dessin corresponds to a Belyĭ function on the resulting Riemann surface. This shows that our previous definitions of dessins (via Belyĭ functions, embedded graphs or permutations) are in fact equivalent. Grothendieck attributed this result to Malgoire and Voisin, and in the meantime there have been many proofs, such as that given by Voevodsky and Shabat, but here we use ideas of Singerman from pre-dessins times, involving maps and triangle groups. The chapter includes an appendix explaining group presentations.

Keywords Belyĭ function • Fuchsian group • Fundamental region • Group presentation • Holomorphic structure • Hyperbolic plane • Klein's quartic curve • Riemann mapping theorem • Riemann surface • Triangle group

3.1 Uniformisation and Fuchsian Groups

This chapter gives an account of some key ideas used in the sequel. As far as the material is classical, proofs are only sketched. Further details can be found in textbooks such as [1, 3, 4, 9, 10, 14, 19].

Gareth A. Jones, J. Wolfart, *Dessins d'Enfants on Riemann Surfaces*, Springer Monographs in Mathematics, DOI 10.1007/978-3-319-24711-3_3

3.1.1 Uniformisation

Theorem 3.1 *Let X be a connected manifold. Then there is a 'universal simply connected covering' $F : Y \to X$, where Y is a simply connected manifold with the following uniqueness property: if F' is any other covering $Y' \to X$ and there exist $p \in X, q \in Y$ and $q' \in Y'$ such that $F(q) = p = F'(q')$, then there is a unique covering map $f : Y \to Y'$ such that $f(q) = q'$ making the diagram*

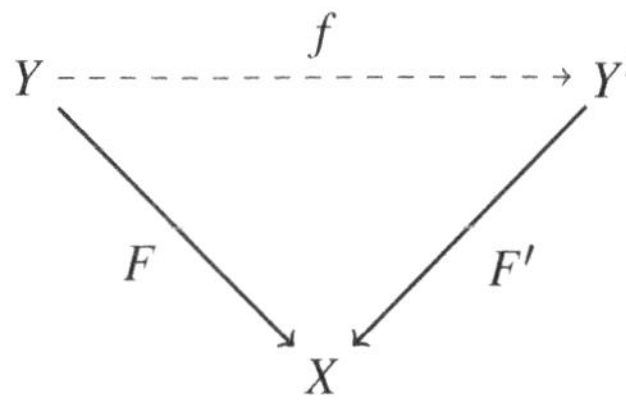

commute, that is, $F = F' \circ f$.

To say that F is a *covering* means that for each $p \in X$ there is a neighbourhood $U = U(p)$ such that $F^{-1}(U)$ is a disjoint union $\dot{\bigcup} V_n$ where each restriction $F|_{V_n} : V_n \to U$ is a homeomorphism. In the case of Riemann surfaces, we can use F to lift all charts from X to Y, so we get the following:

Proposition 3.1 *If X is a Riemann surface, its universal covering space Y is a simply connected Riemann surface, and the covering map F is locally biholomorphic.*

The construction of Y and F is given by homotopy theory. As a set,

$$Y = \{ (p, [\gamma]) \mid p \in X, [\gamma] \in \pi_1(X, p) \}$$

where the symbols $[\gamma]$ denote equivalence classes of closed curves γ modulo homotopy with base point p. One has to define a topology on Y, define F as a projection, and check the required properties, in particular simple connectedness.

Theorem 3.2 (Extended Riemann Mapping Theorem) *If Y is a simply connected Riemann surface, then Y is isomorphic to $\hat{\mathbb{C}}$, $\mathbb{C}$ or $\mathbb{H}$ ($\cong \mathbb{D}$, the open unit disc).*

This is the main theorem of uniformisation theory. Proofs can be found in many books on Riemann surfaces, for example [4]. As a consequence of both theorems and the proposition we get

Theorem 3.3 *Every Riemann surface X is isomorphic to the quotient space $\Gamma \backslash Y$ where Y is the universal covering space of X and Γ is the 'covering group', the group consisting of all 'covering transformations', those $\sigma \in \mathrm{Aut}\, Y$ with $F \circ \sigma = F$; this group permutes the fibres of F transitively.*

Outline Proof The existence of this *covering group* follows from the uniqueness part of Theorem 3.1: for any two $y, z \in Y$ in the same fibre of the universal covering map F there is a covering transformation $\sigma : Y \to Y$ with $\sigma(y) = z$ and $F \circ \sigma = F$, and by transposing y and z it is easy to see that σ is in fact invertible. The group Γ of all covering transformations σ of F acts without fixed points, it is torsion-free (has no non-identity elements of finite order), and it acts discontinuously. Here (properly) *discontinuously* means: for each $q \in Y$ there is a neighbourhood $V = V(q)$ of q with the property that $V \cap \sigma V = \emptyset$ except for finitely many $\sigma \in \Gamma$. In the case of this universal covering space, $\sigma = \mathrm{id}$ is the only exception, so Γ acts fixed point freely. □

Proposition 3.2

(a) *If* $Y = \hat{\mathbb{C}}$, *then* $\mathrm{Aut}\,\hat{\mathbb{C}} = \mathrm{PSL}_2(\mathbb{C})$. *The only covering group acting on* Y *is* $\Gamma = \{\mathrm{id}\}$, *so* $X = \Gamma\backslash Y = \hat{\mathbb{C}}$. *The genus of* X *is* 0.

(b) *If* $Y = \mathbb{C}$ *then* $\mathrm{Aut}\,\mathbb{C} = \mathrm{AGL}_1(\mathbb{C}) = \{z \mapsto az + b \,|\, a \in \mathbb{C}^*, b \in \mathbb{C}\}$. *The covering group* Γ *consists of translations and is either trivial, infinite cyclic, or a lattice* Λ, *so that* $X \cong \mathbb{C}$ *or* $\mathbb{Z}\backslash\mathbb{C}$ *or a torus (elliptic curve)* $\Lambda\backslash\mathbb{C}$, *respectively. For example in the case* $X = \mathbb{Z}\backslash\mathbb{C}$ *we have*

$$F(z) = \exp(z)\,, \quad F : \mathbb{C} \to \mathbb{C}^* = \mathbb{C} \setminus \{0\}$$

and $\exp(z + 2\pi i k) = \exp(z)$ *for all* $k \in \mathbb{Z}$. *The only compact quotients are the tori, which have genus* $g = 1$.

(c) *We have* $Y = \mathbb{H} \cong \mathbb{D}$ *in all other cases, and in particular for compact Riemann surfaces* X *if and only if they have genus* $g > 1$. *Here* Γ *is a subgroup of* $\mathrm{Aut}\,\mathbb{H} = \mathrm{PSL}_2(\mathbb{R})$ *and is called a (torsion-free) 'Fuchsian group'.*

Outline Proof

(a) Observe that any automorphism of the Riemann sphere has to be a rational function of degree 1, and that any Möbius transformation $z \mapsto (az+b)/(cz+d)$ of $\hat{\mathbb{C}}$ has fixed points.

(b) Here the proof uses Exercise 1.7 and the fact that $z \mapsto az+b$ has no fixed points only in the case $a = 1$. The other statements follow from the discontinuity of G. It is not hard to prove that all discontinuous groups of translations of $\mathbb{C}$ are isomorphic to $\{0\}$, $\mathbb{Z}$ or a lattice (see [9, Theorem 3.1.3], for example).

(c) It is clear that $\mathrm{Aut}\,\mathbb{H}$ contains $\mathrm{PSL}_2(\mathbb{R})$. Conversely, if $\gamma \in \mathrm{Aut}\,\mathbb{H}$ then by composing it with an element of $\mathrm{PSL}_2(\mathbb{R})$ (which acts transitively on $\mathbb{H}$) we may assume that γ fixes i. Conjugating with the element $z \mapsto (z-i)/(1-iz)$ of $\mathrm{PSL}_2(\mathbb{C})$, which maps $\mathbb{H}$ to $\mathbb{D}$ and i to 0, we obtain an element $\delta \in \mathrm{Aut}\,\mathbb{D}$ fixing 0. A classical function theoretic lemma by Schwarz says that if a holomorphic function $\delta : \mathbb{D} \to \mathbb{D}$ satisfies $\delta(0) = 0$, then $|\delta(z)| \leq |z|$ for all $z \in \mathbb{D}$, with equality if and only if $\delta(z) = \lambda z$ where $|\lambda| = 1$. Hence δ and its inverse mapping both satisfy $|\delta^{\pm 1}(z)| \leq |z|$, so $\delta(z) = \lambda z$ with $|\lambda| = 1$. Thus δ is an element of $\mathrm{PSL}_2(\mathbb{C})$ preserving $\mathbb{D}$, so γ is an element of $\mathrm{PSL}_2(\mathbb{C})$ preserving $\mathbb{H}$. It follows that γ also preserves the boundary $\hat{\mathbb{R}}$ of $\mathbb{H}$ in $\hat{\mathbb{C}}$, and one easily deduces that $\gamma \in \mathrm{PSL}_2(\mathbb{R})$. □

Exercise 3.1 Fill in the details omitted from parts (a) and (b) of the above proof. (The details for part (c) are more demanding.)

For compact X the covering group Γ is called a *surface group*. Note that surface groups are—again using Theorem 3.1—uniquely determined by their quotient spaces up to conjugation by (biholomorphic) automorphisms of $\mathbb{H}$. The characterisation of these different compact cases by the different genera can be proved by an application of Euler's formula to the fundamental regions to be discussed in Sect. 3.1.3.

In general, discontinuous groups Γ may have fixed points, that is, points $p \in Y$ fixed by a non-identity subgroup

$$\Gamma_p := \{\gamma \in \Gamma \mid \gamma(p) = p\} < \Gamma$$

(necessarily finite, by discontinuity). Using suitable local charts sending p to 0 and the fact that γ is biholomorphic, one can simplify the charts even further to get part (a) of the following:

Theorem 3.4 *For a discontinuous group Γ acting on $Y = \hat{\mathbb{C}}, \mathbb{C}$ or $\mathbb{H}$ the following hold:*

(a) If $p \in Y$, $\gamma \in \Gamma$ and $\gamma(p) = p$ there exists a local chart such that the diagram

$$\begin{array}{ccc} U(p) & \xrightarrow{\gamma} & U(p) \\ z \downarrow & & \downarrow z \\ \mathbb{D} & \longrightarrow & \mathbb{D} : z \longmapsto \zeta_n^k z \end{array}$$

commutes and $z \mapsto z^n : \mathbb{D} \to \mathbb{D}$ induces the quotient map $U \to \Gamma_p \backslash U$ for $\Gamma_p = \langle \gamma \rangle$.

(b) All subgroups with fixed points are finite cyclic.

(c) The points fixed by non-identity elements of Γ form a discrete subset.

(d) The quotient $\Gamma \backslash Y$ has a holomorphic structure as a Riemann surface such that the quotient map $Y \to \Gamma \backslash Y : z \mapsto \Gamma z$ is holomorphic, ramified with multiplicity n at fixed points of order n.

The other parts of the theorem follow from simple topological and function theoretic considerations. From now on we suppose that $Y = \mathbb{H}$ and Γ is a discontinuous subgroup of $\operatorname{Aut} \mathbb{H}$. Such a group is called a *Fuchsian group*.

Exercise 3.2 Prove that a non-identity automorphism of $\mathbb{H}$ has either one fixed point in $\mathbb{H}$ and none in its boundary $\hat{\mathbb{R}} = \mathbb{R} \cup \{\infty\}$, or no fixed points in $\mathbb{H}$ and either one or two in $\hat{\mathbb{R}}$. Such elements are called *elliptic*, *parabolic* and *hyperbolic*, respectively; show that parabolic and hyperbolic elements have infinite order, whereas in a Fuchsian group elliptic elements have finite order.

In addition to its analytic structure as a Riemann surface, $\mathbb{H}$ also has a geometric structure, as a model of the hyperbolic plane, that is, as a metric space with the hyperbolic metric given by

$$ds^2 = \frac{dx^2 + dy^2}{y^2} = \frac{|dz|^2}{y^2}$$

where $z = x + iy \in \mathbb{H}$. These two structures are almost equivalent, in the sense that the groups of orientation-preserving isometries and of Riemann surface automorphisms are the same (see Proposition 3.2(c)):

Theorem 3.5 *The group of orientation-preserving hyperbolic isometries of* $\mathbb{H}$ *is* $\mathrm{PSL}_2(\mathbb{R})$.

Outline Proof The hyperbolic distance between two points in $\mathbb{H}$ is the minimum of the lengths (obtained by integrating ds) of all piecewise differentiable paths between them; a straightforward calculation shows that the elements of $\mathrm{PSL}_2(\mathbb{R})$ preserve these path-lengths, so they act as hyperbolic isometries. Being holomorphic, they preserve orientation. For the converse, since $\mathrm{PSL}_2(\mathbb{R})$ acts transitively on the set of hyperbolic lines, composing any isometry γ with a suitable element $\delta \in \mathrm{PSL}_2(\mathbb{R})$ gives an isometry $\delta\gamma$ which preserves the imaginary line $L = \{z \in \mathbb{H} \mid x = 0\}$; indeed, by composing with a further element of $\mathrm{PSL}_2(\mathbb{R})$ we may assume that $\delta\gamma$ fixes L point-wise. Now each point in $\mathbb{H}$ is uniquely determined (up to reflection in L) by its distances from the points in L, so if γ (and hence $\delta\gamma$) preserves orientation then one easily deduces that $\delta\gamma$ must be the identity, giving $\gamma = \delta^{-1} \in \mathrm{PSL}_2(\mathbb{R})$. □

The great advantage of this result is that it allows one to apply hyperbolic geometry when studying Fuchsian groups. The following result gives a very useful alternative characterisation of these groups:

Theorem 3.6 *A group* $\Gamma < \mathrm{PSL}_2(\mathbb{R})$ *acts discontinuously on* $\mathbb{H}$ *if and only if* Γ *is a discrete subgroup of* $\mathrm{PSL}_2(\mathbb{R})$.

(Here discreteness refers to the quotient topology on $\mathrm{PSL}_2(\mathbb{R})$ obtained from regarding $\mathrm{SL}_2(\mathbb{R})$ as a subset of $\mathbb{R}^4$.)

Outline Proof The idea for proving *discontinuous* $\Rightarrow$ *discrete* is obvious: if Γ is not discrete, take a convergent sequence of elements and prove that for any $z \in \mathbb{H}$ the image points under this sequence must also converge.
Proving *discrete* $\Rightarrow$ *discontinuous* is less easy: suppose one has a sequence of pairwise different elements $\gamma_n \in \Gamma$ and some $z \in \mathbb{H}$ such that $\gamma_n(z)$ converges to $z_0 \in \mathbb{H}$. This point has a compact neighbourhood C such that those $\sigma \in \mathrm{PSL}_2(\mathbb{R})$ with $\sigma(C) \cap C \neq \emptyset$ form a compact subset K of $\mathrm{PSL}_2(\mathbb{R})$. For all sufficiently large n and m we have $\gamma_n\gamma_m^{-1} \in K$, so the elements γ_n have an accumulation point, contradicting the discreteness. □

Exercise 3.3 Prove that every Fuchsian group is countable.

One obvious way of using the above theorem to construct a Fuchsian group is to restrict the coefficients of Möbius transformations to a discrete subring of $\mathbb{R}$, such as $\mathbb{Z}$.

Exercise 3.4 Prove that this construction always yields a discrete (and hence discontinuous) subgroup $\Gamma < \mathrm{PSL}_2(\mathbb{R})$. Classify the discrete subrings of $\mathbb{R}$, and deduce that the only non-identity group arising in this way is the modular group $\mathrm{PSL}_2(\mathbb{Z})$.

Fortunately there are many alternative constructions of Fuchsian groups, as we shall see later in this chapter.

3.1.2 Triangle Groups

There are two general methods for the construction of Fuchsian groups:

- Arithmetic: construct discrete subgroups of $\mathrm{PSL}_2(\mathbb{R})$, such as the modular group $\Gamma = \mathrm{PSL}_2(\mathbb{Z})$, by number theory.
- Geometry (Poincaré): start with a 'suitable' hyperbolic polygon F, and generate a group Γ by side-pairing transformations. Later, F will serve as a *fundamental region* for Γ, that is, a subset of $\mathbb{H}$ with the property that $\mathring{F} \cap \gamma\mathring{F} = \emptyset$ for all $\gamma \in \Gamma \setminus \{1\}$, and $\bigcup_{\gamma\in\Gamma} \gamma\overline{F} = \mathbb{H}$; in other words, $\mathbb{H}$ is tessellated by the Γ-images of F, which overlap only at boundary points.

Example 3.1 Figure 3.1 shows a hyperbolic triangle T with internal angles $\pi/l, \pi/m, \pi/n$ satisfying

$$\frac{1}{l} + \frac{1}{m} + \frac{1}{n} < 1,$$

and a mirror image T' of it. For convenience, they are drawn in the open unit disc $\mathbb{D}$ instead of the upper half plane $\mathbb{H}$, with one vertex at the centre 0 of $\mathbb{D}$. If $l, m, n \in \mathbb{N}\setminus\{0\}$ then the *triangle group* $\Delta(l, m, n)$ with *periods* l, m, n and

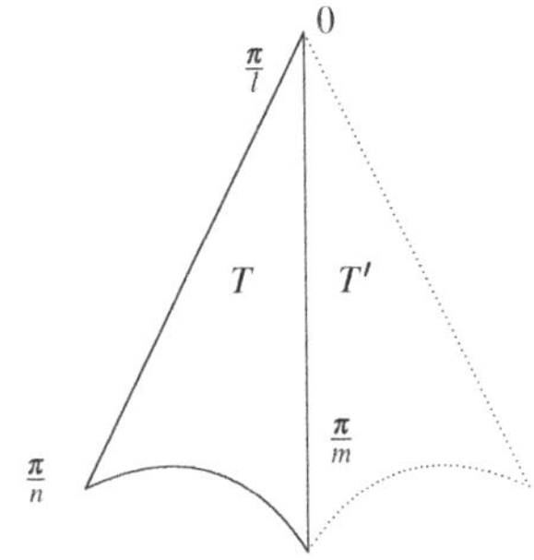

Fig. 3.1 Fundamental region for a triangle group $\Delta(l, m, n)$ acting on $\mathbb{D}$

signature (l, m, n) is the group generated by hyperbolic rotations $\gamma_0, \gamma_1, \gamma_\infty$ around the vertices of T through angles $2\pi/l, 2\pi/m, 2\pi/n$. (The use of the notation γ_∞, rather than the apparently more natural γ_2, for the third generator is motivated by future applications in which the three generators correspond to the critical values 0, 1 and ∞ of a Belyĭ function.) These generators satisfy the relations

$$\gamma_0^l = \gamma_1^m = \gamma_\infty^n = \gamma_\infty\gamma_1\gamma_0 = 1,$$

as is easily seen from the pattern of the images of the fundamental region $F = T \cup T'$ under the action of these generators on the hyperbolic plane. (We will show in the next section that these are, in fact, defining relations for $\Delta(l, m, n)$.)

Exercise 3.5 Prove that, by an appropriate change of generators, a triangle group $\Delta(l, m, n)$ may also be regarded as a triangle group $\Delta(l', m', n')$ where l', m', n' is any permutation of l, m, n.

We note for later use that one is also allowed to include limiting cases of hyperbolic triangles for which one or more of the angles is 0 and each corresponding vertex is a *cusp* on the boundary of the hyperbolic plane (a circle $\mathbb{R} \cup \{\infty\}$ or S^1 as we use $\mathbb{H}$ or $\mathbb{D}$ as a model for this geometry). In this case one has to replace the corresponding period l, m, n with ∞. The corresponding generator γ_j is then a parabolic Möbius transformation of infinite order, so one has to omit the corresponding finite order relation for this generator.

Example 3.2 As an example of this possibility, the modular group $\Gamma = \mathrm{PSL}_2(\mathbb{Z})$ (acting on $\mathbb{H}$) is a triangle group $\Delta(2, 3, \infty)$, where T has vertices at $\zeta_4 = i$, at $\zeta_3 = \omega = e^{2\pi i/3}$ and at ∞, with internal angles $\pi/2$, $\pi/3$ and 0. Taking T' to be the mirror image of T in the imaginary axis, we obtain the fundamental region

$$F = T \cup T' = \{z \in \mathbb{H} \mid -\frac{1}{2} \leq \mathrm{Re}(z) \leq \frac{1}{2}, \ |z| \geq 1\}$$

for Γ. Similarly, as an example of a triangle group $\Delta(\infty, \infty, \infty)$ one can take

$$\Gamma(2) = \{\gamma \in \Gamma = \mathrm{PSL}_2(\mathbb{Z}) \mid \gamma \equiv \pm I \bmod 2\},$$

where I denotes the 2×2 identity matrix and the congruence is read coefficient-wise. In fact, $\Gamma(2)$ is a normal subgroup of Γ, namely the principal congruence subgroup of level 2, with

$$\Gamma/\Gamma(2) \cong \mathrm{PSL}_2(\mathbb{Z}/2\mathbb{Z}) \cong S_3$$

(see Exercise 1.9). In this case one can take T to have vertices at -1, 0 and ∞. Altogether there are 85 types of triangle groups which are 'arithmetically defined' (Takeuchi [20, 21]), including all those with 'small periods', such as $\Delta(2, 3, 7)$. See Sect. 5.1.2 for more information about this topic.

Exercise 3.6 Find generating triples $\gamma_0, \gamma_1, \gamma_\infty$ for Γ and $\Gamma(2)$ corresponding to the above triangles T.

Theorem 3.7

(a) *Each hyperbolic triangle group*

$$\Delta := \Delta(l, m, n) = \langle \gamma_0, \gamma_1, \gamma_\infty \rangle$$

acts discontinuously on $\mathbb{H}$ *with the double triangle* F *as a fundamental region.*

(b) *The generators* γ_0 *and* γ_∞ *act as 'side-pairing' transformations for* F *(from which we can derive the defining relations given above).*

(c) *The quotient* $\Delta\backslash\mathbb{H}$ *is isomorphic to the Riemann sphere* $\hat{\mathbb{C}}$*; more precisely, we have*

(d) *there is a meromorphic function* $j : \mathbb{H} \to \hat{\mathbb{C}}$ *which is* Δ*-invariant, that is, it satisfies* $j(\gamma(z)) = j(z)$ *for all* $z \in \mathbb{H}$ *and* $\gamma \in \Delta$*; it maps the two open triangles in* F *biholomorphically onto* $\mathbb{H}$ *and* $-\mathbb{H}$*, the boundary edges onto* $\hat{\mathbb{R}}$*, and the vertices onto* $0, 1, \infty$ *with multiplicities* l, m, n.

Outline Proof

(b) This is true by construction, and it is rather obvious that locally a neighbourhood of F is covered by a finite number of Δ-images of F, overlapping only at boundary points. Here, the particular choice of the angles plays a key role.

(a) This is a special case of Poincaré's construction, see Theorem 3.8 below. Consider $\Delta \times \overline{F}$ as an infinite number of disjoint closed polygons, glue them together by suitable identifications on the boundaries to obtain a 2-manifold, locally looking like the situation we considered above in part (b), with a clearly discontinuous action of Δ on it and a fundamental domain $\{1\} \times F$. Prove that this manifold is in fact simply connected by writing any closed path as a sum of small closed paths (recall the proof of the global Cauchy integral theorem!) in neighbourhoods like those considered above. Then prove that the function

$$\Delta \times \overline{F} \to \mathbb{H} \ : \ (\gamma, z) \mapsto \gamma(z)$$

induces a covering map of the manifold onto $\mathbb{H}$. Since both are simply connected, this is in fact a homeomorphism, moreover compatible with the action of Δ.

(d) Extending the construction in Remark 1.1 for the modular group, one may deduce the existence of j on the left-hand open part of F from the Riemann mapping theorem; a stronger version asserts that j may be extended continuously to the boundary and normalized so that the vertices of F are mapped to $0, 1, \infty$. Then extend j to the right-hand part of F using Schwarz' reflection principle. It is possible to continue this process by applying the same idea on all boundary lines of all images $\gamma(F)$, $\gamma \in \Delta$: we obtain a well-defined holomorphic function on $\mathbb{H}$ at least outside the Δ-orbits of the vertices of F. At these points, we can continue j by applying Riemann's removable singularity theorem.

Another possibility is to construct j as the inverse function of a Schwarz mapping s from $\mathbb{H}$ onto the left-hand open triangle of F. This function s can be given explicitly as a quotient of two linearly independent solutions of some Gauss hypergeometric differential equation with rational parameters depending on l, m and n, see [24].

(c) This follows directly from (d) in the case where l, m and n are finite. Otherwise, one has to add Δ-orbits of cusps to the quotient space to get a compact Riemann surface. □

Remark 3.1

(1) In the cases

$$\frac{1}{l}+\frac{1}{m}+\frac{1}{n}=1 \quad \text{or} \quad \frac{1}{l}+\frac{1}{m}+\frac{1}{n}>1$$

one obtains euclidean or spherical triangle groups acting on the Gauss plane $\mathbb{C}$ or the Riemann sphere $\hat{\mathbb{C}}$, respectively. The arguments remain valid under the necessary changes. For example, the euclidean triangle groups

$$\Delta(2,4,4)\,, \quad \Delta(3,3,3)\,, \quad \Delta(2,3,6)$$

each contain a translation subgroup isomorphic to $\mathbb{Z}^2$, so the j-functions are special examples of elliptic functions. Among the spherical cases we meet the orientation-preserving symmetry groups

$$\Delta(2,3,3)\cong A_4\,, \quad \Delta(2,3,4)\cong S_4\,, \quad \Delta(2,3,5)\cong A_5$$

of the platonic solids. For the spherical triangle groups the j-functions are, of course, rational functions. In the case $\Delta=\Delta(2,2,n)\cong D_n$, for example, we have

$$j(z)=\frac{1}{4}\left(2+z^n+\frac{1}{z^n}\right).$$

For the other spherical triangle groups see [2].

(2) Observe that $n=\infty$ is allowed in the euclidean case $\Delta(2,2,\infty)$ (an infinite dihedral group D_∞), but not in the spherical case.

(3) For $G=\Delta(\infty,\infty,\infty)$ the j-function is just the universal covering map

$$\mathbb{H}\to\hat{\mathbb{C}}\backslash\{0,1,\infty\}\,.$$

(4) Fuchsian triangle groups are the only *rigid* Fuchsian groups, meaning that they are uniquely determined, up to conjugation in $\mathrm{PSL}_2(\mathbb{R})$, by their signature. This

is because any two triangles in $\mathbb{H}$ with the same internal angles are equivalent under an isometry.

(5) From the fact that their quotient space $\hat{\mathbb{C}}$ has a rational function field (see Exercise 1.3) we may deduce that for Fuchsian triangle groups without cusps, the field of all Δ-automorphic functions is just $\mathbb{C}(j)$.

3.1.3 More General Facts About Fuchsian Groups

We call a Fuchsian group Γ *cocompact* if $\Gamma\backslash\mathbb{H}$ is compact, or equivalently if Γ has a fundamental region $F \subset \mathbb{H}$ such that the closure $\overline{F}$ of F in $\mathbb{H}$ is compact. For instance, a triangle group is cocompact if and only if its periods are all finite. The following construction and theorem make this definition more precise.

Let Γ be a Fuchsian group, and let p be a point in $\mathbb{H}$ not fixed by any non-identity element of Γ. The *Dirichlet region* for Γ, centred at p, is the set

$$F := \{ z \in \mathbb{H} \mid d(p,z) < d(\gamma(p),z) \text{ for all } \gamma \in \Gamma \setminus \{\mathrm{id}\} \}$$

where d denotes the hyperbolic distance in $\mathbb{H}$. It is bounded by edges

$$l_\sigma := \{ z \in \mathbb{H} \mid d(p,z) \le d(\gamma(p),z) \text{ for all } \gamma \in \Gamma \setminus \{\mathrm{id}\}, \\ \text{and } d(p,z) = d(\sigma(p),z)\}$$

for various elements $\sigma \in \Gamma \setminus \{\mathrm{id}\}$.

Exercise 3.7 Prove that if p and p' are distinct points in $\mathbb{H}$ then the set

$$\{z \in \mathbb{H} \mid d(p,z) = d(p',z)\}$$

is a hyperbolic line. (This shows that the edges l_σ of a Dirichlet region are segments of hyperbolic lines.)

Theorem 3.8

(a) F is a fundamental region for Γ.

(b) For each compact $C \subset \mathbb{H}$, $l_\sigma \cap C \neq \emptyset$ for only finitely many $\sigma \in \Gamma$. In particular, $X := \Gamma\backslash\mathbb{H}$ is a compact Riemann surface if and only if $\overline{F}$ is compact, in which case F is a (hyperbolic) convex polygon bounded by finitely many sides.

(c) Γ is generated (finitely in the cocompact case) by 'side-pairing' transformations $\sigma \in \Gamma$ sending $l_{\sigma^{-1}}$ to l_σ, and sending F to a neighbour σF with a common side l_σ.

(d) Loops around vertices of F correspond to 'local' relations between these generators, and these form a presentation of Γ.

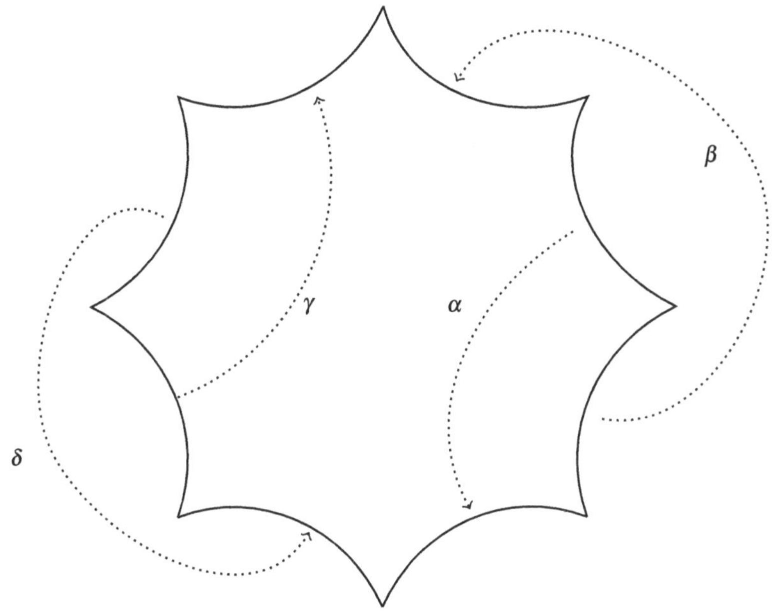

Fig. 3.2 Fundamental region for a Fuchsian group of genus 2

(e) (Poincaré) Conversely, if F is a polygon in $\mathbb{H}$ with side-pairings and some condition on its internal angles guaranteeing that locally around F the images γF do not overlap, then the side-pairing transformations generate a Fuchsian group Γ, and the images $\gamma\overline{F}$ $(\gamma \in \Gamma)$ form a tessellation of $\mathbb{H}$ in the sense that they cover $\mathbb{H}$ but their interiors γF are mutually disjoint.

(Note that parts (a)–(d) refer only to Dirichlet regions for Γ, whereas in (e) F could be any fundamental region.)

Example 3.3 Let F be an 8-sided polygon with side-pairings as indicated in Fig. 3.2 (drawn in $\mathbb{D}$ for convenience), the sum of the internal angles being 2π. These side-pairings generate a Fuchsian group with a presentation

$$\Gamma = \langle \alpha, \beta, \gamma, \delta \mid \alpha\beta\alpha^{-1}\beta^{-1}\gamma\delta\gamma^{-1}\delta^{-1} = 1 \rangle,$$

and $\Gamma \backslash \mathbb{H}$ is a compact Riemann surface of genus $g = 2$. More generally, any compact Riemann surface of genus $g \geq 2$ can be formed in this way from a suitable hyperbolic $4g$-gon, with $2g$ side-pairing generators α_i, β_i $(i = 1, \ldots, g)$ and a single defining relation

$$\alpha_1\beta_1\alpha_1^{-1}\beta_1^{-1} \ldots \alpha_g\beta_g\alpha_g^{-1}\beta_g^{-1} = 1.$$

(In the case $g = 1$ we replace $\mathbb{H}$ with $\mathbb{C}$, and use translations α_1, β_1 to pair opposite sides of a parallelogram.) We call Γ a *surface group* of genus g; it is torsion-free, and is isomorphic to the fundamental group of the surface $\Gamma\backslash\mathbb{H}$.

Example 3.4 Let Δ be a triangle group $\Delta(l, m, n)$, constructed as in Example 3.1, with the union $F = T \cup T'$ of two adjacent triangles as a fundamental region, and two of γ_0, γ_1 and γ_∞ as side-pairing transformations. Then one can use the idea in (d) to show that the relations

$$\gamma_0^l = \gamma_1^m = \gamma_\infty^n = \gamma_\infty\gamma_1\gamma_0 = 1,$$

form a set of defining relations for Δ. (Here we ignore any relation of the form $\gamma^\infty = 1$ corresponding to an infinite period in the signature of Δ.) This implies that if G is any group generated by two elements of orders dividing l and m, with a product of order dividing n, then there is an epimorphism $\theta : \Delta \to G$, so Δ has a normal subgroup $K = \ker\theta$ with quotient isomorphic to G. (Here we regard an infinite period as divisible by all periods.) Moreover, the following exercise shows that if these elements have orders exactly l, m and n then K is torsion-free. These facts will be important in later applications.

Exercise 3.8 Let $\Delta = \langle\gamma_0, \gamma_1, \gamma_\infty\rangle$ be a triangle group and let $\gamma \in \Delta$ with $\gamma \neq 1$. Prove that the following are equivalent:

1. γ has finite order;
2. γ has a fixed point in $\mathbb{H}$;
3. γ is conjugate in Δ to a power of one of the generators $\gamma_0, \gamma_1, \gamma_\infty$ of finite order.

Exercise 3.9 Show that a triangle group is perfect (that is, it has no non-trivial abelian quotients) if and only if its periods are mutually coprime integers.

Exercise 3.10 Show that for each integer $n \geq 2$ the symmetric group S_n is a quotient of the triangle group $\Delta(2, n, n-1)$.

Exercise 3.11 Show that for each integer $n \geq 2$ the group $\mathrm{PSL}_2(\mathbb{Z}/n\mathbb{Z})$ is a quotient of the triangle group $\Delta(2, 3, n)$.

Theorem 3.9

(a) *Suppose that Γ_0 is a Fuchsian group, with F_0 as a fundamental region, and that Γ_1 is a subgroup of finite index in Γ_0, with $\Gamma_0 = \bigcup_k \Gamma_1\gamma_k$. Then $F_1 := \bigcup_k \gamma_k F_0$ is a fundamental region for Γ_1.*

(b) *Let X and X' be Riemann surfaces with surface (universal covering) groups $\Gamma, \Gamma' < \mathrm{Aut}\,\mathbb{H} = \mathrm{PSL}_2(\mathbb{R})$. Then $X \cong X'$ if and only if Γ and Γ' are conjugate in $\mathrm{PSL}_2(\mathbb{R})$.*

Example 3.5 As a special case of (a), if Γ_0 is a triangle group, and as before we take F_0 to be the union of two adjacent triangles, then we obtain a triangulation of the quotient surface $X = \Gamma_1\backslash\mathbb{H}$ by identifying sides of F_1. The number of triangles is twice the index $|\Gamma_0 : \Gamma_1|$ of Γ_1 in Γ_0.

Part (b) is based on the following commutative diagram, where γ is a well-defined lift of the isomorphism if and only if it induces a conjugation $\Gamma \to \Gamma'$.

$$\begin{array}{ccc} \mathbb{H} & \overset{\gamma \in \mathrm{Aut}\,\mathbb{H}}{\longrightarrow} & \mathbb{H} \\ \downarrow & & \downarrow \\ X & \overset{\cong}{\longrightarrow} & X' \end{array}$$

Example 3.6 Unlike triangle groups, which are conjugate if and only if they have the same signatures, surface groups are not rigid. By this we mean that by varying the side-lengths and internal angles of the fundamental $4g$-gon F in Example 3.3 one can construct uncountably many conjugacy classes in $\mathrm{PSL}_2(\mathbb{R})$ of surface groups of a given genus $g \geq 2$, each corresponding by (b) to an isomorphism class of Riemann surfaces of that genus. As a real manifold the group $\mathrm{PSL}_2(\mathbb{R})$ is 3-dimensional (its elements have four real coefficients satisfying $ad - bc = 1$) so we have $6g$ real parameters (or degrees of freedom) in choosing $2g$ generators α_i, β_i; the single defining relation and conjugacy in $\mathrm{PSL}_2(\mathbb{R})$ each reduce this by 3, giving $6g - 6$ independent parameters. Locally, at least, they can serve as coordinates for a *Teichmüller space* parametrizing Riemann surfaces of genus g.

Recall that a compact Riemann surface X is a smooth projective algebraic curve given by some equations. In the case $g > 1, X = \Gamma\backslash\mathbb{H}$ for some Fuchsian group Γ. How are the equations determined by Γ, and conversely? These are important and difficult questions, to be considered later in this book. The next section represents an important first step towards an answer.

3.1.4 Triangle Groups and Belyĭ Functions

Theorem 3.10 *Suppose that X is a (compact) projective smooth algebraic curve. It has a Belyĭ function $\beta : X \to \hat{\mathbb{C}}$ if and only if there is a triangle group $\Delta = \Delta(l, m, n)$ (cocompact) and a finite index subgroup $\Gamma \leq \Delta$ such that $X \cong \Gamma\backslash\mathbb{H}$. Using this isomorphism, the Belyĭ function can be written as*

$$\Gamma\backslash\mathbb{H} \to \Delta\backslash\mathbb{H} \cong \hat{\mathbb{C}} : \Gamma z \mapsto j(z)$$

in terms of the j-function corresponding to Δ.

Proof ($\Leftarrow$) If $X \cong \Gamma\backslash\mathbb{H}$, $\Gamma \leq \Delta$, then the j-function $j : \mathbb{H} \to \hat{\mathbb{C}}$ for $\Delta = \Delta(l, m, n)$ induces a well-defined meromorphic mapping $\beta : \Gamma\backslash\mathbb{H} \to \hat{\mathbb{C}}$, $\Gamma z \mapsto j(z)$. This is ramified only at points $\Gamma p_0, \Gamma p_1, \Gamma p_\infty \in \Gamma\backslash\mathbb{H}$, where p_i is the point in $\mathbb{H}$ fixed by the standard generator γ_i of Δ, so β is a Belyĭ function; the corresponding dessin is the quotient by Γ of the Δ-tessellation of $\mathbb{H}$.

($\Rightarrow$) Given a Belyĭ function $\beta : X \to \hat{\mathbb{C}}$, let l, m and n be the least common multiples of all its multiplicities l', m' and n' above $0, 1$ and ∞ respectively.

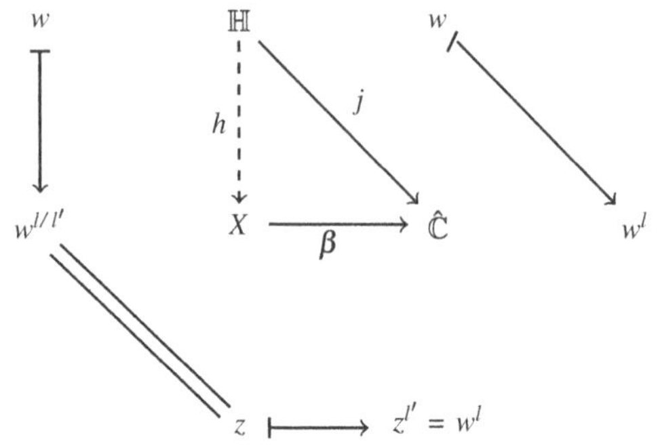

Fig. 3.3 Belyĭ function and triangle function

(Any common multiples will do equally well!) Take $\Delta = \Delta(l, m, n) < \mathrm{PSL}_2(\mathbb{R})$ and let $j : \mathbb{H} \to \hat{\mathbb{C}}$ be its j-function. Now β^{-1} is only locally biholomorphic away from 0, 1 and ∞, but $\beta^{-1} \circ j$ is everywhere locally well-defined and holomorphic, so for suitable local coordinates w and z near zeros of j and β the diagram of Fig. 3.3 commutes if β has multiplicity $l' \mid l$. Since $\mathbb{H}$ is simply connected, by the monodromy theorem $\beta^{-1} \circ j$ can be defined globally as a holomorphic map h with the property that

$$h(z) = h(z') \Rightarrow z \in \Delta z'.$$

Thus $X \cong \Gamma \backslash \mathbb{H}$, where Γ is defined as

$$\{ \gamma \in \Delta \mid h(z) = h(\gamma z) \text{ for all } z \in \mathbb{H} \} .$$

□

Remark 3.2

(1) In general, Γ is not a surface group for X, because it may have elements of finite order. However, if $l' = l$, $m' = m$ and $n' = n$ at all ramification points, that is, if β has the same multiplicity l, m or n at all points above 0, 1 or ∞, then h is the universal covering map, and Γ is the surface group for X, unique up to conjugation in $\mathrm{PSL}_2(\mathbb{R})$. In this case we say that the dessin corresponding to β is *uniform*. In particular, regular dessins are uniform.
(2) The degree $\deg \beta$ of β is equal to the index $|\Delta : \Gamma|$ of Γ in Δ.

3.1.5 Inclusions Between Fuchsian Groups

The following criteria, developed by Singerman [15] for inclusions $\Gamma_1 \leq \Gamma_0$ between Fuchsian groups, turn out to be crucial for the treatment of triangle groups and their subgroups. We restrict our attention to cocompact Fuchsian groups Γ_0.

Recall that these are of signature

$$\langle g\,;\, m_1, \dots, m_r\rangle\ ,$$

for integers $g \geq 0$ and $m_j \geq 1$, meaning that Γ_0 has generators

$$\alpha_i, \beta_i\ (i = 1, \dots, g),\ \gamma_j\ (j = 1, \dots, r)$$

and defining relations

$$\gamma_1^{m_1} = \cdots = \gamma_r^{m_r} = \prod_{i=1}^{g}[\alpha_i, \beta_i]. \prod_{j=1}^{r}\gamma_j = 1.$$

(Note that generators γ_j with $m_j = 1$ can be deleted from the presentation, but it is occasionally useful to allow them.) Here g denotes the genus of the quotient space $\Gamma_0\backslash\mathbb{H}$ and the periods $m_j > 1$ denote the orders of elliptic elements γ_j which generate a maximal system of elliptic elements in the following sense: every elliptic element $\gamma \in \Gamma_0$ is conjugate in Γ_0 to a power of precisely one such γ_j. (See Exercise 3.8 for an example of this in the case where Γ_0 is a triangle group.) The positivity of the normalized covolume

$$M(\Gamma_0)\ :=\ 2g - 2 + \sum_{j=1}^{r}\left(1 - \frac{1}{m_j}\right)\ >\ 0$$

is the only restriction on the possible signatures. In the torsion-free case, where $r = 0$, Γ_0 is a surface group of genus g. For triangle groups, where $g = 0$ and $r = 3$, we will continue to use the earlier notation $\Delta(m_1, m_2, m_3)$ for a group with signature $\langle 0\,;\, m_1, m_2, m_3\rangle$.

Subgroups Γ_1 of finite index s in Γ_0 are again cocompact. The elliptic generators of Γ_1 are powers of conjugates of the generators γ_j, so the signature of Γ_1 may be written in the form

$$\langle g_1\,;\, n_{11}, \dots, n_{1\rho_1},\ \dots\ , n_{r1}, \dots, n_{r\rho_r}\rangle\ ,$$

where the orders n_{ji} belong to generators of Γ_1 conjugate in Γ_0 to powers of γ_j. Clearly, n_{ji} divides m_j for all i and j. If Γ_1 contains no non-identity element conjugate in Γ_0 to a power of γ_j, we put $\rho_j = s/m_j$ and $n_{ji} = 1$ for all i. By [15] we have the following:

Theorem 3.11 *Suppose that Γ_0 is a cocompact Fuchsian group which has signature $\langle g\,;\, m_1, \dots, m_r\rangle$. Then Γ_0 contains a subgroup Γ_1 of finite index s and of signature*

$$\langle g_1\,;\, n_{11}, \dots, n_{1\rho_1},\ \dots\ , n_{r1}, \dots, n_{r\rho_r}\rangle$$

if and only if

(1) the Riemann-Hurwitz formula

$$M(\Gamma_1) \;=\; sM(\Gamma_0)$$

is satisfied, and

(2) there is a transitive permutation group G on s objects and an epimorphism $\Theta : \Gamma_0 \to G$ sending each elliptic generator γ_j of Γ_0 to a product of ρ_j disjoint cycles of lengths

$$\frac{m_j}{n_{j1}}, \ldots, \frac{m_j}{n_{j\rho_j}}.$$

(The cases $n_{ji} = m_j$ are counted as cycles of length 1, so that for each $j = 1, \ldots, r$ the sum of all the corresponding cycle-lengths is s.)

The subgroup Γ_1 has the form $\Theta^{-1}(H)$ where H is the subgroup of G fixing one of the s objects. It is a normal subgroup of Γ_0 if and only if G is a regular permutation group, that is, $H = \{\mathrm{id}\}$ and so $|G| = s$; in this case

$$n_{11} = \ldots = n_{1\rho_1}\,, \;\ldots\;, n_{r1} = \ldots = n_{r\rho_r}\,,$$

$$s \;=\; \rho_j \cdot \frac{m_j}{n_{ji}} \quad \textit{for all} \quad j = 1, \ldots, r\,,\; i = 1, \ldots, \rho_j\,,$$

Γ_1 is the kernel of Θ and $G \cong \Gamma_0/\Gamma_1$. The subgroup Γ_1 is torsion-free if and only if $n_{ji} = 1$ for all i and j. In particular, Γ_1 is a torsion-free normal subgroup of Γ_0 if and only if G is a regular permutation group and Θ preserves the orders of the elliptic generators of Γ_0.

Normal inclusions are particularly important, since if Γ_1 is a normal subgroup of Γ_0, then the group $G = \Gamma_0/\Gamma_1$ acts as a group of automorphisms of the Riemann surface $\Gamma_1\backslash\mathbb{H}$; see Chap. 5 for more details. The last conclusion on normal torsion-free subgroups follows from the general case through the action of G on itself by left multiplication. Then G acts on s objects as a transitive permutation group of order $s = [\Gamma_0 : \Gamma_1]$, and $\Theta(\gamma_j)$ is a product of s/m_j cycles of length m_j for each j.

The permutation group G is the *monodromy group* of the covering

$$\Gamma_1\backslash\mathbb{H} \;\to\; \Gamma_0\backslash\mathbb{H}\,.$$

One can regard it as permuting the sheets of the covering. In the special case of a triangle group $\Delta = \Gamma_0$ and a finite index subgroup $\Gamma = \Gamma_1$, this covering is a Belyĭ function, and the permutation group G can be regarded as the monodromy group of the corresponding dessin (see Sect. 2.1). Equation (2.1), for the genus of a regular dessin, is a particular case of the Riemann-Hurwitz formula, where Γ is a torsion-free normal subgroup of finite index in a triangle group.

For applications to dessins, triangle groups form the most important class of Fuchsian groups. Theorem 3.11 has the following important corollary, also due to

Singerman [16], which classifies the inclusions between triangle groups. This is particularly important, as any Fuchsian group which contains a triangle group must also be a triangle group.

Excluding the rather trivial cases of cyclic and dihedral groups, irrelevant here, Singerman lists the normal and non-normal inclusions between triangle groups. Here we will write $\Delta_1 \lhd_i \Delta_0$ or $\Delta_1 <_i \Delta_0$ to denote that Δ_1 is a normal or non-normal subgroup of index i in Δ_0.

Theorem 3.12 *The normal inclusions between hyperbolic triangle groups have the forms*

(a) $\Delta(s,s,t) \lhd_2 \Delta(2,s,2t)$ *where* $(s-2)(t-1) > 2$, *with quotient group* C_2,
(b) $\Delta(t,t,t) \lhd_3 \Delta(3,3,t)$ *where* $t > 3$, *with quotient group* C_3,
(c) $\Delta(t,t,t) \lhd_6 \Delta(2,3,2t)$ *where* $t > 3$, *with quotient group* S_3.

The non-normal inclusions between hyperbolic triangle groups have the forms

(A) $\Delta(7,7,7) <_{24} \Delta(2,3,7)$, (B) $\Delta(2,7,7) <_9 \Delta(2,3,7)$,
(C) $\Delta(3,3,7) <_8 \Delta(2,3,7)$, (D) $\Delta(4,8,8) <_{12} \Delta(2,3,8)$,
(E) $\Delta(3,8,8) <_{10} \Delta(2,3,8)$, (F) $\Delta(9,9,9) <_{12} \Delta(2,3,9)$,
(G) $\Delta(4,4,5) <_6 \Delta(2,4,5)$, (H) $\Delta(n,4n,4n) <_6 \Delta(2,3,4n)$ *with* $n \geq 2$,
(I) $\Delta(n,2n,2n) <_4 \Delta(2,4,2n)$ *with* $n \geq 3$,
(J) $\Delta(3,n,3n) <_4 \Delta(2,3,3n)$ *with* $n \geq 3$,
(K) $\Delta(2,n,2n) <_3 \Delta(2,3,2n)$ *with* $n \geq 4$.

Lower values of the parameters s,t *and* n *correspond to inclusions between euclidean or spherical triangle groups.*

Recent papers [5–7] by Girondo, Torres-Teigell and Wolfart give useful information about these inclusions and their consequences for dessins. As a simple example, suppose that $\Delta_1 = \Delta(p,q,r)$ is a subgroup of $\Delta_0 = \Delta(l,m,n)$, and that $\mathscr{D}_1$ is a dessin of type (p,q,r). Then $\mathscr{D}_1$ corresponds to a conjugacy class of subgroups $M \leq \Delta_1$. These can also be regarded as subgroups of Δ_0; they all lie in the same conjugacy class in Δ_0, and this corresponds to a dessin $\mathscr{D}_0$ of type (l,m,n). Thus every dessin of type (p,q,r) gives rise to a dessin of type (l,m,n). In fact, we have a functor between the categories of dessins of these two types, since any covering $\mathscr{D}_1 \to \overline{\mathscr{D}}_1$ between dessins of type (p,q,r) corresponds to an inclusion $M \leq \overline{M}$ between map subgroups in Δ_1 and hence in Δ_0, and this gives a covering $\mathscr{D}_0 \to \overline{\mathscr{D}}_0$ between the corresponding dessins of type (l,m,n).

Example 3.7 Inclusion (a) in Theorem 3.12 arises from mapping $\Delta_0 = \Delta(2,s,2t)$ onto $G = \Delta(2,1,2) \cong S_2$ by sending the second generator, of order s, to the identity, so the point-stabiliser Δ_1 in the resulting permutation representation of Δ_0 is a subgroup of index 2 in Δ_0. This generator has two fixed points, the first generator, of order 2, acts freely, and the third generator, of order $2t$, has one orbit of length 2, so Theorem 3.11 implies that Δ_1 has two elliptic generators of order s

and one of order $2t/2 = t$. The Riemann-Hurwitz formula

$$2g + 1 - \frac{2}{s} - \frac{1}{t} = M(\Delta_1) = 2M(\Delta_0) = 2\left(1 - \frac{1}{2} - \frac{1}{s} - \frac{1}{2t}\right)$$

shows that Δ_1 has genus $g = 0$, so it is a triangle group of type (s, s, t); having index 2, it is a normal subgroup of Δ_0. By permuting periods we can regard Δ_0 as the triangle group $\Delta(s, 2, 2t)$. The inclusion $\Delta_1 \leq \Delta_0$ then converts a dessin $\mathcal{D}_1$ of type (s, s, t), regarded as a bipartite map with black and white vertices of valency s, into a map $\mathcal{D}_0$ with vertices of valency s, and faces of the same valency $2t$, simply by ignoring the black and white colouring of the vertices. This 'forgetful functor' can, in fact, be applied to a dessin $\mathcal{D}_1$ of any type (p, q, r), by taking $s = \mathrm{lcm}(p, q)$ and $t = r$, and lifting map subgroups from $\Delta(p, q, r)$ to $\Delta(s, s, t)$ via the natural epimorphism $\Delta(s, s, t) \to \Delta(p, q, r)$.

Exercise 3.12 Find similar interpretations for the normal inclusions (b) and (c).

Example 3.8 Inclusion (K) is obtained by mapping $\Delta_0 = \Delta(2, 3, 2n)$ onto $G = \Delta(2, 3, 2) \cong S_3$, acting naturally with degree 3, so that the point-stabiliser Δ_1 is a non-normal subgroup of index 3. The first and third generators of Δ_0, of orders 2 and $2n$, each have orbits of length 1 and 2, so Δ_1 has elliptic generators of orders $2, n$ and $2n$. Regarded as an embedding of $\Delta_1 = \Delta(n, 2, 2n)$ in $\Delta_0 = \Delta(2n, 2, 3)$, this inclusion induces the stellation of the map $\mathcal{D}_1$: it adds a new vertex to each face, and joins it by an edge to each incident vertex, thus doubling the valency of each vertex and creating triangular faces. Alternatively, we can take duals and regard this inclusion as embedding $\Delta(2n, 2, n)$ in $\Delta(3, 2, 2n)$, so that it truncates maps, replacing each vertex with a small polygon having one vertex on each incident edge. Again, these functors can be applied to dessins of any type by lifting map subgroups.

Exercise 3.13 Find the permutation groups G arising as monodromy groups for the non-normal inclusions (H), (I) and (J).

The monodromy groups G in the remaining cases (A)–(G) are less obvious.

Exercise 3.14 Show that the action of S_5 by conjugation on its six Sylow 5-subgroups (equivalently, that of $\mathrm{PGL}_2(\mathbb{F}_5)$ on the projective line $\mathbb{P}^1(\mathbb{F}_5)$) gives rise to inclusion (G).

Several other inclusions are explained by the example in the next section.

3.1.6 Klein's Quartic Curve

We now give a classic example illustrating many of the topics considered in this chapter.

Klein [11] found a surface group $K \lhd \Delta = \Delta(2, 3, 7)$ of genus 3 and index $s = 168$, with quotient group $\Delta/K \cong \mathrm{PSL}_2(\mathbb{F}_7)$ acting as the automorphism

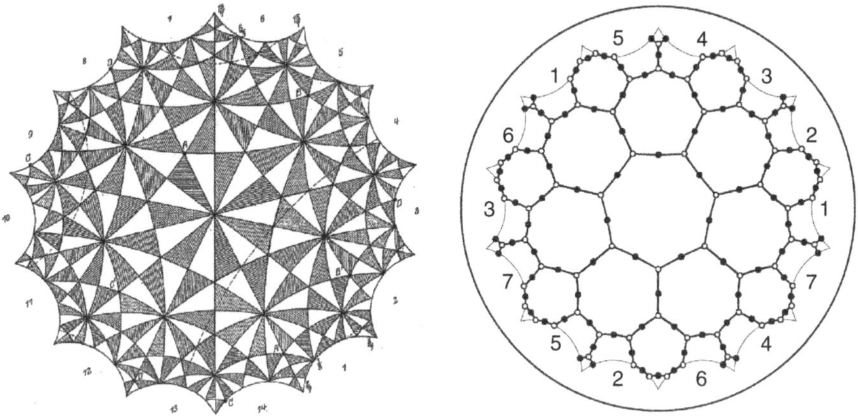

Fig. 3.4 Dessin on Klein's quartic $x^3y + y^3z + z^3x = 0$

group of the Riemann surface $X = K\backslash\mathbb{H}$. The left-hand side of Fig. 3.4 shows his original drawing of its fundamental region, tessellated with 168 fundamental regions of the triangle group Δ. (These are double triangles, as in Theorem 3.7(a).) The j-function for Δ induces on X a Belyĭ function whose dessin is drawn on the right-hand side of the figure, clearly part of the 1-skeleton of the tessellation by the Δ-triangles. The numbers on the boundary lines indicate the identifications necessary to get the surface from the fundamental domain (and induced by the side-pairing transformations generating K). As a projective algebraic curve, X is given by the quartic equation $x^3y+y^3z+z^3x = 0$. The symmetry of this equation reveals an obvious automorphism of order 3, and multiplying coordinates by appropriate 7th roots of 1 gives automorphisms of order 7 (find one!). For (much) more on this beautiful and important example, including an English translation of Klein's paper, see [13].

Exercise 3.15 The automorphisms of orders 3 and 7 mentioned above generate a subgroup of $\mathrm{Aut}\,X$. Show that this group has order 21, and is a quotient of the triangle group $\Delta(3,3,7)$. Deduce that the monodromy group corresponding to inclusion (C) in Theorem 3.12 is isomorphic to $\mathrm{PSL}_2(\mathbb{F}_7)$, acting naturally on the projective line $\mathbb{P}^1(\mathbb{F}_7)$, or equivalently acting by conjugation on its Sylow 7-subgroups.

Exercise 3.16 In Theorem 3.12, composing inclusion (C) with an instance of inclusion (b) gives another triangle group inclusion. Which inclusion is it, and which permutation group is the corresponding monodromy group?

3.2 Appendix: Group Presentations

It is often useful to describe specific examples of groups G by means of presentations $G = \langle X \mid R\rangle$. Here X is a set of elements of G, called *generators*, and R is a set of *relations*, that is, equations involving the generators. Intuitively, the elements of G are the products of powers of the generators, and two such products (often referred to as *words*) represent the same group element if and only if the relations in R, together with the group axioms, force them to be equal. We say that G is *finitely generated* or *finitely presented* if it has a presentation $\langle X \mid R\rangle$ with respectively X or both X and R finite.

For example, let G be the group defined by the presentation

$$G = \langle x \mid x^n = e\rangle,$$

where e represents the identity element x^0 (this element is often denoted by the symbol 1, or by 0 in the case of an abelian group under addition). The elements of G are the powers x^i of x, where $i \in \mathbb{Z}$. If $i \equiv j \bmod n$ then x^i and x^j represent the same element of G, since if $i = j + kn$ for some integer k then

$$x^i = x^j(x^n)^k = x^je^k = x^j.$$

Thus G consists of the elements $e\ (= x^0), x, x^2, \ldots, x^{n-1}$. To see that these are mutually distinct, let G^* be the group of rotations of the plane about the origin by multiples of $2\pi/n$. This is generated by a rotation a through $2\pi/n$ satisfying $a^n = e$. If the relation $x^n = e$ forces $x^i = x^j$ in G then by the same argument, with a replacing x, we see that the relation $a^n = e$ forces $a^i = a^j$ in G^*. However, the rotations a^i $(i = 0, \ldots, n-1)$ are all distinct, so the elements x^i $(i = 0, \ldots, n-1)$ are all distinct. (Equivalently, there is an epimorphism $G \to G^*$, $x^i \mapsto a^i$, and $|G^*| = n$, so $|G| \geq n$.) Thus G consists of n distinct elements, and is a cyclic group C_n of order n.

A less trivial example is that of a *dihedral group* D_n of order $2n$, defined by a presentation

$$G = \langle x, y \mid x^n = y^2 = (xy)^2 = e\rangle.$$

The last relation implies that $yx^i = x^{-i}y$ for all integers i, and from this one easily shows that every element can be written in the form x^iy^j where $i = 0, 1, \ldots, n-1$ and $j = 0$ or 1. To show that these $2n$ elements are all distinct one can imitate the preceding argument, representing x and y as a rotation and a reflection of the plane. Removing the relation $x^n = e$ we obtain the infinite dihedral group D_∞, with presentation

$$G = \langle x, y \mid y^2 = (xy)^2 = e\rangle.$$

Here x and y can be represented as a translation and a reflection of the plane in perpendicular axes, so that xy is a reflection. The elements of G now have the unique form $x^i y^j$ with $i \in \mathbb{Z}$ and $j = 0$ or 1.

In general, any presentation $\langle X \mid R\rangle$ defines a group $G = F/N$ where F is the free group $F(X)$ generated by X and N is the normal subgroup of F generated by the relators. We first form the *free group* $F = F(X)$ generated by X. We can define this to be the set of reduced words $x_1^{e_1} \dots x_m^{e_m}$ $(m \geq 0)$ where each $x_i \in X$, each $e_i \in \mathbb{Z} \setminus \{0\}$, and 'reduced' means that $x_i \neq x_{i+1}$ for $i = 1, \dots, m-1$. We multiply two reduced words $x_1^{e_1} \dots x_m^{e_m}$ and $y_1^{f_1} \dots y_n^{f_n}$ by first concatenating them to form a word $x_1^{e_1} \dots x_m^{e_m} y_1^{f_1} \dots y_n^{f_n}$, and then performing any required reductions: specifically, if $x_m = y_1$ we replace the subword $x_m^{e_m} y_1^{f_1}$ with $x_m^{e_m+f_1}$, removing this if $e_m + f_1 = 0$; iterating this finitely many times eventually yields a reduced word. Under this product the reduced words form a group: the empty word (with $m = 0$) is the identity element, denoted by e, and the inverse of a word $x_1^{e_1} \dots x_m^{e_m}$ is $x_m^{-e_m} \dots x_1^{-e_1}$.

To form N we write each relation in R in the form $r = e$ for some reduced word r, called a *relator*. Then N is the smallest normal subgroup of F containing all the relators, that is, the set of all products of conjugates of relators and their inverses. Finally we define G to be the quotient group F/N. Intuitively, this is the largest group generated by X and satisfying all the relations in R.

In the example $G = C_n$, for instance, $X = \{x\}$, F is the infinite cyclic group generated by x, and N is the (normal) subgroup of F generated by $r = x^n$. Another important example is the *free abelian group* generated by X, where the relations take the form $x_i x_j = x_j x_i$ for all $x_i, x_j \in X$: the elements of G are the words $x_1^{e_1} \dots x_n^{e_n}$ $(n \geq 0)$ where $x_1, \dots, x_n$ are distinct elements of X and $e_1, \dots, e_n \in \mathbb{Z}$; by commutativity, the factors $x_i^{e_i}$ can be written in any order. Alternatively, if we use additive notation then the elements of G are the linear combinations $e_1 x_1 + \dots + e_n x_n$ of elements $x_i \in X$, with integer coefficients e_i. In particular, if $|X| = n \in \mathbb{N}$ then $G \cong \mathbb{Z}^n$.

If G_1 and G_2 are groups with presentations $\langle X_1 \mid R_1\rangle$ and $\langle X_2 \mid R_2\rangle$, where (without loss of generality) X_1 and X_2 are mutually disjoint sets, then the *free product* $G_1 * G_2$ is the group with presentation $\langle X_1 \cup X_2 \mid R_1 \cup R_2\rangle$. One can show that, up to isomorphism, this group is independent of the presentations chosen for G_1 and G_2, and that for each $i = 1, 2$ the subgroup generated by X_i is isomorphic to G_i. For example, the infinite dihedral group D_∞ has a presentation $\langle y, z \mid y^2 = z^2 = e\rangle$ where $z = xy$, so

$$D_\infty = \langle y \mid y^2 = e\rangle * \langle z \mid z^2 = e\rangle \cong C_2 * C_2.$$

A less straightforward example is the modular group

$$\mathrm{PSL}_2(\mathbb{Z}) \cong \Delta(2, 3, \infty) = \langle x, y \mid x^2 = y^3 = e\rangle \cong C_2 * C_3.$$

More generally, $\Delta(p, q, \infty) \cong C_p * C_q$ for any $p, q \in \mathbb{N} \cup \{\infty\}$. The free product should not be confused with the direct product $G_1 \times G_2$, where we add the relations

$x_1x_2 = x_2x_1$ for all $x_i \in X_i$ to those of $G_1 * G_2$, so that the subgroups $\langle X_1 \rangle \cong G_1$ and $\langle X_2 \rangle \cong G_2$ commute.

In simple cases, such as those considered here, one can deduce the structure of G from the presentation, either by hand or by computer. However, there is no systematic way of doing this in general. Indeed, it is an undecidable problem to determine whether two finite presentations define isomorphic groups, or even whether a single finite presentation defines the trivial group. Similarly, there exist finitely presented groups for which the word problem is undecidable, that is, there is no algorithm to determine whether a given word in the generators represents the identity element.

3.3 From Dessins to Holomorphic Structures

Grothendieck [8] made the important observation that even a purely topological definition of a dessin as a graph embedded in a compact oriented surface leads to a unique conformal structure on this surface, so that the dessin comes from a Belyĭ function on the resulting Riemann surface. He attributed this result to Malgoire and Voisin [23], and in the meantime there have been many proofs for it, such as that given by Voevodsky and Shabat in [22], but the ideas to prove it go back to pre dessins times. The first source, written in the language of maps rather than dessins, is probably Singerman's paper [17]. In Sect. 3.3.3 we give a full proof; recall first that by Sect. 2.1.2 we can replace the topological data with the more group theoretic data of algebraic bipartite maps.

3.3.1 Coverings

Let $\mathscr{B} = (G, x, y, E)$ and $\mathscr{B}' = (G', x', y', E')$ be algebraic bipartite maps. Extending the definition of isomorphism in Sect. 2.1.4 (see Exercise 2.9), we define a *morphism* $\gamma : \mathscr{B} \to \mathscr{B}'$ or *covering* to consist of a group-homomorphism $\theta : G \to G'$ and a function $\phi : E \to E'$ such that $x \mapsto x'$ and $y \mapsto y'$ under θ, and $\phi(eg) = \phi(e)\theta(g)$ for all $e \in E$, and for $g = x, y$ (equivalently for all $g \in G$).

Example 3.9 $\mathscr{B}_1 \to \mathscr{B}_2 = C_2 \backslash \mathscr{B}_1$ in Sect. 2.1.1.

More generally, we can take $\mathscr{B} \to \mathscr{B}' := A \backslash \mathscr{B}$, where $A \leq \operatorname{Aut} \mathscr{B}$, and G', x' and y' are the actions of G, x and y on the orbits of A. Coverings induced by automorphisms in this way are *regular*, or *normal* (see Theorem 2.3).

Remark 3.3 The mappings θ and ϕ are surjective since the image of θ contains the generators of G', and since this group acts transitively on E'. The morphism γ is an isomorphism if and only if θ and ϕ are bijections, and it is an automorphism if $\mathscr{B} = \mathscr{B}'$.

The topological analogue of a morphism γ is a branched covering $X \to X'$ of surfaces, preserving orientation, with black vertices, white vertices, edges and faces on X' lifting to the same on X, and branching only at vertices or face-centres. We have a category of topological bipartite maps, and Chap. 2 describes a functor from these to algebraic bipartite maps. We can easily reverse this process, but with more work we can obtain holomorphic, rather than topological structures from algebraic bipartite maps.

3.3.2 Triangle Groups and Bipartite Maps

Consider algebraic bipartite maps of a given type (l, m, n), so in G we have $x^l = y^m = z^n = xyz = 1$. Consider the (abstract) group

$$\Delta = \Delta(l, m, n) = \langle X, Y, Z \,|\, X^l = Y^m = Z^n = XYZ = 1 \rangle .$$

Then G is a quotient of Δ by $X \mapsto x$, etc., and the composition $\Delta \to G \to \mathrm{Sym}(E)$ gives a transitive action of Δ on the edge set E of $\mathscr{B}$. We therefore have a correspondence

$$\text{bipartite maps of type } (l, m, n) \quad \longleftrightarrow \quad \text{transitive actions of } \Delta.$$

(Warning: actions of Δ can give maps of type (l', m', n') where $l'|l$, etc.) These actions correspond to conjugacy classes of subgroups $\Delta_e \leq \Delta$, namely the stabilisers of edges $e \in E$. An algebraic bipartite map $\mathscr{B}$ is finite (corresponding to a compact topological map) if and only if Δ_e has finite index in Δ. Coverings $\mathscr{B} \to \mathscr{B}'$ correspond to inclusions $\Delta_e \leq \Delta_{e'}$ (easy exercise), and regular coverings correspond to normal inclusions. The isomorphism $\mathrm{Aut}\,\mathscr{B} \cong N_\Delta(\Delta_e)/\Delta_e$ (see Theorem 2.2) gives the following:

Theorem 3.13 *$\mathscr{B}$ is regular if and only if $\Delta_e \trianglelefteq \Delta$, in which case*

$$G \cong \mathrm{Aut}\,\mathscr{B} \cong \Delta/\Delta_e.$$

Example 3.10 Let $\mathscr{B}$ correspond to the regular representation of

$$G = C_n \times C_n = \langle x, y \,|\, x^n = y^n = 1, xy = yx \rangle.$$

Then xy has order n, so the type is (n, n, n). Take $\Delta = \Delta(n, n, n)$, $\Delta_e = \ker(\Delta \to G) \trianglelefteq \Delta$. Since G is abelian, Δ_e contains the commutator subgroup Δ' of Δ. Both have index n^2 in Δ, so $\Delta_e = \Delta'$. Here $G \cong \mathrm{Aut}\,\mathscr{B} \cong \Delta/\Delta_e = \Delta^{\mathrm{ab}}$ (the abelianisation of Δ).

The triangle group of type (l, m, n) has the same presentation as Δ (generators γ_0, γ_1, γ_∞ in Sect. 3.1.2). We may therefore identify Δ with this group, X, Y and

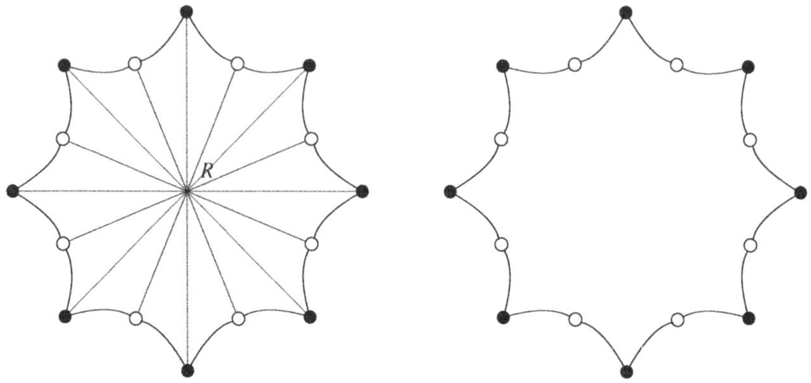

Fig. 3.5 Triangulation and dessin of genus 2

Z denoting the rotations through $2\pi/l$, $2\pi/m$ and $2\pi/n$ about the vertices of a triangle T with internal angles π/l, π/m and π/n. Assume that $\frac{1}{l}+\frac{1}{m}+\frac{1}{n}<1$ (the most typical case); if not, replace $\mathbb{H}$ with $\mathbb{C}$ or $\hat{\mathbb{C}}$. The hyperbolic plane $\mathbb{H}$ is tessellated by the images of T under the extended triangle group $\Delta[l,m,n]$ generated by reflections in the sides of T, and $\Delta=\Delta(l,m,n)$ is the even subgroup of index 2, preserving orientation. This tessellation induces a tessellation on the quotient surface corresponding to any subgroup of the triangle group Δ.

We can colour the vertices white, black or red as they are images of the vertices of T fixed by X, Y or Z. Every triangle has one vertex of each colour. Their valencies are $2l$, $2m$ and $2n$ respectively. For instance, the left hand diagram in Fig. 3.5 shows part of the triangulation induced by the extended triangle group $\Delta[2,8,8]$. For convenience it is drawn in $\mathbb{D}$, with a red vertex R at the centre 0. Identifying opposite sides of the octagon gives a quotient surface of genus 2, corresponding to a normal subgroup of index 8 in $\Delta(2,8,8)$ (see also Fig. 5.1 in Sect. 5.2.1). On the right hand side we see the corresponding dessin on this surface, obtained by omitting the red vertex—the image of R—and its incident edges: the resulting dessin has one face, one black vertex and four white vertices.

More generally, by removing the red vertices and their incident edges, one can convert the triangulation (of $\mathbb{H}$ or $\mathbb{D}$) induced by $\Delta[l,m,n]$ into a bipartite map of type (l,m,n), namely the universal bipartite map $\mathcal{B}_\infty(l,m,n)$. It is a regular map, with automorphism group $\mathrm{Aut}\,\mathcal{B}_\infty(l,m,n)=\Delta(l,m,n)$. The following theorem explains our use of the word '*universal*' in this context:

Theorem 3.14 *Every bipartite map $\mathcal{B}$ of type (l,m,n) is isomorphic to the quotient $A\backslash\mathcal{B}_\infty(l,m,n)$ of $\mathcal{B}_\infty(l,m,n)$ by a subgroup $A\leq\mathrm{Aut}\,\mathcal{B}_\infty(l,m,n)$.*

Exercise 3.17 Take A to consist of the automorphisms of $\mathcal{B}_\infty(l,m,n)$ induced by the subgroup Δ_e of Δ, and check that $\mathcal{B}\cong A\backslash\mathcal{B}_\infty(l,m,n)$.

3.3.3 Holomorphic Structures

The following theorem shows that Definition 3 of 'dessin', in Sect. 2.1.2, describes the same structures as Definitions 1 and 2 in Sect. 1.4.3 and 2.1.1:

Theorem 3.15 *Every bipartite map $\mathcal{B}$ induces on its underlying orientable compact surface a unique conformal structure, depending only on the isomorphism class of $\mathcal{B}$. The resulting Riemann surface is a Belyĭ surface.*

Proof Take the same subgroup A of Aut $\mathcal{B}_\infty(l, m, n)$ as in Exercise 3.17. Then, $A\backslash\mathbb{H}$ has the following holomorphic structure, denoted by $\mathcal{B}^{\text{hol}}$. The subgroup $\Delta_e := A$ acts as a discontinuous group of automorphisms of the Riemann surface $\mathbb{H}$ (since Δ does), so $\mathcal{B}^{\text{hol}}$ is on a Riemann surface $X = A\backslash\mathbb{H}$. Coverings $\mathcal{B} \to \mathcal{B}'$ of bipartite maps correspond to inclusions $\Delta_e \leq \Delta_{e'}$ in Δ, so these induce branched coverings $X \to X'$ of Riemann surfaces. In particular, if we take $\Delta_{e'} = \Delta$, so $|E'| = 1$ corresponding to the trivial bipartite map with one edge, we get a covering $X \to X' = \hat{\mathbb{C}}$ branched only over the vertices 0 and 1 and the face-centre at ∞. This is a Belyĭ function (provided X is compact, that is, $\mathcal{B}$ is finite).

Clearly, this construction gives a unique conformal structure for a fixed triple (l, m, n), for example if the triple describes the type of the bipartite map. But sometimes we also need descriptions of the bipartite map by means of multiples of the type, for example if we pass to quotients as in Theorem 2.3. Also in Theorem 3.1 the triangle group was not uniquely determined by the Belyĭ function. We therefore have to show that the uniqueness remains valid even if $\mathcal{B}$ is not described via a subgroup A of $\mathcal{B}_\infty(l, m, n)$, but also via a subgroup $\tilde{A}$ of $\mathcal{B}_\infty(\tilde{l}, \tilde{m}, \tilde{n})$ where $\tilde{l}, \tilde{m}$ and $\tilde{n}$ denote multiples of the valencies l, m and n, and where $\tilde{A}$ is the pre-image of A under the canonical epimorphism

$$\psi : \Delta(\tilde{l}, \tilde{m}, \tilde{n}) \to \Delta(l, m, n)$$

sending the generators $\tilde{\gamma}_0, \tilde{\gamma}_1, \tilde{\gamma}_\infty$ of $\Delta(\tilde{l}, \tilde{m}, \tilde{n})$ to the corresponding generators of $\Delta(l, m, n)$. To see the uniqueness of the conformal structure we have to prove that $\tilde{A}\backslash\mathbb{H} \cong X = A\backslash\mathbb{H}$, and this can be done by constructing a ramified covering map $\Psi : \mathbb{H} \to \mathbb{H}$ compatible with the group epimorphism ψ, or more precisely with the property that

$$\Psi(\tilde{\gamma}(z)) = \psi(\tilde{\gamma})(\Psi(z)) \quad \text{for all} \quad z \in \mathbb{H} \quad \text{and all} \quad \tilde{\gamma} \in \Delta(\tilde{l}, \tilde{m}, \tilde{n}) .$$

The construction of Ψ was in fact suggested by Klein [12]: take a hyperbolic triangle $\tilde{T}$ with internal angles $\pi/\tilde{l}$, $\pi/\tilde{m}$ and $\pi/\tilde{n}$, and another hyperbolic triangle T with internal angles π/l, π/m and π/n (fundamental domains for the respective extended triangle groups). By Riemann's mapping theorem, there is a biholomorphic mapping Ψ sending the (open) triangle $\tilde{T}$ biholomorphically to the (open) triangle T, respecting the corners. Now extend Ψ continuously to the boundary so that the angle $\pi/\tilde{l}$ is sent to the angle π/l, and so on. One can apply Schwarz's reflection principle to extend Ψ to the neighbouring triangles, and by iteration to all of $\mathbb{H}$. Because $l|\tilde{l}$, $m|\tilde{m}$ and $n|\tilde{n}$, this gives a well-defined mapping $\mathbb{H} \to \mathbb{H}$, holomorphic by Riemann's theorem about removable singularities, ramified only at the fixed points of $\Delta(\tilde{l}, \tilde{m}, \tilde{n})$ and over the fixed points of $\Delta(l, m, n)$ with orders $\tilde{l}/l$, $\tilde{m}/m$ and $\tilde{n}/n$, respectively. The uniqueness of the conformal structure is now given by the following easy exercise. □

Exercise 3.18 Verify that Ψ induces a biholomorphic map $\tilde{A}\backslash\mathbb{H} \to A\backslash\mathbb{H}$ on the quotient spaces.

Then from the direction of Belyĭ's Theorem to be proved in Chap. 4 we may draw the following conclusion:

Proposition 3.3 *If $\mathcal{B}$ is a finite algebraic map, then the Riemann surface X underlying $\mathcal{B}^{\text{hol}}$ is defined, as a smooth projective algebraic curve, over the field $\overline{\mathbb{Q}}$ of algebraic numbers.*

Example 3.11 If $\mathcal{B}$ is as in Example 3.10, the Riemann surface X uniformised by Δ', the commutator subgroup of $\Delta = \Delta(n, n, n)$, is the nth degree Fermat curve $F = F_n$ with affine equation $x^n + y^n = 1$, and with Belyĭ function $\beta : (x, y) \mapsto x^n$. The white vertices are at $v_j = (0, \zeta_n^j)$, $j = 0, 1, \ldots, n-1$, and the black vertices are at $w_k = (\zeta_n^k, 0)$, $k = 0, 1, \ldots, n-1$. The edges (given by $\beta^{-1}([0, 1])$) between vertices v_j and w_k are given by $(r\zeta_n^k, s\zeta_n^j)$ where $r, s \in [0, 1]$ and $r^n + s^n = 1$.

In general,

$$\begin{aligned}
\operatorname{Aut}\mathcal{B} \cong \operatorname{Aut}\mathcal{B}^{\text{hol}} &\cong N_\Delta(\Delta_e)/\Delta_e \\
&\le N_{\mathrm{PSL}_2(\mathbb{R})}(\Delta_e)/\Delta_e \qquad (\text{since } \Delta \le \mathrm{PSL}_2(\mathbb{R})) \\
&\cong \operatorname{Aut}X.
\end{aligned}$$

Thus automorphisms of $\mathcal{B}$ act as automorphisms of the Riemann surface X (equivalently, of the algebraic curve).

If $\mathcal{B}$ is as in this example then $\operatorname{Aut}\mathcal{B} \cong C_n \times C_n$, and this acts on X by multiplying x and y independently by nth roots of 1. In this case, $\operatorname{Aut}\mathcal{B} \neq \operatorname{Aut}X$, since $\operatorname{Aut}X$ is a semidirect product $(C_n \times C_n) \rtimes S_3$ of $\operatorname{Aut}\mathcal{B}$ by a complement S_3. The extra S_3 comes from permuting the three vertex-colours, or equivalently the three coordinates in the projective form $x^n + y^n + z^n = 0$ of X.

Exercise 3.19 Explain this example by describing $N_{\mathrm{PSL}_2(\mathbb{R})}(\Delta_e)$.

3.3.4 Non-cocompact Triangle Groups

Suppose that we want to consider all bipartite maps $\mathscr{B}$ of type $(3, 2, n)$ without restricting n. We take

$$\begin{aligned}\Delta &= \Delta(3, 2, \infty)\\ &= \langle X, Y, Z \mid X^3 = Y^2 = Z^\infty = XYZ = 1 \rangle\\ &= \langle X, Y \mid X^3 = Y^2 = 1 \rangle \qquad \text{(eliminating } Z = (XY)^{-1})\\ &\cong C_3 * C_2.\end{aligned}$$

(See Sect. 3.2 for presentations and free products.) The algebraic theory works as before. Geometrically, we take T to have a white vertex at ζ_3 (with angle $\pi/3$), a black vertex at i (with angle $\pi/2$), and a red vertex at ∞ on $\partial\mathbb{H}$ (with angle $\pi/\infty = 0$). Reflections in the sides of T generate the extended triangle group $\Delta[3, 2, \infty]$, and the images of T tessellate $\mathbb{H}$, with vertices at the images of ζ_3, i and ∞.

Exercise 3.20 Show that $\Delta[3, 2, \infty]$ is the group $\mathrm{PGL}_2(\mathbb{Z})$, consisting of the transformations

$$\tau \mapsto \frac{a\tau + b}{c\tau + d} \qquad (a, \dots, d \in \mathbb{Z},\ ad - bc = 1)$$

and

$$\tau \mapsto \frac{a\bar{\tau} + b}{c\bar{\tau} + d} \qquad (a, \dots, d \in \mathbb{Z},\ ad - bc = -1).$$

The transformations of the first type form the even subgroup $\Gamma = \mathrm{PSL}_2(\mathbb{Z})$ of index 2.

The orbit of ∞ under Γ is $\mathbb{P}^1(\mathbb{Q}) = \mathbb{Q} \cup \{\infty\}$, so this is the set of red vertices. Deleting the red vertices and their incident edges, we get a bipartite map $\mathscr{B}_\infty(3, 2, \infty)$ of type $(3, 2, \infty)$. If Δ_e is a subgroup of finite index in $\Delta = \Gamma$, then $\Delta_e \backslash \mathbb{H}$ is a compact Riemann surface minus finitely many points, one for each orbit of Δ_e on $\mathbb{P}^1(\mathbb{Q})$.

Exercise 3.21 For each prime p, the modular group $\Gamma = \mathrm{PSL}_2(\mathbb{Z}) = \Delta(3, 2, \infty)$ has a natural action as a transitive permutation group on the projective line $\mathbb{P}^1(\mathbb{F}_p) = \mathbb{F}_p \cup \{\infty\}$, via the reduction mod p: $PSL_2(\mathbb{Z}) \to \mathrm{PSL}_2(\mathbb{F}_p)$. Calculate the type and the genus of the corresponding dessin, as a function of p, and draw it for each prime $p \leq 13$.

To deal with bipartite maps $\mathscr{B}$ of all possible types, one can use the triangle group $\Delta(\infty, \infty, \infty) = \Gamma(2)$, the principal congruence subgroup of level 2 in Γ. Here T has three vertices on $\partial\mathbb{H} = \mathbb{R} \cup \{\infty\}$, at 0, 1 and ∞. The group $\Gamma(2)$ is the

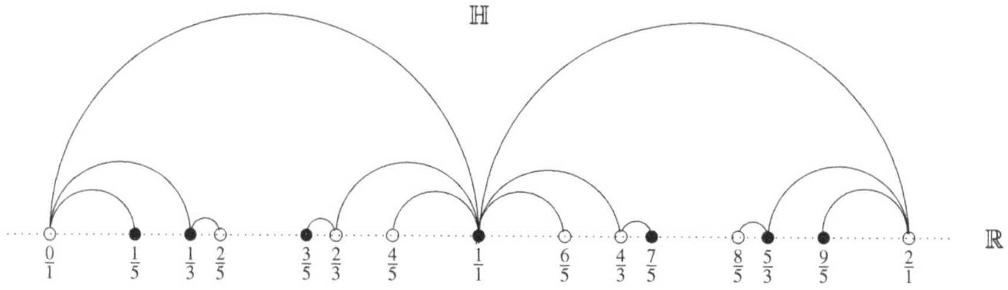

Fig. 3.6 Part of the universal bipartite map $\mathscr{B}_\infty$

even subgroup of $\Delta[\infty, \infty, \infty]$, the group generated by reflections in the sides of T. The images of T tessellate $\mathbb{H}$: the vertices are the elements $p/q \in \mathbb{P}^1(\mathbb{Q})$, coloured white, black or red as p is even and q is odd, or p and q are both odd, or p is odd and q is even (these are the orbits of $\Gamma(2)$, see Exercise 1.9). Deleting the red vertices and their incident edges gives $\mathscr{B}_\infty(\infty, \infty, \infty) = \mathscr{B}_\infty$, the universal bipartite map. Every bipartite map $\mathscr{B}$ is a quotient of $\mathscr{B}_\infty$, see also [18].

Figure 3.6 shows part of $\mathscr{B}_\infty$ for $0 \leq \operatorname{Re} z \leq 2$; this pattern repeats, in both directions, with period 2. The vertices p/q form a dense subset of the real line, so we show only those with $q \leq 5$.

Exercise 3.22 Find the Möbius transformations representing the standard generators X, Y and Z for the triangle group $\Delta(\infty, \infty, \infty)$, fixing the vertices 0, 1 and ∞ of the triangle T and satisfying $XYZ = 1$.

References

1. Beardon, A.F.: The Geometry of Discrete Groups. Springer, Berlin/Heidelberg/New York (1983)
2. Beazley Cohen, P., Itzykson, C., Wolfart, J.: Fuchsian triangle groups and Grothendieck dessins: variations on a theme of Belyi. Commun. Math. Phys. **163**, 605–627 (1994)
3. Farkas, H.M., Kra, I.: Riemann Surfaces. Springer, Berlin/Heidelberg/New York (1991)
4. Forster, O.: Lectures on Riemann Surfaces. Springer, Berlin/Heidelberg/New York (1991)
5. Girondo, E.: Multiply quasiplatonic Riemann surfaces. Exp. Math. **12**, 463–475 (2003)
6. Girondo, E., Torres-Teigell, D., Wolfart, J.: Shimura curves with many uniform dessins. Math. Z. **271**, 757–779 (2012)
7. Girondo, E., Wolfart, J.: Conjugators of Fuchsian groups and quasiplatonic surfaces. Q. J. Math. **56**, 525–540 (2005)
8. Grothendieck, A.: Esquisse d'un programme. In: Schneps, L., Lochak, P. (eds.) Geometric Galois Actions 1. Around Grothendieck's Esquisse d'un Programme. London Mathematical Society Lecture Note Series, vol. 242, pp. 5–48. Cambridge University Press, Cambridge (1997)
9. Jones, G.A., Singerman, D.: Complex Functions, an Algebraic and Geometric Viewpoint. Cambridge University Press, Cambridge (1986)
10. Jost, J.: Compact Riemann Surfaces. An Introduction to Contemporary Mathematics. Springer, Berlin/Heidelberg/New York (1997)

11. Klein, F.: Über die Transformationen siebenter ordnung der elliptischen Functionen. Math. Ann. **14**, 428–497 (1878/1879)
12. Klein, F.: Vorlesungen über die Hypergeometrische Funktion. Springer, Berlin/Heidelberg/New York (1933)
13. Levy, S. (ed.): The Eightfold Way: The Beauty of Klein's Quartic Curve. MSRI Publications, Cambridge University Press, Cambridge (1999)
14. Reyssat, E.: Quelques Aspects des Surfaces de Riemann. Birkhäuser, Boston (1989)
15. Singerman, D.: Subgroups of Fuchsian groups and finite permutation groups. Bull. Lond. Math. Soc. **2**, 319–323 (1970)
16. Singerman, D.: Finitely maximal Fuchsian groups. J. Lond. Math. Soc. (2) **6**, 29–38 (1972)
17. Singerman, D.: Automorphisms of maps, permutation groups and Riemann surfaces. Bull. Lond. Math. Soc. **8**, 65–68 (1976)
18. Singerman, D.: Universal tessellations. Rev. Mat. Complut. **1**, 111–123 (1988)
19. Springer, G.: Introduction to Riemann Surfaces. Addison-Wesley, Reading, MA (1957)
20. Takeuchi, K.: Arithmetic triangle groups. J. Math. Soc. Jpn. **29**, 29–38 (1977)
21. Takeuchi, K.: Commensurability classes of arithmetic triangle groups. J. Fac. Sci. Tokyo Sect. IA Math. **24**, 201–212 (1977)
22. Voevodsky, V.A., Shabat, G.: Equilateral triangulations of Riemann surfaces and curves over algebraic number fields. Soviet Math. Dokl. **39**, 38–41 (1989)
23. Voisin, C., Malgoire, J.: Cartes cellulaires. Cahiers Mathématiques, vol. 12. Université de Montpellier, Montpellier (1977)
24. Yoshida, M.: Fuchsian Differential Equations. Vieweg, Braunschweig/Wiesbaden (1987)

Chapter 4
Galois Actions

Abstract This chapter first collects basic material about Galois theory for finite and infinite field extensions, with examples chosen from number fields and function fields. The latter examples provide a link between Galois groups and covering groups for regular coverings. Another important example is the absolute Galois group $\mathbb{G}$, the automorphism group of the field of all algebraic numbers: as the projective limit of the (finite) Galois groups of the Galois extensions of the rationals, this is a profinite group, with a natural topology, the Krull topology, making it a topological group. Belyĭ's Theorem implies that $\mathbb{G}$ has a natural action on dessins, through its action on the algebraic numbers defining them. As observed by Grothendieck, this action is faithful, so it gives a useful insight into the Galois theory of algebraic number fields.
In the second section, moduli fields of algebraic curves are defined, and we discuss their relation to fields of definition. Weil's cocycle condition is explained. We sketch two proofs of the other direction of Belyĭ's theorem, that a curve can be defined over an algebraic number field if it admits a Belyĭ function. We list some Galois invariants and non-invariants of dessins, which are useful in determining orbits of $\mathbb{G}$, and we give a proof due to Lenstra and Schneps that $\mathbb{G}$ acts faithfully on the set of dessins formed from trees in the plane.

Keywords Absolute Galois group • Algebraic number • Field extension • Field of definition • Galois group • Galois invariant • Grothendieck-Teichmüller tower • Krull topology • Moduli field • Number field • Profinite completion • Projective limit

4.1 Galois Theory

4.1.1 Basic Galois Theory

Every field F has an algebraic closure $\overline{F}$, a minimal extension field of F over which every polynomial $f \in F[x]$ splits into linear factors. This field $\overline{F}$ is:

- unique up to isomorphisms fixing F,

Gareth A. Jones, J. Wolfart, *Dessins d'Enfants on Riemann Surfaces*, Springer Monographs in Mathematics, DOI 10.1007/978-3-319-24711-3_4

- an algebraic extension of F, that is, every $\alpha \in \overline{F}$ is a root of some non-zero $f \in F[x]$, or equivalently $|F(\alpha) : F| < \infty$.

Here, the most important case is

$$\overline{\mathbb{Q}} := \{\alpha \in \mathbb{C} \mid f(\alpha) = 0 \text{ for some non-zero } f \in \mathbb{Q}[x]\},$$

the field of algebraic numbers.

A field extension $K \geq F$ is *normal* (or *Galois*) if every embedding $e : K \hookrightarrow \overline{F}$ (fixing F element-wise) satisfies $e(K) = K$.

(Strictly speaking, "Galois" means "normal and separable", where "separable" means that irreducible polynomials do not have repeated roots; all fields of characteristic 0 are separable, so we will ignore this point by assuming that $\operatorname{char} F = 0$ for every field F mentioned.)

Example 4.1 Let $F = \mathbb{Q}$ and $K = \mathbb{Q}(\zeta_n)$, the nth cyclotomic field, where $\zeta_n = \exp(2\pi i/n)$. Any embedding $e : K \hookrightarrow \overline{\mathbb{Q}}$ sends ζ_n to some $\zeta_n^j \in K$, so $e(K) = K$. Thus this is a Galois extension.

Example 4.2 Let $F = \mathbb{Q}$ and $K = \mathbb{Q}(\alpha)$, where $\alpha = 2^{1/3} \in \mathbb{R}$. There is an embedding $e : K \hookrightarrow \overline{\mathbb{Q}}$ sending α to $\alpha\zeta_3 \notin K$, so this is not a Galois extension.

Theorem 4.1 *$K \geq F$ is a finite Galois extension if and only if K is the splitting field of some $f \in F[x]$.*

The Galois group $\operatorname{Gal} K$ of a field K is the group of all field automorphisms of K. If $H \leq \operatorname{Gal} K$, then $\operatorname{fix} H$ is the subfield fixed point-wise by H. If $F \leq K$ then $\operatorname{Gal} K/F$ is the subgroup of $\operatorname{Gal} K$ fixing F point-wise.

In Theorem 4.1, the group $G = \operatorname{Gal} K/F$ permutes the roots of f faithfully, so we can embed G in the symmetric group S_n where $n = \deg f$ is the number of roots of f, and $|G| = |K : F|$.

Example 4.3 Let $K = \mathbb{Q}(\alpha, \zeta_3)$, with $\alpha = 2^{1/3} \in \mathbb{R}$ as in Example 4.2, and with $F = \mathbb{Q}$. Then K is the splitting field of $f(x) = x^3 - 2$. The degree $|K : F|$ of this extension is 6, and we can take $1, \alpha, \alpha^2, \zeta_3, \alpha\zeta_3, \alpha^2\zeta_3$ as a basis for K over F. The three roots $\alpha_j = \alpha\zeta_3^j$ $(j = 0, 1, 2)$ of f are permuted faithfully by $G = \operatorname{Gal} K/F$, so $G \hookrightarrow S_3$. Since $|G| = |K : F| = 6$ and $|S_3| = 6$, $G \cong S_3$.

Theorem 4.2 (Fundamental Theorem of Galois Theory) *Let $K \geq F$ be a finite Galois extension, with $G = \operatorname{Gal} K/F$. There is an order-reversing bijection $L \mapsto H = \operatorname{Gal} K/L$ between fields L such that $K \geq L \geq F$, and subgroups $H \leq G$. The inverse sends each H to $L = \operatorname{fix} H$. We have $|K : L| = |H|$ and $|L : F| = |G : H|$. The subextension $L \geq F$ is Galois if and only if $H \trianglelefteq G$, in which case $\operatorname{Gal} L/F \cong G/H$.*

$$\begin{array}{ccc} K & \longleftrightarrow & 1 \\ \uparrow & & \downarrow \\ L & \longleftrightarrow & H \\ \uparrow & & \downarrow \\ F & \longleftrightarrow & G \end{array}$$

In Example 4.1,

$$\operatorname{Gal} \mathbb{Q}(\zeta_n)/\mathbb{Q} = \{\, \theta_j : \zeta_n \mapsto \zeta_n^j \mid (j,n) = 1 \,\} \cong U_n = (\mathbb{Z}/n\mathbb{Z})^* \ ,$$

the group of units mod n. This group is abelian, so all its subgroups are normal and hence all subfields of $\mathbb{Q}(\zeta_n)$ are Galois over $\mathbb{Q}$.

In Example 4.3, $S_3 \rhd A_3 \cong C_3$, and the field L corresponding to $H = A_3$ is the Galois extension $\mathbb{Q}(\zeta_3)$ of $\mathbb{Q}$. The subfield $L = \mathbb{Q}(\alpha)$ corresponds to a non-normal subgroup of G, isomorphic to C_2.

Exercise 4.1 Find the splitting field K of $x^n - 2$, describe the Galois group of K, and find the subgroups fixing $2^{1/n} \in \mathbb{R}$ and ζ_n.

4.1.2 The Absolute Galois Group

The *absolute Galois group* of a field F is $\operatorname{Gal} \overline{F}/F$, the group of automorphisms of its algebraic closure $\overline{F}$ fixing F point-wise. The *absolute Galois group* is $\operatorname{Gal} \overline{\mathbb{Q}}/\mathbb{Q}$, denoted by $\mathbb{G}$. Let $\mathscr{K}$ denote the set of all finite Galois extensions K of $\mathbb{Q}$, and let $G_K = \operatorname{Gal} K/\mathbb{Q}$, a finite group of order $|K : \mathbb{Q}|$.

Theorem 4.3

1. *$\overline{\mathbb{Q}}$ is the union of all the fields $K \in \mathscr{K}$.*
2. *Each $K \in \mathscr{K}$ is invariant under $\mathbb{G}$.*

Proof

(1) Each $K \in \mathscr{K}$ is a finite extension of $\mathbb{Q}$, so if $\alpha \in K$ then we have $|\mathbb{Q}(\alpha) : \mathbb{Q}| \leq |K : \mathbb{Q}| < \infty$, so $\alpha \in \overline{\mathbb{Q}}$. Conversely, if $\alpha \in \overline{\mathbb{Q}}$ then $f(\alpha) = 0$ for some non-zero $f \in \mathbb{Q}[x]$, and α is contained in the splitting field K of f.
(2) This follows by definition of “Galois”. □

Thus each $g \in \mathbb{G}$ is uniquely determined by its restrictions $g_K \in G_K$ to the fields $K \in \mathscr{K}$. If $K \geq L$ where $K, L \in \mathscr{K}$ then L is invariant under G_K so there is a restriction homomorphism (in fact, an epimorphism) $\rho_{K,L} : G_K \to G_L$ sending g_K to g_L, that is,

$$\rho_{K,L}(g_K) = g_L$$

whenever $K \geq L$. Conversely if we have elements $g_K \in G_K$ for each $K \in \mathscr{K}$, with $\rho_{K,L}(g_K) = g_L$ whenever $K \geq L$, we can define $g \in \mathbb{G}$ by $g(\alpha) = g_K(\alpha)$ where $\alpha \in K \in \mathscr{K}$. (Check that $g(\alpha)$ does not depend on the choice of K containing α.) We can therefore identify $\mathbb{G} = \operatorname{Gal}\overline{\mathbb{Q}}$ with the group

$$\{ (g_K) \in \Pi := \Pi_{K \in \mathscr{K}} G_K \mid \rho_{K,L}(g_K) = g_L \text{ whenever } K \geq L \text{ in } \mathscr{K} \} ,$$

the subgroup of the cartesian product Π consisting of elements whose coordinates are compatible with the epimorphisms $\rho_{K,L}$.

This is the projective limit $\varprojlim G_K$ of the finite groups G_K and epimorphisms $\rho_{K,L}$. Groups which arise as projective limits of finite groups, via systems of epimorphisms between them, are called *profinite groups*. These are very important in infinite Galois theory, where one deals with field extensions of infinite degree. This is illustrated by the following exercise and example:

Exercise 4.2 Let $K_n := \mathbb{Q}(\zeta_n)$ for each $n \in \mathbb{N}$. Show that the *maximal cyclotomic field*

$$K_\infty := \bigcup_{n \geq 1} K_n$$

is a subfield of $\overline{\mathbb{Q}}$, that K_∞ is an infinite normal extension of $\mathbb{Q}$, and that its Galois group $G_\infty := \operatorname{Gal} K_\infty/\mathbb{Q}$ is an infinite abelian group.

Example 4.4 In fact, we have $K_n \geq K_m$ if and only if m divides n, in which case the restriction mapping $\rho_{n,m}$ from $G_n := \operatorname{Gal} K_n/\mathbb{Q}$ onto G_m corresponds to the natural epimorphism $U_n \to U_m$. It follows that the Galois group G_∞ of $K_\infty/\mathbb{Q}$ is also a profinite group

$$G_\infty := \operatorname{Gal} K_\infty/\mathbb{Q} = \varprojlim G_n \cong \varprojlim U_n .$$

As in the case of $\mathbb{G}$, this group can be identified with the subgroup of $\Pi_{n \geq 1} G_n$ consisting of those elements (g_n) whose components g_n are compatible with the restriction mappings.

Exercise 4.3 Show that $\overline{\mathbb{Q}}$ is countable, whereas G_∞ and hence $\mathbb{G}$ are uncountable.

The cyclotomic fields K_n all have abelian Galois groups, and hence so do all their subfields K. The classical Kronecker-Weber Theorem states that the converse is also true: if K is a Galois extension of $\mathbb{Q}$ and $\operatorname{Gal} K/\mathbb{Q}$ is abelian, then K is a subfield of a cyclotomic field. It follows from this that G_∞ is the largest abelian quotient of $\mathbb{G}$, that is, it is the abelianisation $\mathbb{G}^{ab} = \mathbb{G}/\mathbb{G}'$ of $\mathbb{G}$, where $\mathbb{G}'$ is the commutator subgroup of $\mathbb{G}$. Similarly, the maximal cyclotomic field K_∞ is the largest abelian extension of $\mathbb{Q}$, so it is often denoted by $\mathbb{Q}^{ab}$.

This is an example of the Galois correspondence between subfields of $\overline{\mathbb{Q}}$ and subgroups of $\mathbb{G}$. However, in order to get a bijection between subfields and subgroups, as in the case of finite extensions, we need some topology:

Let us put the discrete topology on each G_K ($K \in \mathscr{K}$), so all subsets are both open and closed. This induces a product topology on the cartesian product Π, the weakest topology such that the projections $\Pi \to G_K$ are continuous. Since $\mathbb{G}$ is contained in Π, it inherits a topology from Π, the *Krull topology*. (Intuitively, elements of $\mathbb{G}$ are "close together" if they agree on a large subfield of $\overline{\mathbb{Q}}$.) Multiplication and inversion are continuous in each G_K, and hence also in Π and $\mathbb{G}$, so these are topological groups.

Exercise 4.4 Show that $\mathbb{G}$ is a closed subgroup of Π, and both Π and $\mathbb{G}$ are compact Hausdorff spaces.

Warning: $\mathbb{G}$ is topologically unpleasant. Although it is a compact topological group, its topology is not that of a Lie group, such as an orthogonal group. It is, in fact, homeomorphic to a Cantor set.

The Fundamental Theorem of Galois Theory, giving an inclusion-reversing correspondence between subgroups and subfields, applies to the extension $\overline{\mathbb{Q}}$ of $\mathbb{Q}$ provided we restrict the correspondence to the *closed* subgroups of $\mathbb{G}$, rather than applying it to all subgroups. (This restriction is also present in the finite case, but we do not notice it because in this situation the topology is discrete and so every subgroup is closed.)

For each subfield L of $\overline{\mathbb{Q}}$, define $\mathbb{G}_L = \operatorname{Gal}\overline{\mathbb{Q}}/L$. The Galois correspondence for the extension $\overline{\mathbb{Q}}/\mathbb{Q}$, the analogue of Theorem 4.2, is the following:

Theorem 4.4 *There is an order-reversing bijection $L \mapsto H = \mathbb{G}_L$ between the subfields L of $\overline{\mathbb{Q}}$ and the closed subgroups H of $\mathbb{G}$. Its inverse is given by $H \mapsto L =$ Fix H. A field L is a finite extension of $\mathbb{Q}$ if and only if the corresponding subgroup H has finite index in $\mathbb{G}$, in which case $|L : \mathbb{Q}| = |\mathbb{G} : H|$. An extension $L \geq \mathbb{Q}$ is normal if and only if the corresponding subgroup H is normal in $\mathbb{G}$, in which case $\mathbb{G}_L \cong \mathbb{G}/H$.*

More generally, if H is an arbitrary subgroup of $\mathbb{G}$, and $L = \operatorname{Fix} H$, then H is a subgroup of its closure $\overline{H} = \mathbb{G}_L$, with equality if and only if H is closed.

Exercise 4.5 Prove that in any topological group, every open subgroup is closed, and every closed subgroup of finite index is open.

It follows from Theorem 4.4 and Exercise 4.5 that the subgroups $H = \mathbb{G}_L$ of $\mathbb{G}$ corresponding to the algebraic number fields L are both open and closed, since they have finite index. In particular, this applies to the normal subgroups $\mathbb{G}_K$ corresponding to the finite Galois extension fields $K \in \mathscr{K}$, with $G_K = \operatorname{Gal} K/\mathbb{Q} \cong \mathbb{G}/\mathbb{G}_K$.

However, not every subgroup of finite index in $\mathbb{G}$ is closed:

Example 4.5 Let P be the set of prime numbers, and let K be the subfield of $\overline{\mathbb{Q}}$ obtained by adjoining $\sqrt{p}$ to $\mathbb{Q}$ for each $p \in P$. This is an infinite normal extension of $\mathbb{Q}$, and the elements of $G_K = \mathrm{Gal}\, K/\mathbb{Q}$ are the automorphisms g_Q, where $Q \subseteq P$, which send $\sqrt{p}$ to $-\sqrt{p}$ or $\sqrt{p}$ as $p \in Q$ or not. Now

$$g_Q \circ g_{Q'} = g_{Q+Q'},$$

where $Q+Q'$ denotes the symmetric difference $(Q\cup Q')\setminus(Q\cap Q')$. It follows that G_K is an abelian group of exponent 2, or equivalently a vector space over the field $\mathbb{F}_2$, isomorphic to the power set $\mathscr{P}(P)$ of P as a group under symmetric difference. Thus G_K is uncountable (by theorems of Euclid and Cantor!), so it has an uncountable basis B. (This means that every element of G_K is a finite linear combination—in this case a finite sum—of elements of B.) If $b \in B$ then $B \setminus \{b\}$ generates a subgroup of index 2 in G_K, and this lifts back under the restriction epimorphism $\rho_K : \mathbb{G} \to G_K$ to a subgroup H_b of index 2 in $\mathbb{G}$. Distinct elements $b \in B$ give rise to distinct subgroups H_b, so $\mathbb{G}$ has uncountably many subgroups of index 2. Now the *closed* subgroups of index 2 correspond to the quadratic extensions of $\mathbb{Q}$. There are only countably many of these, since they are the splitting fields of the countably many irreducible quadratic polynomials over $\mathbb{Q}$, so $\mathbb{G}$ has uncountably many subgroups of index 2 which are not closed.

Exercise 4.6 Show that the commutator subgroup $\mathbb{G}'$ of $\mathbb{G}$ is closed.

4.1.3 Coverings and Galois Groups

In the context of compact Riemann surfaces and algebraic curves, Galois theory plays an important role for the following reason. By analogy with the coverings of algebraic bipartite maps considered in Sect. 3.3.1 and with the infinite coverings of uniformisation theory in Sect. 3.1.1 we now consider non-constant holomorphic functions

$$f : Y \to X$$

between compact Riemann surfaces Y and X. Away from the ramification points f behaves like an (unramified) topological covering, that is, with the property that each non-critical point $x \in X$ has an open neighbourhood $U \subset X$ such that its pre-image $f^{-1}(U)$ is the disjoint union of open subsets $V_j \subset Y$ on which the restriction of f induces a biholomorphic mapping $V_j \to U$. We may therefore consider f as a (possibly) *ramified covering* of X by Y. Many facts already used in Sect. 1.2.5 for the special situation $f : X \to \hat{\mathbb{C}}$ can be proved for general ramified coverings, for example:

Exercise 4.7 Prove that under these hypotheses, the number $|f^{-1}(x)|$ of pre-images of x is constant away from the critical points.

Exercise 4.8 Let $\gamma : \mathscr{B} \to \mathscr{B}'$ be a covering of algebraic bipartite maps as defined in Sect. 3.3.1, and let Y and X be the compact Riemann surfaces with the holomorphic structures induced by $\mathscr{B}$ and $\mathscr{B}'$, respectively. Then γ corresponds to a holomorphic covering $f : Y \to X$.

In this context, too, there is particular interest in *regular* coverings $f : Y \to X$, those for which there is a group G of (automatically holomorphic) automorphisms of Y acting transitively on the fibres of f over the non-critical points. In other words, G is a *covering group* of f, that is, with the property that $f \circ a = f$ for all $a \in G$, but in addition it has the maximal possible size $\deg f$. Such coverings are also called *normal* or *Galois* by a reason which should now become clear:

Theorem 4.5 *Let $f : Y \to X$ be a (possibly ramified) regular holomorphic covering of compact Riemann surfaces Y and X with covering group G, and let $\mathbb{C}(Y)$ and $\mathbb{C}(X)$ be the respective fields of meromorphic functions on Y and X. We consider $\mathbb{C}(X)$ as a subfield of $\mathbb{C}(Y)$ via the embedding $g \mapsto g \circ f$. Then $\mathbb{C}(Y)$ is a Galois extension of $\mathbb{C}(X)$ with Galois group anti-isomorphic to G. The action of $a \in G$ on $\mathbb{C}(Y)$ is defined by*

$$h \mapsto h \circ a \quad .$$

(The meaning of "anti-isomorphic" will become clear in the course of the proof.)

Proof We will use and extend the ideas outlined in Sect. 1.2.5. As already explained in this special case, the subfield $\mathbb{C}(X) \leq \mathbb{C}(Y)$ for the covering $f : Y \to X$ consists of those functions meromorphic on Y and constant on the fibres of f. Each $h \in \mathbb{C}(Y)$ satisfies an algebraic equation of degree at most $n = \deg f$ over the base field $\mathbb{C}(X)$, so $\mathbb{C}(Y)$ is an algebraic extension of $\mathbb{C}(X)$ of degree at most n. In fact we have equality here as a consequence of the Riemann-Roch theorem: for every generic fibre of f there is some $h \in \mathbb{C}(Y)$ taking n pair-wise different values at its n points. This function h generates the algebraic extension $\mathbb{C}(Y)/\mathbb{C}(X)$; by the same arguments as in Sect. 1.2.5, h satisfies an algebraic equation of degree n over the ground field.

Now every automorphism of Y acts as an automorphism of $\mathbb{C}(Y)$ by $h \mapsto h \circ a =: a(h)$, and for all $a, b \in G$ we have

$$(ab)(h) = a(h \circ b) = h \circ b \circ a = h \circ (ba),$$

so we have an anti-homomorphism from G to the automorphism group of the field $\mathbb{C}(Y)$. By regularity, G acts transitively on the fibres of f, so it fixes the subfield $\mathbb{C}(X)$ element-wise and has at least n elements. By the previous section, G therefore has precisely n elements and is anti-isomorphic to $\mathrm{Gal}\,\mathbb{C}(Y)/\mathbb{C}(X)$. □

It follows that many consequences of Galois theory apply to function field extensions corresponding to regular coverings, and this applies in turn to the determination of equations for curves with regular dessins. As an example we consider cyclic coverings: they correspond to cyclic field extensions, and because the ground field contains all roots of unity, Galois theory predicts that they are generated by *pure roots*. This means the following:

Corollary 4.1 *Let $f : Y \to X$ be a cyclic covering of compact Riemann surfaces, that is, with covering group $G \cong C_n$ for some n. Then $\mathbb{C}(Y)$ is generated over $\mathbb{C}(X)$ by an element y satisfying an irreducible equation*

$$y^n = h \quad \textit{for some} \quad h \in \mathbb{C}(X) .$$

If in addition $X = \hat{\mathbb{C}}$, the covering curve Y can be described by the (affine and possibly singular) curve equation $y^n = h(x)$ where $h \in \mathbb{C}(x)$ is a rational function of x. In this model the covering map is described by $f : (x, y) \mapsto x$.

If the covering map is regular, and is given as a Belyĭ function, we can determine h by using ramification properties of f; many examples are given in Chap. 5.

4.2 Moduli Fields and Fields of Definition

The aim of this section is to give a flavour of the *"if"* part of Theorem 1.3:

"If there is a Belyĭ function β on X then X is defined over $\overline{\mathbb{Q}}$."

4.2.1 Basic Facts and Definitions

As already mentioned in Sect. 1.4.1, a field K is a *field of definition* for a compact Riemann surface X if X is isomorphic to a smooth projective algebraic curve in $\mathbb{P}^N(\mathbb{C})$ given by homogeneous equations $p_i(x_0, \dots, x_N) = 0$, with all $p_i \in K[x_0, \dots, x_N]$. If K is a field of definition, then so is any subfield of $\mathbb{C}$ containing K. Is there a minimal field of definition? Is it in $\overline{\mathbb{Q}}$?

Define

$$\mathbb{G}_{\mathbb{C}} := \text{ the group of field automorphisms of } \mathbb{C} .$$

This is a very large group: for instance, it acts transitively on the (uncountable) set of all transcendental numbers. Suppose that X is defined over K by homogenous equations $p_i(x_0, \dots, x_N) = 0$. For any $\sigma \in \mathbb{G}_{\mathbb{C}}$, let X^σ be defined by the equations $p_i^\sigma(x_0, \dots, x_N) = 0$ (apply σ to the coefficients of each p_i), that is,

$$X^\sigma := \{[\sigma(x_0), \dots, \sigma(x_N)] \mid [x_0, \dots, x_N] \in X\}.$$

This is again a smooth curve: singularities can be characterised by vanishing conditions of partial derivatives of certain rational functions formed by means of the defining polynomials, and these conditions remain invariant under the action of σ.

For the same reason, the diagram

$$\begin{array}{ccc} X & \longrightarrow & X' \\ {\scriptstyle\beta}\downarrow & & \downarrow{\scriptstyle\beta^\sigma} \\ \mathbb{P}^1(\mathbb{C}) & \longrightarrow & \mathbb{P}^1(\mathbb{C}) \end{array}$$

commutes, where the horizontal arrows indicate the action of σ on the components, and where β^σ is defined by applying σ to the coefficients of the Belyĭ function β; it remains a Belyĭ function on X^σ, because vanishing of derivatives is preserved under σ, and $\sigma(0) = 0$, $\sigma(1) = 1$, $\sigma(\infty) = \infty$. The list of all multiplicities of β is preserved under σ and the degree of β is equal to that of β^σ. This implies that σ maps the dessin $\mathscr{D}$ for β onto a dessin $\mathscr{D}^\sigma$ of the same type for β^σ on X^σ, with the same number of edges. Now $\mathbb{G}_\mathbb{C}$ acts on the set of all dessins of a given type and with a given number of edges! Translated into the language of triangle groups and their subgroups, this means that all dessins $\mathscr{D}^\sigma$ correspond to subgroups of the same finite index of some fixed triangle group. So, the orbits of $\mathbb{G}_\mathbb{C}$ on the isomorphism classes of dessins are finite, giving point (a) of the following theorem:

Theorem 4.6

(a) The subgroup $\mathbb{G}(\mathscr{D}) := \{\sigma \in \mathbb{G}_\mathbb{C} \mid \mathscr{D} \cong \mathscr{D}^\sigma\}$ *is of finite index in* $\mathbb{G}_\mathbb{C}$*, where isomorphism here means isomorphism as algebraic bipartite maps.*

(b) $\sigma \in \mathbb{G}(\mathscr{D})$ *if and only if there exist (biholomorphic) isomorphisms* $f_\sigma : X \to X^\sigma$ *for which* $\beta^\sigma \circ f_\sigma = \beta$*, that is, the following diagram commutes:*

$$\begin{array}{ccc} X & \xrightarrow{\;f_\sigma\;} & X^\sigma \\ {\scriptstyle\beta}\downarrow & \swarrow{\scriptstyle\beta^\sigma} & \\ \hat{\mathbb{C}} & & \end{array}$$

(c) The "moduli field" $M(\mathscr{D}) := \{\zeta \in \mathbb{C} \mid \sigma(\zeta) = \zeta \text{ for all } \sigma \in \mathbb{G}(\mathscr{D})\}$ *has finite degree* $[M(\mathscr{D}) : \mathbb{Q}]$ *over* $\mathbb{Q}$*, and is therefore a number field.*

Proof Point (b) follows from Theorem 3.15 and its proof. The reason for the last point is that each $\zeta \in M(\mathscr{D})$ has finite orbit length under the action of $\mathbb{G}_\mathbb{C}$, bounded by the index $|\mathbb{G}_\mathbb{C} : \mathbb{G}(\mathscr{D})|$ of $\mathbb{G}(\mathscr{D})$ in $\mathbb{G}_\mathbb{C}$, so

$$[M(\mathscr{D}) : \mathbb{Q}] \le |\mathbb{G}_\mathbb{C} : \mathbb{G}(\mathscr{D})| \,.$$

Consequently we also have

$$\mathbb{G}(X) := \{\sigma \in \mathbb{G}_{\mathbb{C}} \mid \text{there is an isomorphism } f_\sigma : X \to X^\sigma\} \geq \mathbb{G}(\mathscr{D}),$$

and it follows that the corresponding fixed field $M(X)$ of $\mathbb{G}(X)$ is contained in $M(\mathscr{D})$, so we again have a number field. □

Theorem 4.7 *The moduli field $M(X)$ depends only on the isomorphism class of X and is contained in any field of definition for X. Similarly, $M(\mathscr{D})$ is contained in any common field of definition for X and β.*

Proof Suppose that $X \cong X'$, that is, there is an isomorphism $h : X \to X'$, and suppose that $\sigma \in \mathbb{G}(X)$, so there is an isomorphism $f_\sigma : X \to X^\sigma$. We have an isomorphism $h^\sigma : X^\sigma \to X'^\sigma$, and we can construct an isomorphism to make the diagram

$$\begin{array}{ccc} X & \xrightarrow{h} & X' \\ f_\sigma \downarrow & & \downarrow \\ X^\sigma & \xrightarrow{h^\sigma} & X'^\sigma \end{array}$$

commute. Then $h^\sigma \circ f_\sigma \circ h^{-1}$ gives the isomorphism we are looking for, so $\sigma \in \mathbb{G}(X')$. The argument works in the converse direction, so we get the claim by $\mathbb{G}(X) = \mathbb{G}(X')$. □

Theorem 4.8 *$M(X)$ is a field of definition for X if X has genus $g(X) = 0$ or 1.*

Proof We have $g(X) = 0$ if and only if $X \cong \hat{\mathbb{C}} \cong \mathbb{P}^1(\mathbb{C})$, and this is clearly defined over $\mathbb{Q}$.

Similarly $g(X) = 1$ if and only if $X \cong \Lambda\backslash\mathbb{C}$, where $\Lambda = \mathbb{Z} \oplus \mathbb{Z}\tau$ for some $\tau \in \mathbb{H}$. The Weierstrass normal form shows that X is defined over $\mathbb{Q}(g_2(\tau), g_3(\tau)) \geq \mathbb{Q}(J(\tau))$, see Sect. 1.2.4, and it is well known that X can even be defined over $\mathbb{Q}(J(\tau))$, see Exercises 1.10 and 1.11. Its Galois conjugate curve X^σ is defined over $\mathbb{Q}(\sigma(g_2(\tau)), \sigma(g_3(\tau)))$, and even over $\mathbb{Q}(\sigma(J(\tau)))$, where $\sigma \in \mathbb{G}_{\mathbb{C}}$. So if $X \cong X^\sigma$, their absolute invariants (the value of the elliptic modular function on their period quotients) coincide by $\sigma(J(\tau)) = J(\tau)$. Therefore $M(X)$ is generated by $J(\tau)$, and $\mathbb{Q}(J(\tau))$ is a field of definition for X. □

However, in higher genera there are counterexamples, given by Earle [4], Shimura [27], and Dèbes and Emsalem [2], where X cannot be defined over $M(X)$.

Example 4.6 (Earle) Define $\zeta := \zeta_3 = e^{2\pi i/3}$ and let X be the curve of genus 2 with affine model

$$y^2 = x(x-\zeta)(x+\zeta)(x-\zeta^2 t)\left(x+\frac{\zeta^2}{t}\right),$$

defined over $\mathbb{Q}(\zeta)$ where $t \in \mathbb{Q},\ t > 0,\ t \neq 1$. Here we have the following situation.

(1) The curve X cannot by defined over $\mathbb{Q}$. To prove this claim, we have to use the fact that the *hyperelliptic involution* $h : (x,y) \mapsto (x,-y)$ of a hyperelliptic curve (see Example 1.4) is uniquely determined by the complex structure of the curve, and hence so are its fixed points. They are the so-called *Weierstrass points* of the curve, see for example Theorem III.7.3 in [6] (in general, Weierstrass points are defined by vanishing properties of global holomorphic differentials on the curve, see Sect. III.5 of [6]). In our case, the Weierstrass points on X are

$$(\infty,\infty),\ (0,0),\ (\zeta,0),\ (-\zeta,0),\ (\zeta^2 t,0)\ \text{and}\ \left(-\frac{\zeta^2}{t},0\right).$$

Since they are uniquely determined by the conformal structure, their images $\mathscr{P} \subset \mathbb{P}^1(\mathbb{C})$ under the quotient mapping $(x,y) \mapsto x$ (taking the quotient by the group $\{h, \mathrm{id}\}$ of automorphisms) are unique up to transformations in $\mathrm{PSL}_2(\mathbb{C})$. The left-hand side of Fig. 4.1 shows $\mathscr{P}$ without the point ∞, while the right-hand side shows the corresponding set $\overline{\mathscr{P}}$ for the complex conjugate curve $\overline{X}$. If X can be defined over $\mathbb{Q}$, then there is an anti-conformal automorphism of X, permuting the points of $\mathscr{P}$ on $\mathbb{P}^1(\mathbb{C})$: it has to permute the lines or circles containing four points of $\mathscr{P}$, and these permutations have to preserve the cross-ratios of the respective quadruples of points. A straightforward (but lengthy) calculation shows that this is impossible. We leave this calculation as the following exercise:

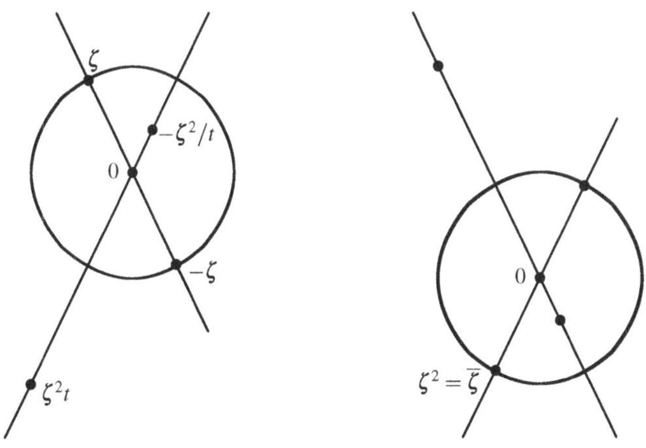

Fig. 4.1 Earle's example: the finite Weierstrass points for X and $\overline{X}$

Exercise 4.9 Verify, by calculating cross-ratios, that the quadruples $\{0,\zeta,-\zeta,\infty\}$, $\{0,\zeta^2 t,-\zeta^2/t,\infty\}$ and $\{\zeta,\zeta^2 t,-\zeta,-\zeta^2/t\}$ lie on lines or circles in $\hat{\mathbb{C}}$ but have different cross–ratios for the parameters t in question.

(2) $M(X) = \mathbb{Q} = \mathbb{R} \cap \mathbb{Q}(\zeta)$, and $X \cong \overline{X}$ because there is a holomorphic isomorphism $X \to \overline{X}$, namely the lift of $\hat{\mathbb{C}} \to \hat{\mathbb{C}} : x \mapsto -\frac{1}{x}$ to

$$X \to \overline{X} \,:\, (x,y) \mapsto (u,v) := \left(-\frac{1}{x}, \frac{iy}{x^3}\right).$$

However, we at least know the following:

Theorem 4.9 *If $M(X) < \overline{\mathbb{Q}}$, then X can be defined over a number field.*

This theorem is often attributed to Weil; proofs in quite different mathematical styles are given in [8, 13] and [19].

Outline Proof [30] Any field of definition K for X contains $M(X)$ because, for a model of X defined over K,

$$\sigma|_K = \mathrm{id} \quad\Rightarrow\quad X = X^\sigma \quad .$$

Such a field K is finitely generated over $\mathbb{Q}$, by the coefficients of the defining polynomials for X. Suppose for simplicity that $K = M(X)(\xi)$ where ξ is transcendental; then given any other transcendental number η, there exists $\sigma \in \mathbb{G}_{\mathbb{C}}$ such that $\sigma|_{M(X)} = \mathrm{id}_{M(X)}$ and $\sigma(\xi) = \eta$. Because $\sigma \in \mathbb{G}(X)$, there exists an isomorphism $f_\sigma : X \to X^\sigma$. Equations $p_i(x) = 0$ for X must be replaced with equations $p_i^\sigma(x) = 0$, where coefficients rational in ξ become coefficients rational in η. Now suppose that we replace η in f_σ with some algebraic number $\alpha \in \overline{\mathbb{Q}}$; then it can be shown that f_σ is still an isomorphism for infinitely many such "specialisations" $\alpha \in \overline{\mathbb{Q}}$. This proves the claim. □

In Sect. 4.3 we will explain another outline proof of this result, based on [21].

In general, it is very difficult to determine explicit equations for a curve from the data of a dessin, and *a fortiori* it is very difficult to determine a minimal field of definition for X or a dessin on X by direct calculation. In contrast to this situation, it is often much easier to determine fields of moduli, so the question arises whether the field of moduli is itself a field of definition. For this question, a result of Weil [29] is extremely useful:

Theorem 4.10 *Let X be defined over a finite extension L of $M := M(X)$. Then X can be defined over M itself if and only if, for each $\sigma \in \mathrm{Gal}\,\overline{M}/M$, there is an isomorphism $f_\sigma : X \to X^\sigma$ such that*

$$f_{\sigma\tau} = f_\sigma^\tau \circ f_\tau \qquad \textit{(Weil's cocycle condition)}$$

for all $\sigma, \tau \in \mathrm{Gal}\,\bar{M}/M$.

(See the beginning of Sect. 4.2.1 for the notation $f_\sigma^\tau := (f_\sigma)^\tau$ in Weil's cocycle condition.) An analogous statement holds for $M(\mathscr{D})$ and the field of definition for X and β, where the isomorphisms must also make the following diagram commute:

$$\begin{array}{ccc} X & \xrightarrow{f_\sigma} & X^\sigma \\ & {\scriptstyle\beta}\searrow \quad \swarrow{\scriptstyle\beta^\sigma} & \\ & \mathbb{P}^1(\mathbb{C}) & \end{array}$$

As a consequence, if $\operatorname{Aut} X = \{\mathrm{id}\}$ then X is defined over $M(X)$, because f_σ is unique. This is in fact the generic case for curves of genus $g > 2$. (Those of genus $g \leq 1$ have infinite automorphism groups, while those of genus 2 have finite automorphism groups of order at least 2.)

Exercise 4.10 Suppose that Weil's cocycle condition in Theorem 4.10 is satisfied by the isomorphisms f_σ, $\sigma \in \operatorname{Gal} \overline{M}/M$. Prove that $f_{\mathrm{id}} = \mathrm{id} : X \to X$.

Exercise 4.11 Show that if σ denotes complex conjugation $\hat{\mathbb{C}} \to \hat{\mathbb{C}}$ then a curve X is definable over $\mathbb{R}$ if and only if there is an isomorphism

$$f := f_\sigma : X \to \overline{X} := X^\sigma \qquad \text{with} \qquad \overline{f} \circ f = \mathrm{id} \; .$$

Exercise 4.12 Using Exercise 4.11, give an alternative proof that Earle's curve in Example 4.6 cannot be defined over $\mathbb{R}$.

4.2.2 *Galois Action: An Example and Some Invariants*

To give an explicit example of the effect of a Galois action on curves, Belyĭ functions and dessins, we return to Example 1.14 and study the effect of the nontrivial Galois action $\sigma : \sqrt[3]{2} \mapsto \zeta_3^{-1} \sqrt[3]{2}$ on it.

Example 4.7 Let X^σ be given by the Legendre model

$$y^2 = x(x-1)\left(x - \frac{\zeta_3}{\sqrt[3]{2}}\right) .$$

As in Example 1.14, the Belyĭ function is $\beta(x, y) = 4x^3(1 - x^3)$, but the dessin now has the form given in Fig. 4.2 (up to homeomorphism, as always):

As before, one has to identify opposite sides of the rectangle to get a dessin on a torus. The Belyĭ function is given by the same formula as for the original dessin because it is a polynomial in $\mathbb{Q}[x]$. The number field is hidden in the equation of the curve; closer inspection of Belyĭ's algorithm shows that the difference between the dessins comes from the different ramifications in the first step $(x, y) \mapsto x$.

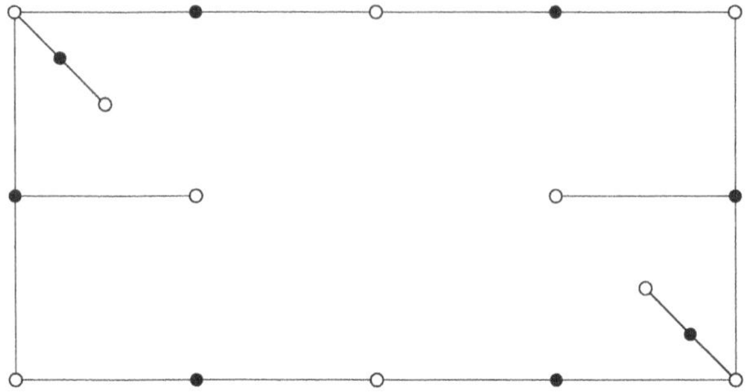

Fig. 4.2 Dessin on a torus, Galois conjugate to that in Example 1.14

A comparison of this dessin with the original dessin in Example 1.14 illustrates some of the combinatorial data invariant under Galois action—most of them already mentioned or obvious consequences of them (the genus invariance follows from Euler's formula). The following is proved in [17]:

Theorem 4.11 *Under Galois actions on dessins the following properties remain invariant:*

- *the list of valencies of white vertices, that is, the zero orders of* β*;*
- *the list of valencies of black vertices, that is, the zero orders of* $1-\beta$*;*
- *the list of valencies of faces, that is, the pole orders of* β*;*
- *the number of edges, that is, the degree of* β*;*
- *the genus of* X*;*
- *the automorphism group (up to isomorphism, of course);*
- *the monodromy group (again, up to isomorphism);*
- *regularity;*
- *uniformity.*

Example 4.8 In both of the dessins illustrated in Examples 1.14 and 4.7, the black vertices have valencies $4, 2, 2, 2, 2$ and the white vertices have valencies $6, 2, 1, 1, 1, 1$. In each case there is a single face, and the automorphism group (preserving orientation!) has order 2, being generated by a half-turn about the centre of that face. Note however that the mirror symmetry in Example 1.14 has disappeared in Example 4.7; this is the subject of the next section. One can also see that these two dessins are not isomorphic by looking at the four white vertices of valency 1: in Fig. 1.3 they are all adjacent to black vertices of valency 2, whereas in Fig. 4.2 only two of them have such neighbours.

Remark 4.1 Using GAP [28], Patrick Reichert (personal communication) was able to determine the monodromy group of these examples as an imprimitive group of degree 12 and order 576. This was confirmed by recent results of Herradón Cueto [14], who also showed that the regular covers of these dessins are defined over the field $\mathbb{Q}(\zeta_3, \sqrt[3]{2})$.

4.2.3 Non-invariants of Galois Action

In contrast to Theorem 4.11, Examples 1.14 and 4.7 show that some properties are not invariant under Galois action:

- Anticonformal symmetry: the mirror image of the dessin in Fig. 4.2 is the corresponding dessin for the complex conjugate curve; these two dessins and curves are not isomorphic, whereas the Galois conjugate dessin and curve in Example 1.14 *are* isomorphic to their complex conjugates.
- The graph: in Example 4.7 the unique black vertex of valency 4 has two white neighbours of valency 1 and one of valency 6, whereas in Example 1.14 the corresponding black vertex has two white neighbours of valencies 2 and 6.

In fact, we will show in Sect. 5.1.2 that there are also *regular* dessins for which Galois conjugation does not induce a graph isomorphism (see [18, Appendix]).

It is useful to have further examples of Galois non-invariants, since it may be possible to use them to show that various pairs of Galois conjugate dessins $\mathcal{D}$ and $\mathcal{D}'$ are non-isomorphic, even in cases where this is not apparent from diagrams. One way is to consider a word $w = w(x, y)$ in the generators x and y of the monodromy group of a dessin. (Recall that these are the permutations of the edges of the bipartite map, following the orientation around their incident white and black vertices respectively.) Any isomorphism $\mathcal{D} \to \mathcal{D}'$ must induce a bijection between the edges of $\mathcal{D}$ and those of $\mathcal{D}'$, commuting with the actions of x and y and hence of w on these two edge-sets, so the cycle-lengths of w must be the same in both cases. If, for some w, they are not, then $\mathcal{D} \not\cong \mathcal{D}'$ as required. It is important to choose words w sensibly: simple words such as x, y or xy will not work, since their cycle-lengths correspond to valencies of white or black vertices, or faces, and these are Galois invariants by Theorem 4.11; similarly, taking powers or conjugates of these words will not work. A better candidate is a more complicated word such as $w = x^i y^j$ with $i, j \neq 0$. Reading this word from left to right, one can think of w as a driving instruction: take the ith exit on the right at the next white vertex, then the jth exit at the next black vertex (with an obvious reinterpretation for negative exponents).

Exercise 4.13 By finding their cycle-lengths, show that the word $w = x^2y$ distinguishes the two dessins illustrated in Examples 1.14 and 4.7, whereas the commutator $w = [x, y] = x^{-1}y^{-1}xy$ does not.

Exercise 4.14 Can the words $w = x^i y^j$ or $[x, y]$ distinguish between non-isomorphic but complex conjugate dessins? Generalise!

4.2.4 Faithful Galois Action on Families of Dessins

Theorem 4.12 *The absolute Galois group $\mathbb{G}$ acts faithfully on the family of all (isomorphism classes of) dessins.*

This statement means that for each $\sigma \in \mathbb{G}$, $\sigma \neq \mathrm{id}$, there is a dessin, in other words a pair (X, β) consisting of an algebraic curve X defined over a number field and a Belyĭ function β on it, with the property that (X^σ, β^σ) is not isomorphic to (X, β).

Probably the easiest *proof* of this statement can be given for dessins of genus 1: suppose that $\sigma \in \mathbb{G}$ sends $J \in \overline{\mathbb{Q}}$ to $J^\sigma \neq J$, and let X be one of the elliptic curves in Exercise 1.11 with invariant J. Any of the algorithms given in Sect. 1.4.4 produces a Belyĭ function β on X also defined over $\mathbb{Q}(J)$, so σ acts in a nontrivial way on genus 1 dessins. □

Girondo and González-Diez [7] have extended this argument to show that $\mathbb{G}$ acts faithfully on dessins for hyperelliptic curves of each genus $g \geq 2$. But even for dessins on the Riemann sphere—where the curve remains invariant under $\mathbb{G}$—we have a faithful action; a sophisticated proof following ideas of Lenstra and written up by Schneps [23] shows moreover:

Theorem 4.13 *The absolute Galois group acts faithfully on the family of dessins given by trees on $\hat{\mathbb{C}}$.*

Lemma 4.1 *Let F be a complex polynomial of degree $n > 0$, let d divide n, let H be a polynomial of degree d with leading coefficient 1 and $H(0) = 0$, and let $G \in \mathbb{C}[z]$ with $F = G \circ H$. Then H and G are uniquely determined.*

Proof Clearly we have $m := \deg G = n/d$ with

$$G(z) = \lambda_m z^m + \ldots + \lambda_0 \quad \text{and} \quad H(z) = z^d + h_{d-1} z^{d-1} + \ldots + h_1 z\,,$$

so

$$F(z) = \lambda_m H^m + \lambda_{m-1} H^{m-1} + \ldots + \lambda_0 = a_n z^n + a_{n-1} z^{n-1} + \ldots + a_0$$

where the coefficients a_j are linear functions of the coefficients λ_k and polynomial functions of the coefficients h_i. Conversely, each coefficient λ_k or h_i can be calculated from the coefficients a_j, since $a_n, a_{n-1}, \ldots, a_{n-d+1}$ uniquely determine λ_m and the coefficients of H by comparison of coefficients:

$$\lambda_m = a_n\,, \quad m h_{d-1} \lambda_m = a_{n-1}\,, \quad \text{and so on.}$$

Similarly, $h_{d-2}, h_{d-3}, \ldots, h_1$ can be recursively calculated from the coefficients of F because a_{n-i} is linear in h_{d-i} and a polynomial in the other coefficients h_{d-i+j}, and likewise $\lambda_{m-1}, \ldots, \lambda_0 = a_0$ can be calculated from the coefficients of F. □

Lemma 4.2 *Let $G, H, \tilde{G}$ and $\tilde{H}$ be polynomials over $\mathbb{C}$ with $\deg H = \deg \tilde{H}$ such that $G \circ H = \tilde{G} \circ \tilde{H}$. Then there are $c, d \in \mathbb{C}$ such that*

$$\tilde{H} \; = \; cH + d.$$

Proof Let μ and $\tilde{\mu}$ be the leading coefficients of H and $\tilde{H}$, and let $\nu = \frac{1}{\mu}H(0)$ and $\tilde{\nu} = \frac{1}{\tilde{\mu}}\tilde{H}(0)$. Then there are polynomials G_1 and G_2 with

$$G \circ H = G_1 \circ \left(\frac{1}{\mu}H - \nu\right) = \tilde{G} \circ \tilde{H} = G_2 \circ \left(\frac{1}{\tilde{\mu}}\tilde{H} - \tilde{\nu}\right).$$

By Lemma 4.1 we have $\frac{1}{\mu}H - \nu = \frac{1}{\tilde{\mu}}\tilde{H} - \tilde{\nu}$, as required. □

Proof of Theorem 4.13 If the graph of the dessin $\mathscr{D}$ on $\hat{\mathbb{C}}$ is a tree, its Belyĭ function β has only one pole—the face centre—and we can suppose this is at ∞, in other words that β is a polynomial. Then each Galois conjugate β^σ of β is also a polynomial.

Now suppose that $\sigma \in \mathbb{G}(\mathscr{D})$, that is, σ is a field automorphism of $\mathbb{C}$ such that there is a (Riemann surface) isomorphism $f_\sigma : \hat{\mathbb{C}} \to \hat{\mathbb{C}}$ satisfying $\beta^\sigma \circ f_\sigma = \beta$. This isomorphism must fix ∞, so $f_\sigma^{-1}(z) = az + b$ for some $a, b \in \mathbb{C}$, $a \neq 0$, see Exercise 1.7. To prove the theorem it is therefore sufficient to show that for each $\sigma \neq \mathrm{id}$ there is a polynomial β with

$$\beta^\sigma(z) \neq \beta(az + b) \quad \text{for all} \quad a, b \in \mathbb{C}\,.$$

To do this, take some $\alpha \in \overline{\mathbb{Q}}$ such that $\gamma := \sigma(\alpha) \neq \alpha$, and some $f_\alpha \in \mathbb{Q}(\alpha)[z]$ such that

$$f_\alpha'(z) = z^3(z-1)^2(z-\alpha)\,.$$

Belyĭ's algorithm produces a polynomial $f \in \mathbb{Q}[z]$ such that $\beta := f \circ f_\alpha$ is a Belyĭ function—and is itself a polynomial. Moreover,

$$\beta^\sigma = f \circ f_{\sigma(\alpha)} = f \circ f_\gamma\,.$$

If there were $a, b \in \mathbb{C}$, $a \neq 0$ with $\beta^\sigma(z) = \beta(az + b)$, we would have

$$f \circ f_\alpha(az + b) = f \circ f_\gamma(z)\,,$$

so that $f_\alpha(az+b) = cf_\gamma(z) + d$ by Lemma 4.2. To show that this is impossible, look at the critical points on the right-hand side (where $cf_\gamma'(z) = 0$): they are 0, 1 and γ, with multiplicities $4, 3$ and 2. They have to coincide with the critical points on the left-hand side. However, the critical points of f_α can be calculated in the same way, giving a system of conditions

$$a \cdot 0 + b = 0$$

$$a \cdot 1 + b = 1$$

$$a \cdot \gamma + b = \alpha$$

which can be satisfied only for $b = 0$, $a = 1$, $\alpha = \gamma$, a contradiction. □

Remark 4.2 González-Diez and Jaikin-Zapirain [9] have recently answered an old open question (see also the proof by Guillot in [11, Sect. 5] and a different access by Bauer, Catanese and the late Fritz Grunewald in [1]): they have proved that $\mathbb{G}$ also acts faithfully on *regular* dessins, and even on their underlying surfaces, called quasiplatonic surfaces. (These dessins and surfaces form the subject-matter of the second part of this book.) A proof of this important theorem, involving some delicate results from profinite group theory, would be beyond the scope of an introductory book such as this one. We will simply note that a major ingredient in the proof is the transfer, from the context of free groups to that of triangle groups, of a theorem of Jarden [16] concerning automorphisms of profinite groups. This allows the authors to prove that $\mathbb{G}$ acts faithfully, even when restricted to relatively small classes of regular dessins, such as those of a given hyperbolic type. We will return briefly to this topic in Sect. 11.3.3, where we consider the action of $\mathbb{G}$ on certain structures called Beauville surfaces, which are constructed from pairs of regular dessins.

At this point, it is appropriate to mention some far-reaching developments and conjectures, also well beyond the scope of this book, which are based on statements, suggestions and speculations in Grothendieck's *Esquisse d'un Programme* [10].

One can regard a dessin as corresponding to an unbranched finite covering of $\mathbb{P}^1(\mathbb{C}) \setminus \{0, 1, \infty\}$, or equivalently to a conjugacy class of subgroups of finite index in its topological fundamental group $\pi_1(\mathbb{P}^1(\mathbb{C}) \setminus \{0, 1, \infty\})$, which is a free group F_2 of rank 2—equivalently, one can think of this as the triangle group $\Delta(\infty, \infty, \infty)$, playing the role described in Sect. 3.3.4. The action of the absolute Galois group $\mathbb{G}$ on dessins induces an action as a group of automorphisms of the algebraic fundamental group $\pi_1^{\mathrm{alg}}(\mathbb{P}^1(\mathbb{C}) \setminus \{0, 1, \infty\})$ of this surface: this is the profinite completion

$$\hat{F}_2 = \varprojlim F_2/N$$

of F_2, the projective limit of its quotients by all its normal subgroups N of finite index. (Such normal subgroups correspond to the finite unbranched regular coverings of $\mathbb{P}^1(\mathbb{C}) \setminus \{0, 1, \infty\}$, with covering groups F_2/N.) The fact that $\mathbb{G}$ acts faithfully on dessins means that it is embedded in $\mathrm{Aut}\,\hat{F}_2$, which is a much larger and more complicated group that the well-understood group $\mathrm{Aut}\,F_2$ (an extension of $\mathrm{Inn}\,F_2 \cong F_2$ by $\mathrm{Out}\,F_2 \cong \mathrm{GL}_2(\mathbb{Z})$—see Sect. 8.3).

One can regard $\mathbb{P}^1(\mathbb{C}) \setminus \{0, 1, \infty\}$ as the moduli space $\mathscr{M}_{0,4}$ of all Riemann surfaces of genus 0 with four marked points, that is, it parametrizes the isomorphism classes of all such objects. The reason is that there is, up to isomorphism, a unique Riemann surface of genus 0, namely $\hat{\mathbb{C}} = \mathbb{P}^1(\mathbb{C})$; its automorphism group $\mathrm{PGL}_2(\mathbb{C})$ acts sharply 3-transitively, so that the first three marked points can be mapped to $0, 1$ and ∞, in that order, by a unique automorphism, and the fourth point is then mapped to a unique point in $\mathbb{P}^1(\mathbb{C}) \setminus \{0, 1, \infty\}$ (namely the cross-ratio of the four points), which serves as a parameter for this 4-tuple.

Grothendieck suggested that one should study the action of $\mathbb{G}$ on what is now called the *Grothendieck-Teichmüller tower*: this is the set of moduli spaces $\mathscr{M}_{g,k}$ of Riemann surfaces of genus g with k marked points, for all $g, k \geq 0$, together with their algebraic fundamental groupoids $\hat{T}_{g,k}$, linked by certain natural connecting homomorphisms. (Fundamental groupoids differ from fundamental groups in having more than one base-point.)

In [3], Drinfel'd used deformations of quasi-Hopf algebras to construct a subgroup $\widehat{GT}$ of $\mathrm{Aut}\,\hat{F}_2$, subsequently reinterpreted by Ihara [15] in terms of braid groups: these describe how finite sets of points can move continuously around a surface, or more precisely they are the fundamental groups of configuration spaces of such points in surfaces. In particular, Ihara showed that $\mathbb{G}$ is naturally embedded as a subgroup of $\widehat{GT}$. There are many deep open questions about these two groups, in particular, whether or not they are equal. Good starting points for learning about these developments would be the survey by Schneps [25] in [26], the papers by Ihara [15], with an appendix by Ensalem and Lochak [5], and by Lochak and Schneps [20] in [24], and the survey by Oesterlé [22]. In [11], Guillot gives a very accessible and rather more explicit approach to a slightly larger group $\widehat{GT}_0$, also first defined by Drinfel'd in [3]; this explicit approach is developed further in [12].

4.3 Appendix: Another Proof of Theorem 4.9

We describe here the main ideas of another way to prove that Belyĭ curves and their Belyĭ functions can be defined over number fields. We already know that their moduli fields are number fields (Theorem 4.6(c)), so it is sufficient to show that Belyĭ curves and Belyĭ functions can be defined over finite extensions of their moduli fields. The idea described here is inspired by the much more general Proposition 2.1 of [21], where the reader can also find sources for even further generalisations. We concentrate first on the fields of definition of the Belyĭ curves, then we add some final remarks about the fields of definition of the Belyĭ functions.

1. Let X be a Belyĭ curve. As already used implicitly in Sect. 1.1, X is uniquely determined by its function field $\mathbb{C}(X)$ (for the terminology, see Sect. 4.1.3), a finite algebraic extension $\mathbb{C}(f, t)$ of the rational function field $\mathbb{C}(t)$, $t = \beta$, where β denotes a Belyĭ function on X. The minimal polynomial $P(u, t) \in \mathbb{C}[u, t]$ of f over $\mathbb{C}(t)$ gives an affine model $P(u, t) = 0$ for X. In general, this model is singular, but by methods similar to those used in Sect. 1.2.5 one may pass to

a nonsingular model, also defined over $\overline{\mathbb{Q}}$ if P has algebraic coefficients. Since P has finitely many coefficients, it is sufficient to find a function f generating the algebraic extension $\mathbb{C}(X)/\mathbb{C}(t)$ such that P has coefficients in $\overline{\mathbb{Q}}$, in other words to prove that there is an algebraic field extension K contained in $\overline{\mathbb{Q}(t)}$, the algebraic closure of $\overline{\mathbb{Q}}(t)$ (and even of $\mathbb{Q}(t)$), of finite degree over $\overline{\mathbb{Q}}(t)$, and unramified outside $t = 0, 1, \infty$, such that

$$\mathbb{C}(X) = K \otimes_{\overline{\mathbb{Q}}} \mathbb{C}.$$

2. We can restrict our attention to the cases where X has genus $g > 1$ since we already know by Theorem 4.8 that Belyĭ curves of genera 0 and 1 can be defined over their moduli fields. As another simplification, we can also assume that X is quasiplatonic and that the dessin is regular, or in other words that $\beta = t$ defines a regular covering, or equivalently (see Sect. 4.1.3) that the field extension $\mathbb{C}(X)/\mathbb{C}(t)$ is a Galois extension: if not, we replace it with its normalisation $\mathbb{C}(Y)/\mathbb{C}(t)$ and write X as a quotient of Y by a subgroup $U \leq \mathrm{Gal}(\mathbb{C}(Y)/\mathbb{C}(t)) \leq \mathrm{Aut}(Y)$. Namely, if we know that Y is defined over $\overline{\mathbb{Q}}$, that is—see above—$\mathbb{C}(Y) = K_Y \otimes_{\overline{\mathbb{Q}}} \mathbb{C}$ for a finite Galois extension K_Y of $\overline{\mathbb{Q}}(t)$, the elements of $\mathrm{Aut}(Y)$ are also defined over $\overline{\mathbb{Q}}$: otherwise $\mathrm{Aut}(Y)$ would be infinite (apply suitable automorphisms of $\mathbb{C}/\overline{\mathbb{Q}}$ to the coefficients of the automorphisms) in contradiction to the finiteness of the automorphism group in genera $g > 1$ (see the Hurwitz bound in Sect. 5.1.2). Therefore $\mathrm{Gal}(\mathbb{C}(Y)/\mathbb{C}(t)) \cong \mathrm{Gal}(K_Y/\overline{\mathbb{Q}}(t))$, so K can be defined as the fixed field of $U \leq \mathrm{Gal}(K_Y/\overline{\mathbb{Q}}(t))$, a function field with constants in $\overline{\mathbb{Q}}$.
3. Let $T = \mathrm{Aut}(\mathbb{C}/\overline{\mathbb{Q}})$ denote the group of field automorphisms of $\mathbb{C}$ fixing $\overline{\mathbb{Q}}$ element-wise, and $T^* = \mathrm{Aut}(\overline{\mathbb{C}(t)}/\overline{\mathbb{Q}}(t))$, extending the action of T to the field $\overline{\mathbb{C}(t)}$ of all algebraic functions, and fixing $\overline{\mathbb{Q}}(t)$ element-wise. For a given degree, there are only finitely many Galois extensions $\mathbb{C}(X)/\mathbb{C}(t)$ unramified outside $t = 0, 1, \infty$, so the subgroup $S \leq T^*$ preserving the field $\mathbb{C}(X)$ (not necessarily fixing it element-wise, of course) has finite index in T^*. For later use, we note that T acts faithfully on the set of all transcendental numbers, so we have fixed point sets

$$\mathbb{C}^T = \overline{\mathbb{Q}} \quad \text{and} \quad \overline{\mathbb{C}(t)}^{T^*} = \overline{\mathbb{Q}}(t).$$

4. Now suppose that $a \in \overline{\mathbb{Q}}$, $a \neq 0, 1, \infty$ and let $b \in t^{-1}(a) \subset X$ be one of its pre-images under t. Recall that S acts on $\mathbb{C}$ and on X itself (via the set of $\mathfrak{p}$-adic valuations of $\mathbb{C}(X)$ given by the orders of the functions at the points of X, for example); S fixes a and t, so there is a finite index subgroup $S_b \leq S$ fixing b. Using the Riemann-Roch Theorem, or even better the Weierstrass gap sequence in b (see for example Sects. III.5.9 and III.5.10 of [6]), we can choose $p > 1$ as the first non-gap for the Weierstrass sequence, and define $L(b)$ to be the $\mathbb{C}$-vector space of all meromorphic functions $m \in \mathbb{C}(X)$ holomorphic at all points except b, where a single pole of order at most $p \leq g + 1$ is allowed (here g is the genus

of X). This space $L(b)$ has dimension 2 and is generated by the constant 1 and a function f. Without loss of generality, we can take $t-a$ as a local coordinate in a neighbourhood of b and assume that f is normalised—and is uniquely determined by this normalisation—by the coefficients $a_{-p} = 1$ and $a_0 = 0$ of its Laurent expansion at b:

$$f(t) = (t-a)^{-p} + \sum_{1 \leq j < p} a_{-j}(t-a)^{-j} + \sum_{j>0} a_j(t-a)^j .$$

We even have $\mathbb{C}(X) = \mathbb{C}(t,f)$ because the subgroup $V \leq \mathrm{Gal}(\mathbb{C}(X)/\mathbb{C}(t)) \leq \mathrm{Aut}(X)$ fixing t and f acts trivially on a neighbourhood of b, so $V = \{\mathrm{id}\}$.

5. How does $\sigma \in S_b$ act on f and its Laurent coefficients at b? We may consider these Laurent series of f and f^σ as elements of the $\mathfrak{p}$-adic completion of $\mathbb{C}(X)$ for the valuation given by the orders of the functions at b. By hypothesis, σ fixes the function field $\mathbb{C}(X)$ and the elements $t, a, b, 0, 1$. Also $\mathfrak{p}$ is invariant, and hence so is the pole order of f. As in Galois actions on $\mathfrak{p}$-adic fields, σ induces an action on the residue class field (here $\mathbb{C}$; recall that T^* extends the action of T on $\mathbb{C}$), so we have a Laurent expansion

$$f^\sigma(t) = (t-a)^{-p} + \sum_{1 \leq j < p} a_{-j}^\sigma(t-a)^{-j} + \sum_{j>0} a_j^\sigma(t-a)^j .$$

By the uniqueness of the normalised function $f \in L(b)$, this must coincide with the expansion of f, so we have

$$a_j^\sigma = a_j \quad \text{for all} \quad j \quad \text{and all} \quad \sigma \in S_b .$$

6. Since S_b has finite index in T^*, each coefficient a_j has only finitely many conjugates under T, so it is an algebraic number. One can determine the coefficients of the minimal polynomial $P(u,t)$ recursively by replacing u with the Laurent expansion of f and inserting it in the equation $P(f,t) = 0$. This gives, for each coefficient of P, a polynomial equation with coefficients in $\overline{\mathbb{Q}}$, thus proving Theorem 4.9.
7. Since the Belyĭ function for the model $P(u,t) = 0$ for X is given by a projection $(u,t) \mapsto t$, it also has algebraic coefficients. This remains true if we pass to a nonsingular model for X, with the difference that we have to consider projections

$$(u_1, \ldots, u_n, t) \mapsto t$$

where the coordinates depend on each other by more than one polynomial equation with algebraic coefficients. Because t belongs to the ground field of the extension $\mathbb{C}(X)/\mathbb{C}(t)$ it is also contained in all the intermediate function fields, also playing the role of the Belyĭ function for the quotients of X, so the claim does not depend on the hypothesis that X is quasiplatonic.

8. For genera 0 and 1 the situation is less straightforward since in such cases we may construct Belyĭ functions with transcendental coefficients by composing a Belyĭ function defined over $\overline{\mathbb{Q}}$ with an automorphism of the curve *not* defined over $\overline{\mathbb{Q}}$. However, this is only a question of taking a better model: for the same reason as in higher genera, Belyĭ functions in genera 0 and 1 can also be defined over number fields since their dessins are quotients of regular dessins of genera $g > 1$.

References

1. Bauer, I., Catanese, F., Grunewald, F.: Faithful actions of the absolute Galois group on connected components of moduli spaces. Invent. Math. **199**, 859–888 (2015)
2. Dèbes, P., Emsalem, M.: On fields of moduli of curves. J. Algebra **211**, 42–56 (1999)
3. Drinfel'd, V.G.: On quasitriangular quasi-Hopf algebras and a group closely connected with Gal $(\overline{\mathbf{Q}}/\mathbf{Q})$. Leningrad Math. J. **2**, 829–860 (1991)
4. Earle, C.J.: On the moduli of closed Riemann surfaces with symmetries. In: Ahlfors, L.V., et al. (eds.) Advances in the Theory of Riemann Surfaces. Annals of Mathematics Studies, vol. 6, pp. 119–130. Princeton University Press, Princeton (1971)
5. Ensalem, M., Lochak, P.: Appendix: the action of the absolute Galois group on the moduli spaces of spheres with four marked points. In: Schneps, L. (ed.) The Grothendieck Theory of Dessins d'Enfants. London Mathematical Society Lecture Note Series, vol. 200, pp. 307–321. Cambridge University Press, Cambridge (1994)
6. Farkas, H.M., Kra, I.: Riemann Surfaces. Springer, Berlin/Heidelberg/New York (1991)
7. Girondo, E., González-Diez, G.: A note on the action of the absolute Galois group on dessins. Bull. Lond. Math. Soc. **39**, 721–723 (2007)
8. González-Diez, G.: Variations on Belyi's theorem. Q. J. Math. **57**, 339–354 (2006)
9. González-Diez, G., Jaikin-Zapirain, A.: The absolute Galois group acts faithfully on regular dessins and on Beauville surfaces. Proc. Lond. Math. Soc. (3) **111**(4), 775–796 (2015)
10. Grothendieck, A.: Esquisse d'un Programme. In: Schneps, L., Lochak, P. (eds.) Geometric Galois Actions 1. Around Grothendieck's Esquisse d'un Programme, London Mathematical Society Lecture Note Series, vol. 242, pp. 5–48. Cambridge University Press, Cambridge (1997)
11. Guillot, P.: An elementary approach to dessins d'enfants and the Grothendieck-Teichmüller group. Enseign. Math. **60**(3–4), 293–375 (2014)
12. Guillot, P.: Some computations with the Grothendieck-Teichmüller group and equivariant dessins d'enfants (2014). arXiv:1407:3112 [math.GR]
13. Hammer, H., Herrlich, F.: A remark on the moduli field of a curve. Arch. Math. (Basel) **81**, 5–10 (2003)
14. Herradón Cueto, M.: The field of moduli and fields of definition of dessins d'enfants (2014). arXiv:1409.7736 [math.AG]. Accessed 20 Jan 2015
15. Ihara, Y.: On the embedding of Gal $(\overline{\mathbf{Q}}/\mathbf{Q})$ into $\widehat{GT}$. In: Schneps, L. (ed.) The Grothendieck Theory of Dessins d'Enfants. London Mathematical Society Lecture Note Series, vol. 200, pp. 289–305. Cambridge University Press, Cambridge (1994)
16. Jarden, M.: Normal automorphisms of free profinite groups. J. Algebra **62**, 118–123 (1980)
17. Jones, G.A., Streit, M.: Galois groups, monodromy groups and cartographic groups. In: Schneps, L., Lochak, P. (eds.) Geometric Galois Actions 2. The Inverse Galois Problem, Moduli Spaces and Mapping Class Groups. London Mathematical Society Lecture Note Series, vol. 243, pp. 25–65. Cambridge University Press, Cambridge (1997)

18. Jones, G.A., Streit, M., Wolfart, J.: Wilson's map operations on regular dessins and cyclotomic fields of definition. Proc. Lond. Math. Soc. **100**, 510–532 (2010)
19. Koeck, B.: Belyi's theorem revisited. Beiträge Algebra Geom. **45**, 253–275 (2004)
20. Lochak, P., Schneps, L. The Grothendieck-Teichmüller group and automorphisms of braid groups. In: Schneps, L. (ed.) The Grothendieck Theory of Dessins d'Enfants. London Mathematical Society Lecture Note Series, vol. 200, pp. 323–358. Cambridge University Press, Cambridge (1994)
21. Malle, G., Matzat, B.H.: Inverse Galois Theory. Springer, Berlin/Heidelberg/New York (1999)
22. Oesterlé, J.: Dessins d'enfants. Astérisque (Sém. Bourbaki 2001/02, Exp. 907) **290**, 285–305 (2003)
23. Schneps, L.: Dessins d'enfants on the Riemann sphere. In: Schneps, L. (ed.) The Grothendieck Theory of Dessins d'Enfants. London Mathematical Society Lecture Note Series, vol. 200, pp. 47–77. Cambridge University Press, Cambridge (1994)
24. Schneps, L. (ed.): The Grothendieck Theory of Dessins d'Enfants. London Mathematical Society Lecture Note Series, vol. 200. Cambridge University Press, Cambridge (1994)
25. Schneps, L.: The Grothendieck-Teichmüller group $\widehat{GT}$: a survey. In: Schneps, L., Lochak, P. (eds.) Geometric Galois Actions 1. London Mathematical Society Lecture Note Series, vol. 242, pp. 183–203. Cambridge University Press, Cambridge (1997)
26. Schneps, L., Lochak, P. (eds.): Geometric Galois Actions 1, 2. London Mathematical Society Lecture Note Series, vols. 242, 243. Cambridge University Press, Cambridge (1997)
27. Shimura, G.: On the field of rationality of an abelian variety. Nagoya Math. J. **45**, 167–178 (1972)
28. The GAP Group: GAP – Groups, Algorithms, and Programming. Version 4.7.6 (2014). http://www.gap-system.org. Accessed 20 Jan 2015
29. Weil, A.: The field of definition of a variety. Am. J. Math. **78**, 509–524 (1956)
30. Wolfart, J.: The 'Obvious' part of Belyi's Theorem and Riemann surfaces with many automorphisms. In: Schneps, L., Lochak, P. (eds.) Geometric Galois Actions 1. Around Grothendieck's Esquisse d'un Programme. London Mathematical Society Lecture Note Series, vol. 242, pp. 97–112. Cambridge University Press, Cambridge (1997)

Part II
Regular Dessins

By Belyĭ's Theorem, all properties of algebraic curves defined over number fields should be encoded somehow in the dessins lying on these curves; in particular it should be possible to obtain explicit equations for models of these curves from their dessins. Unfortunately, we are still far from being able to decode this information in general. As almost always in mathematics, the situation is better if sufficiently large automorphism groups are available—in this case, if the dessins are regular. (In principle there is no great loss of generality in restricting attention to regular dessins, since every dessin is the quotient of a regular dessin, of the same type, by some group of automorphisms; however, the explicit passage between a dessin and its regular cover is not always easy to implement.) The following chapters are therefore devoted to fundamental properties of regular dessins, their underlying quasiplatonic curves and their automorphism groups. In particular, we will consider the enumeration and classification of all examples of low genera, and the construction of infinite families of regular dessins by using families of finite groups as automorphism groups or by embedding families of highly symmetric graphs, such as complete graphs or complete bipartite graphs. We will also consider how certain map and hypermap operations, introduced by Wilson and James, are related to the action of the absolute Galois group on dessins, together with the consequences of this for the determination of Galois orbits, fields of definition and explicit equations.

Chapter 5
Quasiplatonic Surfaces, and Automorphisms

Abstract Quasiplatonic Riemann surfaces or algebraic curves, sometimes also called *curves with many automorphisms* or *triangle curves*, can be characterised in many equivalent ways, for example as those curves having a regular dessin, one with the greatest possible degree of symmetry. The sphere and the torus each support infinitely many regular dessins, easily described in both cases. For each genus $g > 1$ there are, up to isomorphism, only finitely many regular dessins; this chapter gives complete lists for genera $2, 3$ and 4, and discusses methods for counting and classifying them. These methods often involve counting generating triples for finite groups, in some cases with the aid of character theory (which we briefly summarise) and Möbius inversion. We present several important infinite families of quasiplatonic curves, such as Hurwitz and Macbeath-Hurwitz curves, Lefschetz and Accola-Maclachlan curves. We prove that like their counterparts, the curves with trivial automorphism group, quasiplatonic curves can be defined over their field of moduli. Many of the automorphism groups appearing in this chapter are 2-dimensional linear or projective groups over finite fields, so we summarise their most relevant properties in the final section.

Keywords Accola-Maclachlan curve • Arithmetic group • Automorphism group • Character table • Character theory • Field of definition • Hurwitz group • Kulkarni curve • Linear group • Low genus map • Möbius function • Möbius inversion • Moduli field • Projective group • Quasiplatonic curve • Quasiplatonic surface • Regular dessin • Shimura curve • Triangle curve • Wiman curve

5.1 Quasiplatonic Surfaces: Construction and Counting

5.1.1 Definitions and Properties

As we have already seen, any compact Riemann surface X of genus $g > 1$ can be uniformised by an essentially unique (up to conjugation in $\mathrm{PSL}_2(\mathbb{R})$) torsion-free Fuchsian group K, isomorphic to its fundamental group $\pi_1 X$. Isomorphisms $X \to X'$ are induced by isometries of $\mathbb{H}$ conjugating K to K'. Taking $X = X'$ we see that automorphisms of X are induced by isometries normalising K. Since K acts

Gareth A. Jones, J. Wolfart, *Dessins d'Enfants on Riemann Surfaces*, Springer Monographs in Mathematics, DOI 10.1007/978-3-319-24711-3_5

trivially on $K\backslash\mathbb{H}$, we get

$$\mathrm{Aut}\,X \cong N(K)/K$$

where N denotes the normaliser in $\mathrm{PSL}_2(\mathbb{R})$. If $g = 1$, we must replace $\mathbb{H}$ with $\mathbb{C}$, and replace K with a lattice Λ, unique up to similarity—see Sects. 1.2.3 and 1.2.4. In the case $g > 1$, we need more information about Fuchsian groups than that given in Sect. 3.1.

First we need a simple exercise:

Exercise 5.1 Two elements of $\mathrm{PSL}_2(\mathbb{R})$ commute if and only if their fixed-point sets in $\hat{\mathbb{C}}$ are the same.

Lemma 5.1 *If K is a non-abelian Fuchsian group, the normaliser $N(K)$ is also a Fuchsian group.*

Proof If $N(K)$ is not Fuchsian, there is a sequence (f_n) of non-identity elements of $N(K)$ converging to 1 as $n \to \infty$. Since K is non-abelian, Exercise 5.1 implies that there exist non-identity elements $g_1, g_2 \in K$ with different fixed-point sets in $\hat{\mathbb{C}}$. Now $f_n \in N(K)$ and $f_n \to 1$, so the elements $f_n g_i f_n^{-1}$ lie in K and converge to g_i for each $i = 1, 2$; since K is discrete, we have $f_n g_i f_n^{-1} = g_i$ for all sufficiently large n, so f_n commutes with each g_i and hence has the same fixed-point set in $\hat{\mathbb{C}}$ as g_i by Exercise 5.1. By our choice of g_1 and g_2 this is impossible, so $N(K)$ is a Fuchsian group. □

If K is a Fuchsian group with a fundamental region of finite hyperbolic area, then it is non-abelian, so $N(K)$ also has this property; moreover, K has finite index in $N(K)$ since the index is the ratio of the areas of the fundamental regions for K and $N(K)$.

We say that X (compact, of genus $g > 1$) is *quasiplatonic* if X is uniformised by a subgroup K as above, with K a normal subgroup of a triangle group. (The term *quasiplatonic* was introduced by Singerman in [50]; in older literature these surfaces were called *'with many automorphisms'*, see for example [41, 43]. In the literature about Beauville surfaces—to be treated in Part III of this volume—quasiplatonic curves are often called *triangle curves.*)

Theorem 5.1 *If X is a compact Riemann surface of genus $g > 1$ with surface group K, then the following are equivalent:*

a) X is quasiplatonic;
b) $N(K)$ is a triangle group;
c) X has a Belyĭ function $\beta : X \to \hat{\mathbb{C}}$ which is a regular covering;
d) (X, β) corresponds to a regular dessin.

Proof

a) ⇒ b): $N(K)$ is a Fuchsian group and it contains a triangle group. Any Fuchsian group containing a triangle group must be a triangle group (by Theorem 3.11 or by Teichmüller theory—triangle groups are the only rigid Fuchsian groups).

b) $\Rightarrow$ c): The inclusion $K \leq N(K)$ induces a Belyĭ function

$$X \cong K\backslash\mathbb{H} \rightarrow N(K)\backslash\mathbb{H} \cong \hat{\mathbb{C}}\,, \; Kz \mapsto N(K)z\,.$$

Since $K \trianglelefteq N(K)$, this is a regular covering.

c) $\Rightarrow$ d): Use β to lift the trivial dessin ($[0, 1] \sim \circ\!\!-\!\!-\!\!-\!\!\bullet$) on $\hat{\mathbb{C}}$ to X; since β is regular we get a regular dessin on X.

d) $\Rightarrow$ a): If X corresponds to a regular dessin $\mathscr{B}$, then K is normal in the corresponding triangle group. □

Example 5.1 The nth degree Fermat curve ($n > 3$) corresponds to a regular dessin, and is uniformised by the commutator subgroup of $\Delta(n, n, n)$ which is normal [even in $\Delta(2, 3, 2n)$], see Figs. 1.2 and 9.1.

Exercise 5.2 For genus $g = 1$, what are the analogues of the quasiplatonic surfaces?

One can characterise the quasiplatonic surfaces X as the isolated local maxima for $|\operatorname{Aut} X|$, in the sense that, within the Teichmüller space of all compact Riemann surfaces of genus g, every other surface sufficiently close to X has fewer automorphisms than X (see [41, 43] for an interpretation in terms of singularities of moduli spaces).

5.1.2 *Hurwitz Groups and Surfaces*

Here we look for global maxima of $|\operatorname{Aut} X|$.

Problem Given $g \geq 2$, what are the most symmetric Riemann surfaces of genus g?

We have $\operatorname{Aut} X \cong N(K)/K$, with $N(K)$ Fuchsian, where K is a surface group uniformising X. The index $|N(K) : K|$ is finite, equal to the ratio of the areas of the fundamental regions of these two groups. For K this area is always $4\pi(g-1)$, so maximising $|\operatorname{Aut} X|$ is equivalent to minimising the area for $N(K)$. One can show (Siegel [47]) that among all Fuchsian groups, this area is minimised when $N(K)$ is the triangle group $\Delta(3, 2, 7) = \Delta(2, 3, 7)$, given by

$$\Delta = \langle X, Y, Z \,|\, X^3 = Y^2 = Z^7 = XYZ = 1 \rangle\,.$$

Exercise 5.3 Prove that Δ has a fundamental region of area $\pi/21$, and that this is the minimum among all hyperbolic triangle groups. (Use the Gauss-Bonnet formula $\pi - \alpha - \beta - \gamma$ for the area of a hyperbolic triangle with internal angles α, β and γ.)

This gives us the *Hurwitz bound*

$$|\operatorname{Aut} X| \leq \frac{4\pi(g-1)}{\pi/21} = 84(g-1)$$

for surfaces of genus $g \geq 2$, attained if and only if $X \cong K \backslash \mathbb{H}$ where K is a normal subgroup of finite index in $\Delta = \Delta(3, 2, 7)$.

Exercise 5.4 Prove that if X has genus $g \geq 2$ and $|\operatorname{Aut} X| < 84(g-1)$ then $|\operatorname{Aut} X| \leq 48(g-1)$; find examples attaining this bound.

Exercise 5.5 Show that if l, m and n are three distinct primes, then every proper normal subgroup K of $\Delta(l, m, n)$ is torsion-free.

These surfaces $X \cong K \backslash \mathbb{H}$, where $K \lhd \Delta(3, 2, 7)$, together with their automorphism groups $G = \operatorname{Aut} X$, are called *Hurwitz surfaces* and *Hurwitz groups*. These surfaces are all quasiplatonic.

Example 5.2 The modular group $\Gamma = \mathrm{PSL}_2(\mathbb{Z}) = \Delta(3, 2, \infty)$ can be mapped onto $G = \mathrm{PSL}_2(7) := \mathrm{PSL}_2(\mathbb{F}_7)$ by reducing coefficients mod 7. (Here and in the following we often use the common short notations $\mathrm{SL}_2(q)$, $\mathrm{PSL}_2(q)$ and so on instead of $\mathrm{SL}_2(\mathbb{F}_q)$ and $\mathrm{PSL}_2(\mathbb{F}_q)$, respectively.) The generator $Z : \tau \mapsto \tau + 1$ is mapped to an element z of order 7 in G, so G is a quotient Δ/K of $\Delta = \Delta(3, 2, 7)$. We have

$$|G| = 168 \left(= \frac{7(7^2 - 1)}{2} \right),$$

(explain this formula, or see Sect. 5.5), so the surface $X = K \backslash \mathbb{H}$ has genus $g = 1 + \frac{168}{84} = 3$. This is Klein's quartic curve, given in projective coordinates by

$$x^3 y + y^3 z + z^3 x = 0,$$

with $\operatorname{Aut} X \cong \mathrm{PSL}_2(7)$, see Fig. 3.4 in Sect. 3.1.6.

Exercise 5.6 Prove that there is no Hurwitz group of genus 2.

Example 5.3 Macbeath [37] extended Example 5.2 by proving that the group $G = \mathrm{PSL}_2(q) := \mathrm{PSL}_2(\mathbb{F}_q)$ is a Hurwitz group if and only if one of the following holds:

- $q = 7$, as above;
- $q = p$ for some prime $p \equiv \pm 1 \bmod 7$;
- $q = p^3$ for some prime $p \equiv \pm 2$ or $\pm 3 \bmod 7$.

In the first and third cases, as we will show later in this chapter, the triangle group Δ has a unique normal subgroup K with $\Delta/K \cong G$, so the Hurwitz surface X is unique up to isomorphism. In the second case, however, there are three such normal subgroups K, giving three non-isomorphic surfaces X. These are distinguished by the generator of order 7 lying in one of the three conjugacy classes of elements of that order in G (see Sect. 5.5). Streit [51] showed that the corresponding dessins are defined over the cubic field $k := \mathbb{Q}(\cos 2\pi/7) < \mathbb{Q}(\zeta_7)$ and are conjugate under the absolute Galois group $\mathbb{G}$. In fact, not only are these three dessins mutually non-isomorphic, but so are the embedded graphs, as shown in the Appendix of [31].

Digression on Arithmetic Fuchsian Groups and Shimura Curves Another possible approach to Macbeath's series of Hurwitz groups has been mentioned by Magnus [39], and has been verified more recently by Džambić [17] (and for dessins of more general type by Feierabend [18]): it is based on the fact that the triangle group $\Delta = \Delta(3,2,7)$ and all its normal subgroups of finite index are *arithmetically defined* Fuchsian groups. This means that they are commensurable with a norm one group of units in a maximal order (= ring of integers) in a quaternion algebra with centre field k. In our case, Δ is precisely such a norm one group, and Γ is called *commensurable* with Δ if $\Gamma \cap \Delta$ has finite index in both groups. To be arithmetic, the centre field k of the quaternion algebra (in our case, it is just the cubic field $\mathbb{Q}(\cos\frac{2\pi}{7})$ generated by the traces of Δ) has to be totally real; it has $|k:\mathbb{Q}|$ (= 3 in our case) inequivalent embeddings $k \hookrightarrow \mathbb{R}$, extending to embeddings of the quaternion algebra into either the Hamiltonians or the matrix algebra $M_2(\mathbb{R})$. For arithmetic Fuchsian groups, only one of these embeddings goes to $M_2(\mathbb{R})$. Quotients $\Gamma\backslash\mathbb{H}$ of the hyperbolic plane by arithmetic Fuchsian groups Γ are called *Shimura curves* and play an important role as moduli spaces for abelian varieties, comparable to the quotient of $\mathbb{H}$ by the elliptic modular group (also an arithmetic triangle group) for the family of all elliptic curves.

This is a topic outside the scope of our book; however, arithmeticity is a key ingredient for a deeper understanding of the Macbeath-Hurwitz curves (and of Bring's curve later in Sect. 5.2.3): the surface groups $K \lhd \Delta = \Delta(3,2,7)$ leading to quotients $\Delta/K \cong G = \mathrm{PSL}_2(\mathbb{F}_q)$ turn out to be *principal congruence subgroups* $\Delta(\mathfrak{p})$ of Δ corresponding to certain ideals $\mathfrak{p}$ in the ring of integers of the centre field k; as for the congruence subgroups of the elliptic modular group, we define these subgroups by the condition that

$$\gamma \in \Delta\,, \quad \gamma \equiv \pm I \bmod \mathfrak{p}\,,$$

where $\mathfrak{p}$ is a prime ideal in k containing p and of norm q, I denotes the identity matrix, and the congruence has to be understood component-wise. In fact, algebraic number theory predicts that the prime $p = 7$ ramifies in k, each prime $p \not\equiv \pm 1$ mod 7 remains prime in k (that is, it is *inert*, hence with norm p^3), and each prime $p \equiv \pm 1$ mod 7 splits into three different prime ideals of norm p, leading to three different (and even non-conjugate) principal congruence subgroups $\Delta(\mathfrak{p})$. It is remarkable—and it fits very well into the more general framework of *Shimura curves* and their canonical models—that the absolute Galois group $\mathbb{G}$ acts on these prime ideals in the same way as it acts on the corresponding quotient curves.

If we regard Δ as $\Delta(7,2,3)$ then each Hurwitz surface carries a regular dessin of type $(7,2,3)$, that is, a 7-valent regular triangular map. In any map, a *Petrie polygon* is a closed zig-zag path in the embedded graph, turning alternately first left and first right, according to the local orientation, at successive vertices. In the monodromy group, where x represents rotation of darts around vertices and y represents reversal of darts along edges, this corresponds to iterated applications of the commutator word $w = [x,y] = x^{-1}y^{-1}xy$. In a regular map, all Petrie polygons have the same length, called the *Petrie length*, equal to twice the order of w (draw a diagram to see why!). Unlike some other numerical parameters, such as valencies of vertices and faces, Petrie length is not a Galois invariant, so this word w can be used as in Sect. 4.2.3 to distinguish Galois conjugate but non-isomorphic dessins.

Example 5.4 We illustrate this by returning to Example 5.3. For primes $p \equiv \pm 1 \mod 7$ the three regular dessins described there all have monodromy group and automorphism group $G \cong \mathrm{PSL}_2(\mathbb{F}_p)$. In the case $p = 13$ the three conjugacy classes of elements x of order 7 in G are represented by matrices in $\mathrm{SL}_2(\mathbb{F}_{13})$ with traces ± 3, ± 5 and ± 6. Simple calculations with matrices show that the commutator $w = [x, y]$ has order 7, 13 or 6 respectively. The three dessins therefore have Petrie lengths 14, 26 and 12, so they are mutually non-isomorphic.

When $p = 29$ the elements of order 7 in G correspond to matrices with traces ± 3, ± 7 and ± 11, and $w = [x, y]$ has order 15, 14 or 15 respectively. Thus in this case the word $[x, y]$ will distinguish the second dessin from the other two, but it will not distinguish the first from the third. For this we can use another word, such as x^4yx^2y, which has order 29 and 14 for these two dessins.

Exercise 5.7 Find three generating triples for $G = \mathrm{PSL}_2(\mathbb{F}_{13})$ corresponding to the three dessins described in Example 5.4, and verify that the Petrie lengths are as claimed.

Remark 5.1 A number of other classes of finite simple groups, such as the Ree groups $Re(3^e)$ and the Fischer-Griess monster group M, have been shown to be Hurwitz groups; see Conder's survey [6] for a recent account.

Remark 5.2 For many decades—until Macbeath [36] found a series of Hurwitz surfaces in the early 1960s—Klein's quartic seemed to be the only example of a Riemann surface attaining the Hurwitz bound. The next example—of genus 7 and with automorphism group $\mathrm{PSL}_2(\mathbb{F}_8)$ of order 504—had however already been considered in 1899 by Fricke [19], who also mentioned the congruence subgroup property of its surface group $\Delta(2)$, though not Hurwitz's upper bound. Shortly before his paper, in the same volume of the Mathematische Annalen, Burnside [4] had already given a purely algebraic proof that $\mathrm{PSL}_2(\mathbb{F}_8)$ is an epimorphic image of $\Delta(2,3,7)$, but without even mentioning Riemann surfaces. It seems reasonable to call this second Hurwitz surface the *Fricke-Macbeath surface*, and the entire series of Hurwitz surfaces with automorphism groups $\mathrm{PSL}_2(\mathbb{F}_q)$ the *Macbeath-Hurwitz surfaces*.

A simple equation for the Fricke-Macbeath curve has been given by Bradley Brock: it is mentioned in [26], a paper concerned with systematic attempts to find such equations.

Remark 5.3 According to [5] and [6], the lowest genera of Hurwitz surfaces are $3, 7, 14, 17$ and 28. Among all positive integers the genera of Hurwitz surfaces are in fact as rare as cubic numbers, in a sense which can be made precise, see [35]. However this does not mean that the number of non-isomorphic Hurwitz surfaces of genera $g \leq x$ grows slowly with x: one may deduce from [44] that this number grows at least as fast as $x^{c \log x}$ for some (very small) constant $c > 0$ and for sufficiently large x (there is also a similar upper bound). The point is that for certain (rather rare) values of g there are many non-isomorphic Hurwitz surfaces of genus g. Recent work of González-Diez and Jaikin-Zapirain [23] (mentioned in Remark 4.2) and of Kucharczyk [32] shows that the Hurwitz surfaces are sufficiently abundant and rich in structure that the absolute Galois group acts faithfully on them.

In the next few sections we will consider more general techniques for constructing and enumerating generating triples of any given type in a given group G. This is important, because it allows us to construct and enumerate regular dessins of that type with automorphism group G.

5.1.3 Kernels and Epimorphisms

Rephrasing the problem posed at the end of the preceding section, it is useful to classify the normal subgroups K of a triangle group Δ with a given quotient group $G \cong \Delta/K$. This is because such normal subgroups correspond to regular dessins with automorphism group G, or equivalently to generating triples for G satisfying the defining relations of Δ; two such triples correspond to the same normal subgroup (or isomorphic dessins) if and only if they are equivalent under automorphisms of G, so classifying and counting can be done entirely within the group G.

In some cases, such as the following, simple arguments can be used.

Example 5.5 Let us look for regular dessins with automorphism group $G = C_n \times C_n$. It is easy to see that any generating triple for G must have type (n, n, n), so we take $\Delta = \Delta(n, n, n)$. We need $\Delta/K \cong G$, so because G is abelian, K must contain the commutator subgroup Δ' of Δ. Abelianising Δ, by adding the relation $XY = YX$ for instance, we find that $\Delta/\Delta' \cong C_n \times C_n$, so $|\Delta : \Delta'| = n^2$. Since K also has index n^2 in Δ we therefore have $K = \Delta'$. Thus there is, up to isomorphism, just one regular dessin with automorphism group $G = C_n \times C_n$. It is, of course, the Fermat dessin, as in Example 5.1. Its uniqueness corresponds to the fact that any two generating triples for G are equivalent under an automorphism of this group.

Exercise 5.8 Show that any regular dessin with an abelian automorphism group is a quotient of a Fermat dessin.

Example 5.6 Among the simplest examples of quasiplatonic curves are the *Lefschetz* curves, which arise as quotients of Fermat curves. Let X have an affine model

$$y^p = x^u(x-1),$$

where p is prime and $u \in \{1, 2, \ldots, p-2\}$. There is a Belyĭ function

$$\beta : (x, y) \mapsto x$$

which realises X as a p-sheeted cover of $\mathbb{P}^1(\mathbb{C})$, branched over 0, 1 and ∞. This is a regular covering, induced by the group $G \cong C_p$ of automorphisms of X generated by

$$\alpha : (x, y) \mapsto (x, \zeta_p y).$$

The multiplicity of β over each of 0, 1 and ∞ is p, so the Riemann-Hurwitz formula shows that X has genus

$$g = 1 - p + \frac{3}{2}(p-1) = \frac{1}{2}(p-1).$$

The existence of β shows that X is uniformised by a torsion-free normal subgroup K of index p in the triangle group $\Delta = \Delta(p,p,p)$. Since $\Delta/K \cong G \cong C_p$ is abelian, K contains the commutator subgroup Δ', so X is the quotient by K/Δ' ($\cong C_p$) of the Fermat curve of degree p, which is uniformised by Δ'. Similarly, the regular dessin of type (p,p,p) corresponding to β is the quotient by K/Δ' of the corresponding Fermat dessin.

Since $\Delta/\Delta' \cong C_p \times C_p$, there are $p+1$ normal subgroups K of index p in Δ, all containing Δ'. Three of these are not torsion-free, each containing one of the three canonical generators of Δ. The remaining $p-2$ are torsion-free, and they correspond to the different choices for u in the equation defining X. This gives $p-2$ dessins, which are mutually non-isomorphic since they correspond to distinct normal subgroups K of Δ. However, there are isomorphisms among the underlying curves: Δ is a normal subgroup of index 6 in the maximal triangle group $\tilde{\Delta} = \Delta(2,3,2p)$ (see [49] and Sect. 3.1.5), and pairs of curves X are isomorphic if the corresponding subgroups K are conjugate in $\tilde{\Delta}$. Moreover, for some K there are automorphisms of X, in addition to those in G, induced by elements of the normaliser $N(K) > \Delta$, and this normaliser is not necessarily contained in $\tilde{\Delta}$. For $p = 7$ we have $N(K) = \Delta(2,3,7)$, see also the last point of Lemma 9.2.

For more general situations, and in particular for enumeration, we need the following result:

Proposition 5.1 *If Δ is any finitely generated group, and G is any finite group, the number $n_\Delta(G)$ of $K \trianglelefteq \Delta$ with $\Delta/K \cong G$ is given by*

$$n_\Delta(G) = \frac{|\operatorname{Epi}(\Delta, G)|}{|\operatorname{Aut} G|},$$

where $\operatorname{Epi}(\Delta, G)$ *is the set of all epimorphisms $\theta : \Delta \to G$.*

Proof These normal subgroups K are the kernels of the epimorphisms $\theta : \Delta \to G$ (compare this with Theorem 3.11), and $\ker\theta = \ker\theta'$ if and only if $\theta' = \alpha \circ \theta$ for some $\alpha \in \operatorname{Aut} G$. Hence the kernels correspond to the orbits of $\operatorname{Aut} G$ acting by composition on $\operatorname{Epi}(\Delta, G)$.

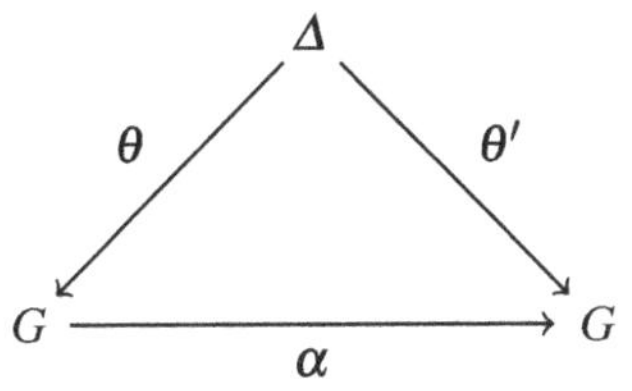

Since $\operatorname{Aut} G$ acts semiregularly (that is, if $\alpha \circ \theta = \theta$ then $\alpha = \mathrm{id}$), its orbits all have size $|\operatorname{Aut} G|$. By the hypotheses, $\operatorname{Epi}(\Delta, G)$ is finite, so the result follows. □

For many finite groups G, $|\operatorname{Aut} G|$ is known or easily found, so we should concentrate on counting epimorphisms. If Δ is a triangle group

$$\Delta(l, m, n) = \langle X, Y, Z \mid X^l = Y^m = Z^n = XYZ = 1 \rangle ,$$

then finding epimorphisms $\Delta \to G$ is equivalent to finding triples $x, y, z \in G$ such that

(a) we have

$$x^l = y^m = z^n = xyz = 1$$

(so there is a homomorphism $h : \Delta \to G$, $X \mapsto x$ etc.),

(b) G is generated by x, y and z (or by any two of them), so that h is an epimorphism. The torsion elements of K are those conjugate to powers of one of the generators of Δ, see Exercise 3.8. Such non-identity elements are therefore contained in the kernel of h only if h does not preserve the order of each generator. So, if we want K to be torsion-free, we also require:

(c) x, y and z must have orders exactly l, m and n.

5.1.4 Direct Counting

Example 5.7 Let $\Delta = \Delta(5, 2, \infty)$ and $G = A_5$, so we count normal subgroups $K \trianglelefteq \Delta$ with $\Delta/K \cong A_5$. This is equivalent to counting regular maps $\mathcal{M}$ $(m = 2)$ with valency 5 $(l = 5)$ and $\operatorname{Aut}\mathcal{M} \cong A_5$. Now A_5 has 24 elements x of order 5 (the 5-cycles), and 15 elements of order 2 (the double transpositions $(ab)(cd)$) giving $24 \times 15 = 360$ pairs x, y satisfying the relations of Δ. The subgroup $H = \langle x, y \rangle$ has order divisible by 10, so from knowledge of the subgroups of A_5—or by using Proposition 5.3 and the isomorphism $A_5 \cong \mathrm{PSL}_2(\mathbb{F}_5)$—we see that $H = A_5$ or H is isomorphic to the dihedral group D_5 of order 10. There are six subgroups $H \cong D_5$, each generated by $4 \times 5 = 20$ pairs x, y, so 120 pairs do not generate A_5. Hence $360 - 120 = 240$ pairs do generate A_5, so $|\operatorname{Epi}(\Delta, G)| = 240$. Now $\operatorname{Aut} A_5 = S_5$

(acting by conjugation) of order 120, so $n_\Delta(G) = 240/120 = 2$. Thus Δ has two normal subgroups K with $\Delta/K \cong A_5$, so there are two 5-valent regular maps $\mathcal{M}$ with $\mathrm{Aut}\,\mathcal{M} \cong A_5$. One is the icosahedron, represented by

$$\theta : X \mapsto x = (1,2,3,4,5), \ Y \mapsto y = (1,2)(3,4), \ Z \mapsto z = (2,5,4).$$

The other map $\mathcal{M}$ is the great dodecahedron; this has the vertices and edges of the icosahedron, but the triangular faces are replaced with 12 pentagons, each incident with the five neighbours of a single vertex. It is represented by

$$\theta : X \mapsto x = (1,2,3,4,5), \ Y \mapsto y = (1,3)(2,4), \ Z \mapsto z = (1,2,3,5,4).$$

This map $\mathcal{M}$ has genus $g = 1 + \frac{N}{2}(1 - \frac{1}{l} - \frac{1}{m} - \frac{1}{n}) = 4$ (where in this case $N = 60$, $l = 5$, $m = 2$ and $n = 5$). The underlying algebraic curve X is Bring's curve, given in $\mathbb{P}^4(\mathbb{C})$ by the equations

$$x_1^k + x_2^k + x_3^k + x_4^k + x_5^k = 0 \qquad (k = 1,2,3)\ .$$

This curve has automorphism group $\mathrm{Aut}\,X \cong S_5$ (permuting the coordinates), and the subgroup A_5 corresponds to the automorphism group of the map; the odd permutations send $\mathcal{M}$ to its dual map $\mathcal{M}^* \cong \mathcal{M}$.

Example 5.8 Let us confirm the statement in Example 5.3 that if $G = \mathrm{PSL}_2(p) := \mathrm{PSL}_2(\mathbb{F}_p)$ for some prime $p \equiv \pm 1 \bmod 7$ then the triangle group $\Delta = \Delta(2,3,7)$ has three normal subgroups K with $\Delta/K \cong G$. For this we need to count the orbits of $\mathrm{Aut}\,G$ on generating triples x, y, z for G satisfying $x^2 = y^3 = z^7 = xyz = 1$. Any non-identity triple satisfying these relations generates a perfect subgroup of G (see Exercise 3.9); by Dickson's description [13, Chap. XII] of the subgroups of G (or by Proposition 5.3), the only perfect proper subgroups are isomorphic to A_5, which has no elements of order 7, so the triple generates G.

For convenience of exposition, let us assume that $p \equiv 1 \bmod 7$ (see Example 5.11 for the case $p \equiv -1 \bmod 7$). Up to conjugacy in $\mathrm{Aut}\,G = \mathrm{PGL}_2(\mathbb{F}_p)$ we may then assume that the element $z \in G$ is represented by a matrix

$$Z = \begin{pmatrix} s & 0 \\ 0 & s^{-1} \end{pmatrix} \in \mathrm{SL}_2(\mathbb{F}_p)$$

where s is a primitive 7th root of 1 in $\mathbb{F}_p$; replacing s with s^2 or s^4 gives representatives of the other two conjugacy classes. Any element $y \in G$ of order 3 is represented by a matrix

$$Y = \begin{pmatrix} a & b \\ c & d \end{pmatrix} \in \mathrm{SL}_2(\mathbb{F}_p)$$

with determinant $ad - bc = 1$ and trace $a + d = 1$. The product

$$YZ = \begin{pmatrix} as & bs^{-1} \\ cs & ds^{-1} \end{pmatrix}$$

represents an element $x = x^{-1} = yz \in G$ of order 2 if and only if it has trace $as + ds^{-1} = 0$. Solving for a and d gives

$$a = \frac{1}{1 - s^2} \quad \text{and} \quad d = \frac{s^2}{s^2 - 1},$$

so we need b and c to satisfy

$$bc = ad - 1 = -\frac{s^4 - s^2 + 1}{(s^2 - 1)^2}\ .$$

Since $s^4 - s^2 + 1 \neq 0$ (why?), this equation has $p - 1$ solutions b, c. The resulting $p - 1$ elements $y \in G$ are all equivalent under conjugation by the centraliser of z in $\mathrm{PGL}_2(\mathbb{F}_p)$, represented by the diagonal matrices in $\mathrm{GL}_2(\mathbb{F}_p)$, so up to automorphisms our choice of the conjugacy class of z gives a unique generating triple. The three possible choices for this conjugacy class give three orbits of $\mathrm{Aut}\,G$ on generating triples, and hence three normal subgroups K of Δ with $\Delta/K \cong G$.

The calculations are similar (though a little more complicated) in the other cases of Macbeath's theorem. When $p \equiv \pm 2$ or $\pm 3 \bmod 7$ one needs to take $q = p^3$ in order that G can contain elements of order 7 and be generated by suitable triples; then $\mathrm{Aut}\,G = \mathrm{P\Gamma L}_2(\mathbb{F}_q)$, an extension of $\mathrm{PGL}_2(\mathbb{F}_q)$ by $\mathrm{Gal}\,\mathbb{F}_q \cong C_3$. These extra automorphisms permute the three conjugacy classes of elements of order 7 in G transitively, so $\mathrm{Aut}\,G$ has only one orbit on generating triples and hence we obtain only one normal subgroup K.

The fact mentioned in the Digression following Example 5.3, that the principal congruence subgroups of $\Delta(2, 3, 7)$ give the correct number of surface groups, relies on the above counting procedures: a priori, it is not evident that there are no non-congruence subgroups leading to the same quotients.

5.1.5 *Counting by Character Theory*

Here we very briefly summarise the basic facts of character theory, as needed for further counting techniques. For full details, see any book on representation theory, such as [20], or [27].

A (complex) *representation* of a group G is a homomorphism $\rho : G \to \mathrm{GL}(V)$, where V is a vector space over $\mathbb{C}$; this gives an action of G on V by invertible linear transformations. Representations $\rho : G \to \mathrm{GL}(V)$ and $\rho' : G \to \mathrm{GL}(V')$ are *equivalent* if there is an isomorphism $V \to V'$ commuting with the actions of

Table 5.1 The character table of A_5

1	(..)(..)	(...)	$(.....)^+$	$(.....)^-$
1	1	1	1	1
3	1	0	λ	μ
3	1	0	μ	λ
4	0	1	-1	-1
5	1	-1	0	0

G on V and V'. A representation ρ is *irreducible* if V has no G-invariant linear subspaces other than 0 and V. A finite group G has c irreducible representations, up to isomorphism, where c is the number of conjugacy classes in G. The *character* of a representation ρ is the function $\chi : G \to \mathbb{C}$ sending each $g \in G$ to the trace of $\rho(g)$; this function is constant on the conjugacy classes of G. The *character table* of G is a $c \times c$ array, with rows and columns indexed by the irreducible characters and the conjugacy classes of G; the entry in the ith row and jth column is the value of the ith irreducible character on the elements of the jth conjugacy class.

Example 5.9 In A_5 there are $c = 5$ conjugacy classes: the identity, 15 double transpositions, twenty 3-cycles, and two classes of twelve 5-cycles (mutually conjugate in S_5). Hence there are five irreducible characters and the character table is as in Table 5.1, where $\lambda, \mu = (1 \pm \sqrt{5})/2$; the columns correspond to the five conjugacy classes, indicated by their cycle-structures, with $(.....)^{\pm}$ indicating the two classes of 5-cycles.

The first row corresponds to the principal representation, obtained (for any group G) by sending each element to the identity matrix in $\mathrm{GL}_1(\mathbb{C})$. The second row corresponds to the irreducible representation of G as the rotation group of an icosahedron, extended in the obvious way to $\mathbb{R}^3$ and then to $\mathbb{C}^3$. This representation is defined over the field $\mathbb{Q}(\sqrt{5})$, and the third row corresponds to the irreducible representation obtained from the preceding one by applying the field automorphism $\sqrt{5} \mapsto -\sqrt{5}$, or equivalently, composing it with an outer automorphism of G (transposing the two classes of 5-cycles). The last two rows are obtained from doubly transitive permutation representations of G of degree 5 (the natural representation) and 6 (as $\mathrm{PSL}_2(\mathbb{F}_5)$): if G is any permutation group of degree n, the corresponding $n \times n$ permutation matrices give a representation of G on $V = \mathbb{C}^n$ with character $\pi(g) = |\operatorname{fix}(g)|$, the number of points fixed by g; the vectors with coordinate-sum 0 form a G-invariant subspace with character $\pi - 1$, and one can show that this is irreducible if and only if G is doubly transitive.

Proposition 5.2 *If $\mathscr{X}$, $\mathscr{Y}$ and $\mathscr{Z}$ are conjugacy classes in a finite group G, then the number of solutions of $xyz = 1$ in G with $x \in \mathscr{X}$, $y \in \mathscr{Y}$, $z \in \mathscr{Z}$ is equal to*

$$\frac{|\mathscr{X}| \cdot |\mathscr{Y}| \cdot |\mathscr{Z}|}{|G|} \cdot \sum_{\chi} \frac{\chi(x)\chi(y)\chi(z)}{\chi(1)},$$

where χ ranges over the irreducible characters of G.

(For this and similar but more general results, see [45, Chap. 7].) Heuristically, if products were evenly distributed over G one would expect about $|\mathscr{X}| \cdot |\mathscr{Y}| \cdot |\mathscr{Z}|/|G|$ solutions; the summation term, which in many cases takes values close to 1, represents a correction, recognising that products are not evenly distributed and ensuring that the formula always produces a non-negative integer.

Example 5.10 Take $\Delta = \Delta(3,3,5), G = A_5$ and count all $K \trianglelefteq \Delta$ with $\Delta/K \cong A_5$. There is only one choice for the classes $\mathscr{X}$ and $\mathscr{Y}$ of elements x and y of order 3, and there are two choices for the class $\mathscr{Z}$ containing the element z of order 5. In each case Proposition 5.2 gives 60 triples x, y, z in these classes satisfying $xyz = 1$, giving a total of 120 triples. They all generate A_5, since this group has no proper subgroups of order divisible by 15. Since Aut $A_5 \cong S_5$, the number of normal subgroups K is $120/120 = 1$. Thus there is a single regular bipartite map $\mathscr{B}$ of type $(3,3,5)$ with Aut $\mathscr{B} \cong A_5$. By the formula (2.1) in Sect. 2.1.2, it has genus

$$g = 1 + \frac{|G|}{2}\left(1 - \frac{1}{l} - \frac{1}{m} - \frac{1}{n}\right) = 5\,.$$

It is a double covering of the dodecahedron branched over the 12 face–centres, with vertices alternately coloured black and white.

Exercise 5.9 Confirm the result of Example 5.7, classifying the 5-valent regular maps with automorphism group A_5, by using character theory.

Exercise 5.10 Show that there are, up to isomorphism, two regular dessins of type $(5,5,3)$ with automorphism group A_5. Find their genus, and describe how to construct them from an icosahedron.

Example 5.11 One can also apply this technique to other groups G. For instance, in Example 5.3 we stated that if $G = \mathrm{PSL}_2(\mathbb{F}_p)$ for some prime $p \equiv \pm 1 \bmod 7$ then the triangle group $\Delta = \Delta(2,3,7)$ has three normal subgroups K with $\Delta/K \cong G$, and in Example 5.8 we proved this in the case $p \equiv 1 \bmod 7$ by directly constructing and counting appropriate generating triples for G. Now let us consider the case $p \equiv -1 \bmod 7$. As before, there are unique conjugacy classes $\mathscr{X}$ and $\mathscr{Y}$ of elements of orders 2 and 3 in G, with $|\mathscr{X}| = p(p+\delta)/2$ and $|\mathscr{Y}| = p(p+\varepsilon)$ where $\delta := \pm 1 \equiv p \bmod 4$ and $\varepsilon := \pm 1 \equiv p \bmod 3$, and there are three classes $\mathscr{Z}$ of elements of order 7, each with $|\mathscr{Z}| = p(p-1)$. Let us assume that $\delta = 1$, that is, $p \equiv 1 \bmod 4$. Putting $q = p$ in the character table of $\mathrm{PSL}_2(q)$ in Table 5.5 (see Sect. 5.5), we see that the irreducible characters of degree $p-1$ all vanish on $\mathscr{X}$, while those of degrees $(p+1)/2$ or $p+1$ all vanish on $\mathscr{Z}$, so the only characters χ making non-zero contributions to the sum in Proposition 5.2 are the two of degrees 1 and p. Since the latter character takes the values 1, ε and -1 on $\mathscr{X}$, $\mathscr{Y}$ and $\mathscr{Z}$, it follows that for each choice of $\mathscr{Z}$ the number of triples (x, y, z) is

$$\frac{p(p+1)/2 \cdot p(p+\varepsilon) \cdot p(p-1)}{p(p^2-1)/2}\left(1 + \frac{1 \cdot \varepsilon \cdot (-1)}{p}\right) = p(p^2-1).$$

As in Example 5.8, each such triple generates G, so $\mathrm{Aut}\,G$ permutes them semiregularly. Since $\mathrm{Aut}\,G \cong \mathrm{PGL}_2(\mathbb{F}_p)$, of order $p(p^2-1)$, it does so transitively, so we obtain one normal subgroup K in Δ for each of the three choices of $\mathscr{X}$.

Exercise 5.11 Use Table 5.6 in Sect. 5.5 to deal with the case in Example 5.11 where $\delta = -1$, that is, $p \equiv -1 \bmod 4$. (Warning: if $\varepsilon = -1$, that is, $p \equiv -1 \bmod 3$, then the irreducible characters of degree $p-1$ and $(p-1)/2$ all contribute to the sum in Proposition 5.2.)

This technique has been applied to other finite groups, for instance in [28] to count the Hurwitz surfaces associated with the Ree groups $R(3^e)$, and in [30] to count the quotients of $\Delta(2,4,5)$ isomorphic to the Suzuki groups $Sz(2^e)$; see Sect. 2.2.3 for these families of finite simple groups.

5.1.6 Counting by Möbius Inversion

We saw in Proposition 5.1 that the number of normal subgroups of a finitely generated group Δ, with a given finite group G as their quotient, is given by the formula

$$n_\Delta(G) = \frac{|\,\mathrm{Epi}(\Delta, G)|}{|\,\mathrm{Aut}\,G|}\,.$$

If we take Δ to be a triangle group, then this is useful in giving the number of regular dessins or maps of a given type, with a given automorphism group G. In many cases, $|\,\mathrm{Aut}\,G|$ is either known or easily found, and the difficulty in applying this result lies in counting the epimorphisms $\Delta \to G$. Given a presentation for Δ, one has to look for sets of elements of G (images of generators of Δ) which satisfy the defining relations of Δ, so that they give a homomorphism $\theta : \Delta \to G$, and one then has to determine whether these elements generate G, so that θ is an epimorphism.

This second step can be troublesome, but in 1936, Hall [24] introduced a method which allows one to by-pass it, and merely count *homomorphisms*, rather than epimorphisms; the cost of this is that one has to consider homomorphisms to *arbitrary* subgroups of G, rather than just G itself. The starting point is the obvious fact that each homomorphism $\Delta \to G$ is an epimorphism to a unique subgroup $H \le G$, so we have

$$|\,\mathrm{Hom}(\Delta, G)| = \sum_{H \le G} |\,\mathrm{Epi}(\Delta, H)|\,.$$

Now one would like to interchange the roles of the operators Hom and Epi in this equation, so that $|\,\mathrm{Epi}(\Delta, G)|$ (which we need) is expressed in terms of $|\,\mathrm{Hom}(\Delta, H)|$ (which is easier to calculate). A technique called Möbius inversion allows one to do this.

Let Λ be the set of all subgroups of G. The Möbius function for G is the function $\mu_G : \Lambda \to \mathbb{Z}$ defined recursively for $H \in \Lambda$ by

$$\sum_{K \geq H} \mu_G(K) = \delta_{H,G},$$

where δ is the Kronecker delta function, so that

$$\mu_G(G) = 1 \quad \text{and} \quad \mu_G(H) = -\sum_{K>H} \mu_G(K) \quad \text{for all} \quad H < G\,.$$

Given a finite group G, with sufficient knowledge of its subgroup lattice Λ one can calculate the values of μ_G by working through Λ from the top downwards. We have $\mu_G(G) = 1$, so $\mu_G(H) = -1$ for each maximal subgroup H of G. Continuing, if the values of $\mu_G(K)$ are known for all subgroups $K > H$ then $\mu_G(H)$ is minus their sum, so after a finite number of iterations, all values of μ_G are known.

Example 5.12 Let $G = D_p$ for some prime $p > 2$. The proper subgroups H of G are: one subgroup $H \cong C_p$, p subgroups $H \cong C_2$, and the identity subgroup 1. All except 1 are maximal, so they satisfy $\mu_G(H) = -1$. The subgroup 1 is contained in G, with $\mu_G(G) = 1$, and in $p + 1$ subgroups H with $\mu_G(H) = -1$, so $\mu_G(1) = -(1 + (p + 1).(-1)) = p$.

Exercise 5.12 Find the values of μ_G for the dihedral and quaternion groups $G = D_4$ and Q_8 of order 8, and the alternating and symmetric groups A_4 and S_4 of degree 4.

Exercise 5.13 Show that $\mu_G(H) = 0$ unless H is an intersection of maximal subgroups of G.

Exercise 5.14 Show that if $G = C_n$ then $\mu_G(H) = \mu(n/m)$, where H is the unique subgroup of G of order m (dividing n), and μ is the Möbius function on $\mathbb{N}$, defined by $\sum_{m|n} \mu(m) = \delta_{n,1}$.

Hall [24] proved the following:

Theorem 5.2 *If Δ is a finitely generated group and G is a finite group, then*

$$|\operatorname{Epi}(\Delta, G)| = \sum_{H \leq G} \mu_G(H) |\operatorname{Hom}(\Delta, H)|\,. \qquad \square$$

The proof is an easy exercise (do it!), using the recursive definition of μ_G. One can regard this result as a generalisation of the Möbius Inversion Formula from elementary number theory, which states that if functions $f, g : \mathbb{N} \to \mathbb{C}$ satisfy

$$f(n) = \sum_{m|n} g(m)$$

for all $n \in \mathbb{N}$, then

$$g(n) = \sum_{m|n} \mu(n/m)f(m) = \sum_{m|n} \mu(m)f(n/m)$$

for all n. Indeed, μ can be regarded as the Möbius function for the group $\mathbb{Z}$, restricted to its subgroups $n\mathbb{Z}$ of finite index.

If we take $\Delta = \Delta(\infty, \infty, \infty)$, a free group F_2 of rank 2, then $|\operatorname{Hom}(\Delta, H)| = |H|^2$ for each finite group H, since any pair of elements of H can serve as images of the two free generators of Δ, so Theorem 5.2 gives

$$|\operatorname{Epi}(\Delta, G)| = \sum_{H \leq G} \mu_G(H)|H|^2 .$$

Now the regular dessins with automorphism group G correspond to the normal subgroups of $\Delta = F_2$ with quotient group G, so by Proposition 5.1 the number $r(G)$ of these dessins (up to isomorphism) is given by

$$r(G) = \frac{1}{|\operatorname{Aut} G|} \sum_{H \leq G} \mu_G(H)|H|^2 .$$

Example 5.13 If $G = D_p$ for some prime $p > 2$ then by using the values of $\mu_G(H)$ given in Example 5.12, and the fact that $|\operatorname{Aut} D_p| = p(p-1)$ (exercise!), we find that

$$r(D_p) = \frac{1 \cdot p - p \cdot 2^2 - 1 \cdot p^2 + 1 \cdot (2p)^2}{p(p-1)} = 3 ,$$

so there are three regular dessins with automorphism group D_p.

Exercise 5.15 Find these three dessins, giving the type and genus of each.

Exercise 5.16 Show that the groups $G = D_4, Q_8, A_4$ and S_4 have $|\operatorname{Aut} G| = 8, 6, 24$ and 24 respectively, and hence find the number of regular dessins with automorphism group G in each case.

Example 5.14 In [24], Hall determined the values of μ_G for a number of groups G, including $\mathrm{PSL}_2(p)$ for all primes p. When $p = 5$ we have $G \cong A_5$, and in this case Hall found the following values:

- $\mu_G(G) = 1$;
- there are five subgroups $H \cong A_4$, with $\mu_G(H) = -1$;
- there are six subgroups $H \cong D_5$, with $\mu_G(H) = -1$;
- there are ten subgroups $H \cong S_3$, with $\mu_G(H) = -1$;
- there are five subgroups $H \cong V_4 \cong C_2 \times C_2$, with $\mu_G(H) = 0$;
- there are ten subgroups $H \cong C_3$, with $\mu_G(H) = 2$;
- there are fifteen subgroups $H \cong C_2$, with $\mu_G(H) = 4$;
- there is the identity subgroup $H = 1$, with $\mu_G(H) = -60$.

Using this, he showed that $r(G) = 19$, so there are 19 regular dessins with automorphism group $G \cong A_5$, corresponding to the orbits of $\mathrm{Aut}\,A_5 = S_5$ on generating pairs for A_5. They are described, as hypermaps, by Breda and Jones in [3].

Exercise 5.17 Confirm Hall's calculation of $r(A_5)$, and determine the type and genus of each of these 19 regular dessins.

Since Hall's 1936 paper, the values of μ_G have been determined and applied to enumeration for a number of other groups G, including the groups $\mathrm{PSL}_2(q)$ and $\mathrm{PGL}_2(q)$ for all prime powers q by Downs [14] (see [15] for applications), the Suzuki groups $Sz(2^e)$ by Downs and Jones [16], and the 'small' Ree groups $R(3^e)$ by Pierro [40]. In [8], Connor and Leemans have an online atlas of subgroup lattices of almost simple finite groups, including their Möbius functions.

5.2 Low Genera

For any fixed genus $g \geq 2$, the Hurwitz bound $|\,\mathrm{Aut}\,X| \leq 84(g-1)$ (see Sect. 5.1.2) shows that only finitely many groups G can arise as automorphism groups of regular dessins of that genus. Each such group has only finitely many generating triples, so it corresponds to only finitely many regular dessins. It is therefore feasible, at least in principle, to classify all the regular dessins of a given genus $g \geq 2$.

In this section we will describe the isomorphism classes of quasiplatonic curves of genus $g = 2, 3$ and 4 (see Sect. 6.2 for genera 0 and 1, and for further classification methods not used here). The classification is based in part on the classification of regular maps given in [11, 21] and [46], that is, the classification of all torsion-free normal subgroups of triangle groups $\Delta(2, q, r)$, and in part on a classification of canonical representations by Kuribayashi and Kuribayashi (see [34] and the references given there). The information given below can all be obtained from a case-by-case application of Singerman's Theorem 3.11 [48, 49], supported by Takeuchi's inclusion relations [54], and in many cases the equations are easy to guess from the ramification data of the Belyĭ functions.

5.2.1 Genus 2

For genus $g = 2$ all compact Riemann surfaces are hyperelliptic. In this case one obtains three non-isomorphic quasiplatonic curves X, with affine models

$$y^2 = x^6 - x \tag{5.1}$$

$$y^2 = x^6 - 1 \tag{5.2}$$

$$y^2 = x^5 - x\,. \tag{5.3}$$

Table 5.2 Regular dessins of genus 2

Equation	p	q	r	$\Delta(p,q,r)/N$	$\|\Delta(p,q,r)/N\|$
(5.1)	2	5	10	C_{10}	10
	5	5	5	C_5	5
(5.2)	2	4	6	$(C_3 \times C_2 \times C_2) \rtimes C_2$	24
	2	6	6	$C_3 \times C_2 \times C_2$	12
	3	6	6	$C_3 \times C_2$	6
	3	4	4	$C_3 \rtimes C_4$	12
(5.3)	2	3	8	$\mathrm{GL}_2(3) \cong Q \rtimes S_3$	48
	3	3	4	$\mathrm{SL}_2(3) \cong Q \rtimes C_3$	24
	2	4	8	$Q \rtimes C_2 \cong C_8 \rtimes C_2$	16
	2	8	8	C_8	8
	4	4	4	Q	8

The corresponding covering groups N are normal subgroups of the triangle groups $\Delta = \Delta(p,q,r)$ listed in Table 5.2. For genus 2 these groups are all arithmetically defined (there are 85 arithmetic triangle groups, classified by Takeuchi [53]). Each normal inclusion $N \trianglelefteq \Delta(p,q,r)$ corresponds to a regular dessin of type (p,q,r) with automorphism group $G = \Delta/N \leq \mathrm{Aut}\,X$; it is the full automorphism group $\mathrm{Aut}\,X$ if we take Δ to be a maximal triangle group containing N as a normal subgroup.

Here C_m denotes the cyclic group of order m. The automorphism group G of curve (5.2), with $\Delta = \Delta(2,4,6)$, is a semidirect product: the complement C_2 acts by conjugation on the normal subgroup $C_3 \times C_2 \times C_2$ by inverting the first factor and transposing the second and third. The other three rows for (5.2) refer to subgroups of G, images of triangle groups contained in $\Delta(2,4,6)$, and thus automorphism groups of regular dessins on this curve.

For the curve (5.3), the isomorphism $G \cong \Delta(2,3,8)/N \cong \mathrm{GL}_2(\mathbb{F}_3)$ is provided by

$$g_0 = \begin{pmatrix} -1 & 0 \\ 1 & 1 \end{pmatrix}, \; g_1 = \begin{pmatrix} 1 & 1 \\ 0 & 1 \end{pmatrix}, \; g_\infty = \begin{pmatrix} 1 & -1 \\ 1 & 1 \end{pmatrix},$$

where we denote the generators of G induced by the generators of Δ by g_0, g_1 and g_∞. Here Q denotes the quaternion group of order 8, generated by i, j, k with $i^2 = j^2 = k^2 = -1$ and $ijk = 1$; the action of S_3 on Q is defined by

$$(12) \; : \; i \mapsto j \,, \; j \mapsto i \,, \; k \mapsto -k$$

$$(23) \; : \; i \mapsto -i \,, \; j \mapsto k \,, \; k \mapsto j \,.$$

The centre $Z \cong C_2$ of G is generated by the hyperelliptic involution g_∞^4, and the factor group G/Z is isomorphic to $S_4 \cong \mathrm{PGL}_2(\mathbb{F}_3)$ acting on $\mathbb{P}^1(\mathbb{C}) \cong X/Z$. The fixed points of g_∞ are the face-centres of a regular cube, or the vertices of a regular octahedron in $\hat{\mathbb{C}}$.

The equations are classical (Bolza [2]) and can be determined, as for all hyperelliptic quasiplatonic curves, by using the fact that the polynomial $p \in \mathbb{C}[x]$ on the right-hand side of the equation $y^2 = p(x)$ has to be invariant under the action of G/Z on $\hat{\mathbb{C}}$. In determining these polynomials it is (in particular for higher genera) more useful to work, not with the maximal automorphism group of the curve, but with a subgroup which is easier to handle. In this example C_8 is a suitable subgroup, visible in the symmetry of Fig. 5.1. This shows a fundamental region for N as a subgroup of $\Delta(2,8,8)$, drawn in the unit disc instead of $\mathbb{H}$; opposite sides of the octagon are identified to form the surface. The boundary lines of the fundamental region are the edges of the corresponding dessin: this has four white vertices $P_1, \ldots, P_4$ of valency 2, one black vertex Q of valency 8, eight edges and one face. (This dessin also appears as an example in Sect. 3.3.2—see Fig. 3.5.) The 16 triangles shown within the face are fundamental regions for the extended triangle group $\Delta[2,8,8]$, with a red vertex at R; adjacent pairs of them form fundamental regions for $\Delta(2,8,8)$.

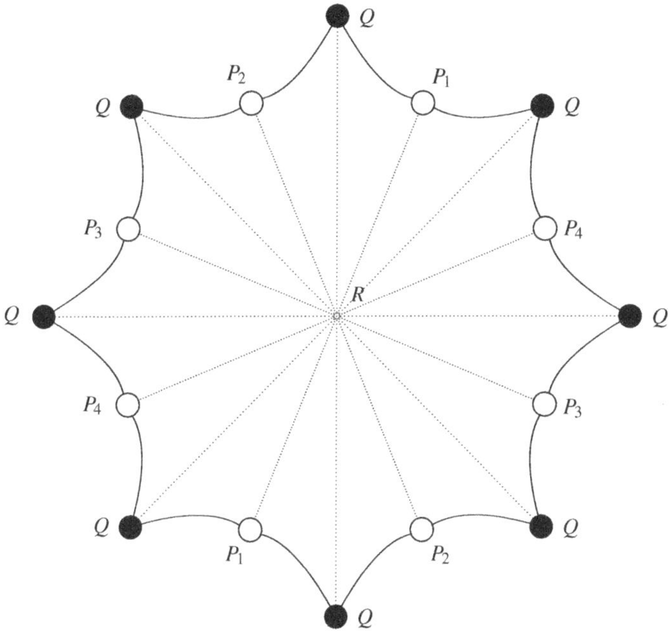

Fig. 5.1 Fundamental region for N in Eq. (5.3)

5.2.2 Genus 3

Here we have eight non-isomorphic quasiplatonic curves, which can be described by the models

$$y^2 = x^8 - x \tag{5.4}$$

$$y^2 = x^7 - x \tag{5.5}$$

$$y^2 = x^8 - 1 \tag{5.6}$$

$$y^2 = x^8 - 14x^4 + 1 \tag{5.7}$$

$$y^3 = x(1 - x^3) \tag{5.8}$$

$$y^4 + x^3 = 1 \tag{5.9}$$

$$y^4 + x^4 = 1 \tag{5.10}$$

$$x^3y + y^3z + z^3x = 0\,. \tag{5.11}$$

Their universal covering groups N are normal subgroups of the triangle groups $\Delta(p, q, r)$ listed with their quotients and their indices in Table 5.3.

The first four curves are hyperelliptic, the last two are the Fermat curve F_4 and Klein's quartic. With the exception of the curve (5.7), all the triangle groups involved are arithmetically defined. The semidirect products are explained in the comments below.

Comments For the curve (5.5), $\Delta(4, 4, 6)/N \cong C_3 \rtimes C_4$ is isomorphic to the group $\Delta(3, 4, 4)/N$ in case (5.2) for genus 2, but with different generators.

In the case of the curve (5.6), if α and β generate the direct factors of the normal subgroup $C_8 \times C_2$ of $G = \Delta(2, 4, 8)/N$, then the generator γ of the complement C_2 acts on this subgroup by

$$\gamma^{-1}\alpha\gamma = \alpha^{-1}\beta\,, \quad \gamma^{-1}\beta\gamma = \beta\,,$$

and we may choose $g_0 = \gamma$, $g_1 = \alpha\gamma$, $g_\infty = \beta\alpha$. The subgroup $\Delta(4, 4, 4)/N$ is generated by α^2 and $\alpha\gamma$.

The polynomial on the right-hand side of (5.7) is chosen so that its zeros form the vertices of a regular cube, or the face-centres of a regular octahedron on $\hat{\mathbb{C}}$. The equation shows that the curve is a double cover of an elliptic curve with equation

$$y^2 = x^4 - 14x^2 + 1\,.$$

To get Eq. (5.8), one can consider the curve X as a three-fold regular cover $X \to \hat{\mathbb{C}}$ with covering group $G/C_3 \cong C_3$, because between the triangle group and N one has by [48] a normal subgroup of signature $\langle 0; 3, 3, 3, 3, 3\rangle$ (see Theorem 3.11). The covering group fixes two of the five fixed points (0 and ∞, say) and permutes the three others (the cube roots of unity ζ_3^k, say). Then the automorphism group of

Table 5.3 Regular dessins of genus 3

Equation	p	q	r	$\Delta(p,q,r)/N$	$\lvert\Delta(p,q,r)/N\rvert$
(5.4)	2	7	14	C_{14}	14
	7	7	7	C_7	7
(5.5)	2	4	12	$S_3 \times C_4$	24
	2	12	12	C_{12}	12
	4	4	6	$C_3 \rtimes C_4$	12
(5.6)	2	4	8	$(C_8 \times C_2) \rtimes C_2$	32
	2	8	8	$C_8 \times C_2$	16
	4	8	8	C_8	8
	4	4	4	$C_4 \rtimes C_4$	16
(5.7)	2	4	6	$S_4 \times C_2$	48
	2	6	6	$A_4 \times C_2$	24
	3	4	4	S_4	24
(5.8)	3	9	9	C_9	9
(5.9)	2	3	12	$SL_2(3) \circ C_4$	48
	3	3	6	$\mathrm{SL}_2(3)$	24
	3	4	12	C_{12}	12
(5.10)	2	3	8	$(C_4 \times C_4) \rtimes S_3$	96
	3	3	4	$(C_4 \times C_4) \rtimes C_3$	48
	2	4	8	$(C_4 \times C_4) \rtimes C_2$	32
	4	4	4	$C_4 \times C_4$	16
	2	8	8	$C_8 \rtimes C_2$	16
	4	8	8	C_8	8
(5.11)	2	3	7	$\mathrm{PSL}_2(7)$	168
	3	3	7	$C_7 \rtimes C_3$	21
	7	7	7	C_7	7

X is generated by

$$(x, y) \mapsto (\zeta_3 x, \zeta_9 y)$$

and the Belyĭ function can be given as $(x, y) \mapsto y^3/x = 1 - x^3$.

The automorphism group $G \cong \Delta(2,3,12)/N$ of the curve (5.9) is the central product $\mathrm{SL}_2(\mathbb{F}_3) \circ C_4 \cong (\mathrm{SL}_2(\mathbb{F}_3) \times C_4)/C_2$ of $\mathrm{SL}_2(\mathbb{F}_3)$ and C_4, amalgamating their central subgroups of order 2. The centre $Z = \langle g_\infty^3 \rangle \cong C_4$ of G has quotient $G/Z \cong A_4$. In G we have an index 2 subgroup $\Delta(3,3,6)/N$, isomorphic to $\mathrm{SL}_2(\mathbb{F}_3)$ via

$$g_0 = \begin{pmatrix} -1 & -1 \\ 1 & 0 \end{pmatrix}, \quad g_1 = \begin{pmatrix} 1 & 0 \\ 1 & 1 \end{pmatrix}, \quad g_\infty = \begin{pmatrix} 0 & 1 \\ -1 & 1 \end{pmatrix}.$$

(Recall that the centre of $\mathrm{SL}_2(\mathbb{F}_3)$ has order 2, with quotient $\mathrm{PSL}_2(\mathbb{F}_3) \cong A_4$.) In order to determine the equation, the automorphism group $C_{12} \cong \Delta(3,4,12)/N$ is more useful, since it allows one to consider the curve as a three- or four-fold cover of $\hat{\mathbb{C}}$ ramified over suitable roots of unity.

For the automorphism group of the Fermat curve (5.10), write C_4 additively, and let $C_4 \times C_4$ be given by

$$\{ (\xi, \eta, \zeta) \in C_4^3 \mid \xi + \eta + \zeta = 0 \}$$

with S_3 permuting the coordinates. For the quotients of the other triangle groups we mention only the presentation

$$\Delta(2,8,8)/N = \langle g_0, g_1 \mid g_0^2 = g_1^8 = 1 \, , \, g_0 g_1 g_0 = g_1^5 \rangle \cong C_8 \rtimes C_2 \, .$$

Finally, (5.11) is Klein's quartic. The last quotient $\Delta(7,7,7)/N = \langle \alpha \rangle \cong C_7$ in the table leads to

$$y^7 = x\,(x-1)^2$$

as another useful model and the (known) conclusion that it is a quotient of the Fermat curve F_7. The generators

$$g_0 = \alpha \, , \qquad g_1 = \alpha^2 \, , \qquad g_\infty = \alpha^4$$

for C_7 used for this model of (5.11) cannot be transformed by an automorphism of C_7 into the generators

$$g_0 = \alpha \, , \qquad g_1 = \alpha \, , \qquad g_\infty = \alpha^5$$

for the curve (5.4), so these two epimorphisms $\Delta(7,7,7) \to C_7$ have different kernels (which are not even conjugate in $\mathrm{PSL}_2(\mathbb{R})$, so the curves are not isomorphic).

5.2.3 Genus 4

Here we have eleven isomorphism classes of quasiplatonic curves, given by the models

$$y^2 = x^9 - 1 \tag{5.12}$$

$$y^2 = x(3x^4 + 1)(3x^4 + 6x^2 - 1) \tag{5.13}$$

$$y^2 = x^9 - x \tag{5.14}$$

$$y^2 = x^{10} - 1 \tag{5.15}$$

$$y^{10} = x^2\,(x-1) \tag{5.16}$$

$$y^{12} = x^3(x-1)^2 \tag{5.17}$$

$$y^{15} = x^5(x-1)^3 \tag{5.18}$$

$$y^3 = 1 - x^6 \tag{5.19}$$

$$y^{12} = x^4 - x^5 \tag{5.20}$$

$$1 = \frac{4}{27}\frac{(x^2-x+1)^3}{x^2(x-1)^2} + \frac{4}{27}\frac{(y^2-y+1)^3}{y^2(y-1)^2} \tag{5.21}$$

$$x_1^n + \ldots + x_5^n = 0 \quad \text{for } n = 1, 2, 3\,. \tag{5.22}$$

With D_m denoting a dihedral group of order $2m$, we give in Table 5.4 the corresponding triangle groups containing their universal covering groups N as

Table 5.4 Regular dessins of genus 4

Equation	p	q	r	$\Delta(p,q,r)/N$	$\lvert\Delta(p,q,r)/N\rvert$
(5.12)	2	9	18	C_{18}	18
	9	9	9	C_9	9
(5.13)	3	4	6	$\mathrm{SL}_2(3)$	24
(5.14)	2	4	16	$C_{16} \rtimes C_2$	32
	2	16	16	C_{16}	16
	4	4	8	$C_8\,.\,C_2$	16
(5.15)	2	4	10	$(C_{10} \times C_2) \rtimes C_2$	40
	2	10	10	$C_{10} \times C_2$	20
	5	10	10	C_{10}	10
	4	4	5	$C_5 \rtimes C_4$	20
(5.16)	5	10	10	C_{10}	10
(5.17)	4	6	12	C_{12}	12
(5.18)	3	5	15	C_{15}	15
(5.19)	2	6	6	$S_3 \times C_6$	36
	3	6	6	$C_3 \times C_6$	18
	3	6	6	$S_3 \times C_3$	18
(5.20)	2	3	12	$C_3 \times S_4$	72
	3	3	6	$C_3 \times A_4$	36
	2	6	12	$C_3 \times D_4$	24
	3	12	12	C_{12}	12
	6	6	6	$C_3 \times C_2 \times C_2$	12
(5.21)	2	4	6	$S_3 \wr C_2$	72
	2	6	6	$S_3 \times S_3$	36
	3	6	6	$C_3 \times S_3$	18
	3	4	4	$(S_3 \wr C_2) \cap A_6$	36
(5.22)	2	4	5	S_5	120
	2	5	5	A_5	60
	4	4	5	$C_5 \rtimes C_4$	20

normal subgroups of genus 4. The curves (5.12)–(5.15) are hyperelliptic. All the associated triangle groups, with the exception of those corresponding to (5.14), (5.17) and (5.18), are arithmetically defined.

Comments In (5.13), the point ∞ and the zeros of the right-hand side form the vertices and the midpoints of the edges of a regular tetrahedron. The automorphism group must therefore have a quotient isomorphic to A_4. In fact, $\Delta(3,4,6)/N \cong \mathrm{SL}_2(\mathbb{F}_3)$ via

$$g_0 = \begin{pmatrix} 1 & 1 \\ 0 & 1 \end{pmatrix}, \; g_1 = \begin{pmatrix} 0 & -1 \\ 1 & 0 \end{pmatrix}, \; g_\infty = \begin{pmatrix} 0 & 1 \\ -1 & 1 \end{pmatrix};$$

recall again that $\mathrm{PSL}_2(\mathbb{F}_3) \cong A_4$.

In (5.14), the generator β of C_2 acts on the generator α of C_{16} by $\beta^{-1}\alpha\beta = \alpha^7$. The group in the third line is a non-split extension of a normal subgroup C_8 by C_2.

In (5.15), if α and β generate the direct factors of the normal subgroup $C_{10} \times C_2$ of $\mathrm{Aut}\,X$, the generator γ of the complement C_2 acts on them by

$$\gamma^{-1}\alpha\gamma \;=\; \alpha^{-1}\beta \quad , \qquad \gamma^{-1}\beta\gamma \;=\; \beta \,.$$

Note that $\delta := \alpha\gamma$ has order 4 and acts on $\lambda := \alpha^2 = (\alpha\beta)^2$ by $\delta^{-1}\lambda\delta = \lambda^{-1}$. This gives the definition of $\Delta(4,4,5)/N = C_5 \rtimes C_4$ (not isomorphic to the corresponding quotient for the curve (5.22)!). The group $\Delta(5,10,10)/N = \langle\alpha\rangle \cong C_{10}$ is isomorphic to the automorphism group of (5.16), but here with generators $g_0 = \alpha^8, g_1 = g_\infty = \alpha$ and for (5.16) with generators $g_0 = \alpha^6, g_1 = \alpha, g_\infty = \alpha^3$.

The curve (5.19) is a 3-fold cover of the curve described in Eq. (5.2). Between N and $\Delta(2,6,6)$ there are two triangle groups of type $\Delta(3,6,6)$, conjugate in $\Delta(2,4,6)$ but not in $\Delta(2,6,6)$. In fact, quasiplatonic curves can have two different regular dessins of the same signature but with different automorphism groups only in cases of signature $\Delta(n,2n,2n)$, see [22]. There is an infinite family of such cases with highly interesting geometric properties [29].

The automorphism group of (5.21), of order 72, is a wreath product of S_3 by C_2, that is, a semidirect product $(S_3 \times S_3) \rtimes C_2$ where the complement C_2 acts by transposing the two factors S_3. We may read these two factors as permutation groups of $\{1,2,3\}$ and $\{4,5,6\}$, respectively, and C_2 as transposing these two sets. Thus the automorphism group is a subgroup of the symmetric group S_6. With $\sigma := (1\,4)(2\,5)(3\,6) \in C_2$ the permutations of orders $2,4,6$

$$(5\,6)\,, \quad (1\,5)(2\,4\,3\,6) = (4\,5)(1\,3\,2)\sigma \quad \text{and} \quad (1\,5\,3\,4\,2\,6) = \sigma(1\,2\,3)(4\,5\,6)$$

generate the full automorphism group of the dessin. Its subgroup of even permutations is the automorphism group of the dessin of type $(3,4,4)$. Manfred Streit determined the affine equation (5.21) as a model, using the facts that the quotient of X by each factor S_3 has genus 0, and that one can easily determine the Belyĭ function induced by the triangle group $\langle 2,6,6\rangle$ on each quotient; compare this with

Eq. (1.2). Both Belyĭ functions generate the function field of X and are related by $\beta_1 + \beta_2 = 1$, giving the equation for the curve.

Finally, there is Bring's curve (5.22), given by three homogeneous equations in $\mathbb{P}^4(\mathbb{C})$ (see Example 5.7). Here, the subgroup $\Delta(4,4,5)/N$ of the automorphism group S_5 is generated by the permutations $\alpha = (12345)$ and $\beta = (2354)$, so that

$$\beta^{-1}\alpha\beta = \alpha^{-2} ;$$

this defines a semidirect product $C_5 \rtimes C_4$ not isomorphic to that for the curve (5.15). Like Klein's quartic and the Macbeath-Hurwitz curves discussed in Sect. 5.1.2, Bring's curve is a Shimura curve, that is, it has an arithmetically defined surface group. In fact, $\Delta(2,5,5)$ is the norm one group of a quaternion algebra with centre $k := \mathbb{Q}(\sqrt{5})$, and the surface group N is its principal congruence subgroup of level $\sqrt{5}$. The prime ideal $\langle\sqrt{5}\rangle$ is unramified in the quaternion algebra, but it is ramified in $k/\mathbb{Q}$ and has norm $q = 5$, so we have a quotient group $\Delta/N \cong \mathrm{PSL}_2(\mathbb{F}_5) \cong A_5$. The other triangle groups $\Delta(2,4,5)$ and $\Delta(4,4,5)$ also have arithmetic interpretations. The prime 2 is inert in k and ramifies in the algebra (see [54]), so it normalises the norm one group and therefore generates an index 2 extension of it; this is the maximal triangle group $\Delta(2,4,5)$, with quotient $\Delta(2,4,5)/N \cong \mathrm{PGL}_2(\mathbb{F}_5)$, whereas $\Delta(4,4,5)/N$ is isomorphic to $\mathrm{AGL}_1(\mathbb{F}_5)$, the stabiliser in $\mathrm{PGL}_2(\mathbb{F}_5)$ of $\infty \in \mathbb{P}^1(\mathbb{F}_5)$.

It is remarkable that each of these low-genus quasiplatonic curves X is uniquely determined up to isomorphism by $\mathrm{Aut}\,X$ and its maximal triangle group Δ. According to a recent investigation [7], based on Conder's list of regular dessins [5], this is also true for genus 5. It fails first in genus 6 for some non-isomorphic curves of the form

$$y^{14} = x^m(1-x)^n$$

with automorphism group C_{14} and with universal covering groups normal in the triangle group $\Delta(7,14,14)$.

5.3 Infinite Families

As g increases, it becomes increasingly laborious to study and understand the quasiplatonic curves of genus g, though Conder's lists [5] now give valuable basic data for g up to about 300—see also the census of rotary (= regular, in our sense) maps compiled by Potočnik [42]. (In fact, it is not so much the size of g as the number of factors of $g-1$ which causes the problem: many quasiplatonic curves X arise naturally as unbranched coverings of curves Y of smaller genus, and in this case the Euler characteristic $\chi = 2(1-g)$ of X is a multiple of that of Y.) However, there are several infinite families of quasiplatonic curves which have interesting and important properties, and which arise either for all values of g, or at least for infinitely many g satisfying some number-theoretic condition (see Example 5.6 for the Lefschetz curves, for instance). In many cases, the curves in a particular family arise as coverings, usually branched, of some basic curves of small genus.

Example 5.15 The *Accola-Maclachlan curves* form one of the most important families. We saw in Sect. 5.1.2 that the automorphism group of a curve of genus $g \geq 2$ has order at most $84(g-1)$, and that this upper bound is attained (by the Hurwitz curves) for infinitely many values of g. However, there are also infinitely many values for which the bound is not attained, and indeed many for which there is a much smaller upper bound. Accola [1] and Maclachlan [38], independently and more or less simultaneously, showed that the upper bound is at least $8(g+1)$ and that it takes this value for infinitely many values of g. Here we will prove the first assertion, by describing a curve with $8(g+1)$ automorphisms for each g, though we will not prove the second assertion, that there are infinitely many values of g for which there is no curve with a larger automorphism group.

It is easy to check that the curve X with affine model

$$y^2 = x^{2k} - 1$$

has automorphisms $\alpha : (x, y) \mapsto (\zeta_{2k}x, y)$ and $\beta : (x, y) \mapsto (x^{-1}, iyx^{-k})$, and that these satisfy $\alpha^{2k} = \beta^4 = (\alpha\beta)^2 = 1$. It follows that X is uniformised by a normal subgroup K of the triangle group $\Delta = \Delta(2k, 2, 4)$ with $\Delta/K \cong G := \langle\alpha, \beta\rangle$. The $2k$ points $(x, 0) \in X$ with $x^{2k} = 1$ form an orbit of G of length $2k$, and the stabiliser of $(1, 0)$ has order divisible by 4 since it contains β, so by the orbit-stabiliser theorem $|G|$ is divisible by their product $2k \cdot 4 = 8k$. One can check that β^2 commutes with α and therefore generates a central subgroup of order 2; then $G/\langle\beta^2\rangle$ is a quotient of $\Delta(2k, 2, 2) \cong D_{2k}$, so $|G|$ divides $8k$. Thus $|G| = 8k$ and $G/\langle\beta^2\rangle \cong D_{2k}$. Moreover, G has a presentation

$$G = \langle \alpha, \beta, \gamma \mid \alpha^{2k} = \beta^2 = \gamma^4 = \alpha\beta\gamma = 1, (\alpha^2)^\beta = \alpha^{-2} \rangle$$

where we use the notation $\delta^\beta := \beta^{-1}\delta\beta$. The normal inclusion of K in Δ shows that X is a quasiplatonic curve, carrying a regular dessin $\mathcal{D}$ of type $(2k, 2, 4)$ with $\operatorname{Aut}\mathcal{D} \cong G$. The inverse image of $\langle\beta^2\rangle$ in Δ is a normal subgroup L, containing K with index 2, with $\Delta/L \cong D_{2k}$; it corresponds to the unique regular dessin $\overline{\mathcal{D}}$ of type $(2k, 2, 2)$ on the sphere. (Like other regular dessins of type $(l, 2, n)$, $\mathcal{D}$ and $\overline{\mathcal{D}}$ may be regarded as regular maps of type $\{4, 2k\}$ and $\{2, 2k\}$ in the notation used by Coxeter and Moser in [11, Chap. 8], as we will explain in Chap. 6.) The dessin $\mathcal{D}$ is a double cover of $\overline{\mathcal{D}}$, branched over its k face centres. Applying the Riemann-Hurwitz formula to this covering, we see that X has genus

$$g = 1 - 2 + \frac{2k}{2} = k - 1,$$

so $|G| = 8(g+1)$.

It follows from Singerman's classification of inclusions between triangle groups (Theorem 3.12) that Δ is a maximal Fuchsian group, so it is the normaliser of K in $PSL_2(\mathbb{R})$ and hence Aut $X = G$. Thus X has $8(g+1)$ automorphisms, as claimed.

There is another useful construction for X and $\mathcal{D}$. The easily-verified relation $(\alpha^2)^\beta = \alpha^{-2}$ shows that α^2 generates a normal subgroup of G; it is isomorphic to C_k, with $G/\langle\alpha^2\rangle \cong \Delta(2,2,4) \cong D_4$. Thus $\mathcal{D}$ is a k-sheeted cyclic cover of the regular spherical dessin of type $(2,2,4)$ (the regular map of type $\{4,2\}$), branched over its four white vertices.

Example 5.16 Kulkarni [33] later found another family of quasiplatonic curves with properties very similar to those of the Accola-Maclachlan curves: for instance, each curve has $8(g+1)$ automorphisms, where g is its genus. However, these curves exist only for $g \equiv 3 \bmod 4$. The following exercise shows their construction.

Exercise 5.18 Show that if $0 < k = 2l \equiv 0 \bmod 4$ then the group

$$G := \langle \alpha, \beta, \gamma \mid \alpha^{2k} = \beta^2 = \gamma^4 = \alpha\beta\gamma = 1, (\alpha^2)^\beta = \alpha^{2(l-1)} \rangle$$

of order $8k$ is the automorphism group of a quasiplatonic curve of genus $g = k - 1$, carrying a regular dessin of type $(2k, 2, 4)$ (or regular map of type $\{4, 2k\}$) which is a k-sheeted cyclic cover of the regular spherical dessin of type $(2,2,4)$, branched over its white vertices.

Turbek [56] has given the equation

$$y^{2g+2} = (x-1)x^{g-1}(x+1)^{g+2}$$

for this curve, and has also given explicit formulae for generators of G. (An automorphism of order $2g+2$ is obvious.)

Accola [1] and Maclachlan [38] used a similar construction to describe another family of curves, this time of genus g divisible by 3, with a slightly larger automorphism group, of order $8(g+3)$, as follows:

Exercise 5.19 Show that for each $k > 1$ the group

$$G = \langle \alpha, \beta, \gamma \mid \alpha^{3k} = \beta^2 = \gamma^4 = \alpha\beta\gamma = 1, (\alpha^3)^\beta = \alpha^{-3} \rangle$$

of order $24k$ is the automorphism group of a quasiplatonic curve of genus $3(k-1)$, carrying a regular dessin of type $(3k, 2, 4)$ (or regular map of type $\{4, 3k\}$) which is a k-sheeted cyclic covering of the unique regular spherical dessin of type $(3,2,4)$ (the cube), branched over its eight white vertices.

The Accola-Maclachlan curve of genus g has an affine model given by

$$y^2 = x^n - 1$$

where $n = 2(g+1)$ is even. If we take an odd value $n = 2g+1$ we obtain the *Wiman curve* [58] of genus g. This attains the upper bound $4g+2$ for the order of a single automorphism of a Riemann surface of genus $g > 1$ (see also [25]):

Exercise 5.20 Show that the curve X given by $y^2 = x^n - 1$, with $n = 2g+1$, has genus g and has an automorphism of order $4g+2$. Show that X is a quasiplatonic curve, carrying a regular dessin of type $(n, 2, 2n)$ (or map of type $\{2n, n\}$). Show how this dessin can be formed by suitably identifying sides of a $2n$-gon.

5.4 Fields of Definition Again

In the next chapters we will see that for many quasiplatonic curves it is possible to determine the field of moduli. Strangely enough, as for curves with trivial automorphism group, their field of moduli can be used as a (minimal!) field of definition. We begin with the analogous statement for regular dessins due to a more general result of Coombes and Harbater [10].

Theorem 5.3 *Regular dessins can be defined over their field of moduli.*

Proof Recall that a dessin $\mathscr{D}$ on a curve X is regular if and only if its Belyĭ function β defines a regular covering of the projective line $\mathbb{P}^1(\mathbb{C})$. For each element g of the absolute Galois group $\mathbb{G}$, the Belyĭ function β^g is again a regular covering $X^g \to \mathbb{P}^1(\mathbb{C})$ since g conjugates all covering transformations of β into covering transformations of β^g. By definition of 'moduli field' we know that, if $g \in \mathbb{G}(\mathscr{D})$, that is if g fixes $M(\mathscr{D})$, there is an isomorphism $f_g : X \to X^g$ such that $\beta = \beta^g \circ f_g$. A priori, there are as many choices for f_g as the degree of β (and β^g, of course). If we can replace this with a *unique* choice forcing the different f_g to satisfy Weil's cocycle condition in Theorem 4.10, the moduli field can serve as a field of definition (as in the case of curves with trivial automorphism group).

The trick is to choose a rational point $r \neq 0, 1, \infty$ in $\mathbb{P}^1(\mathbb{C})$ and some $x \in \beta^{-1}(r) \subset X$. Then $g(r) = r$ and $g(x) \in (\beta^g)^{-1}(r) \subset X^g$; since the covering group acts transitively on the fibres of β^g there is a unique choice of f_g such that $f_g(x) = g(x)$ making the following diagram

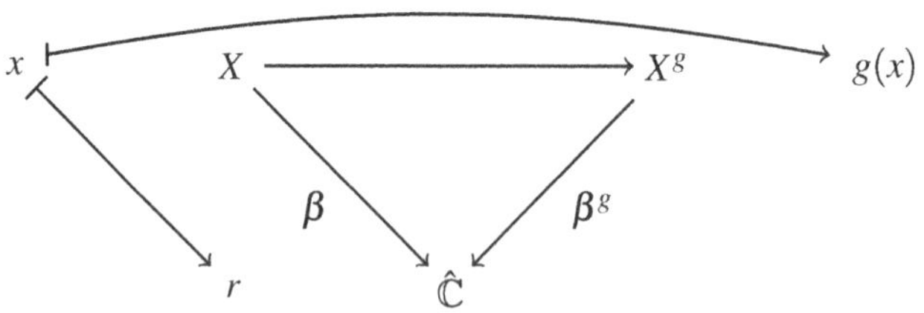

commute. With this choice of the isomorphisms f_g, Weil's cocycle condition is automatically satisfied. □

Theorem 5.4 *Quasiplatonic curves X can be defined over their moduli fields M(X).*

Proof (We follow [59, Theorem 5]; another possible proof can be deduced from [12].) The canonical projection $X \to (\operatorname{Aut} X)\backslash X \cong \mathbb{P}^1(\mathbb{C})$ is a Belyĭ function β; assume first that the critical points are $0, 1, \infty$ and that the ramification orders of β over $0, 1, \infty$ are pairwise distinct. Now let $g \in \mathbb{G}(X)$, that is, let g fix $M(X)$ elementwise. By definition, there is an isomorphism $f_g : X \to X^g$. Since the ramification orders of β are g-invariant by Theorem 4.11, and since

$$\operatorname{mult}_{g(x)} \beta^g = \operatorname{mult}_x \beta \qquad \text{and} \qquad \operatorname{mult}_{f_g(x)} \beta^g = \operatorname{mult}_x \beta ,$$

the composed map $\beta^g \circ f_g$ must also be a canonical projection

$$X \to (\operatorname{Aut} X)\backslash X \cong \mathbb{P}^1(\mathbb{C})$$

like β, so it coincides with β up to a Möbius transformation m_g satisfying

$$\beta^g \circ f_g = m_g \circ \beta . \tag{5.23}$$

On the other hand, m_g has to fix $0, 1, \infty$ elementwise, since the multiplicities are pairwise distinct and hence determine the critical values. Therefore m_g is the identity, so $g \in \mathbb{G}(\mathscr{D})$ and hence $\mathbb{G}(X) \leq \mathbb{G}(\mathscr{D})$. Thus $M(\mathscr{D}) \leq M(X)$, and because the reverse inclusion is obvious, both moduli fields coincide, and X can be defined over $M(X)$ by Theorem 5.3.

This argument needs to be modified (as we will do later in this section) if some multiplicities coincide, that is, if the regular dessin has type $\langle p, p, q \rangle$ or $\langle p, p, p \rangle$. In such cases the zeros and poles of the canonical projection may be transposed, so we can no longer expect that each Möbius transformation m_g is the identity. Thus $M(\mathscr{D})$ could be a proper extension of $M(X)$. We will first prove, by comparing the groups $\mathbb{G}(X)$ and $\mathbb{G}(\mathscr{D})$, that in this case the extension has degree $2, 3$ or 6 (we currently know no examples of degree 6):

Lemma 5.2 *Let X be a quasiplatonic curve with canonical Belyĭ function β and let $g \in \mathbb{G}(X)$. Fix an identification of $(\operatorname{Aut} X)\backslash X$ with $\mathbb{P}^1(\mathbb{C})$. Then*

1. *the projection $X \to (\operatorname{Aut} X)\backslash X \cong \mathbb{P}^1(\mathbb{C})$ is uniquely determined, so by Eq. (5.23) $m_g := \beta^g \circ f_g \circ \beta^{-1}$ is well-defined;*
2. *m_g depends only on $g \in \mathbb{G}(X)$, not on the choice of f_g;*
3. *if S_3 denotes the symmetric group on $\{0, 1, \infty\}$, the map*

$$\mathbb{G}(X) \to S_3 \ : \ g \mapsto m_g|_{\{0,1,\infty\}}$$

defines an antihomomorphism with kernel $\mathbb{G}(\mathscr{D})$;

4. *$M(\mathscr{D})/M(X)$ is a Galois extension;*

5. *m_g depends only on the restriction of g to $M(\mathscr{D})$, and it defines an injective antihomomorphism*

$$\operatorname{Gal} M(\mathscr{D})/M(X) \to S_3 \ : \ g \mapsto m_g|_{\{0,1,\infty\}} \ .$$

Proof Part (1) is clear, and (2) follows from the fact that for another choice f_g' of the isomorphism, $f_g^{-1} \circ f_g'$ is an automorphism of X fixing the fibres of β over $0, 1$ and ∞.

For (3) recall that each m_g is a Möbius transformation with rational coefficients. For all $g, h \in \mathbb{G}(X)$ we therefore have

$$m_{gh} \ = \ m_h^g \circ m_g \ = \ m_h \circ m_g \ ,$$

and by definition $m_g \ = \ \mathrm{id}$ if and only if $g \ \in \ \mathbb{G}(\mathscr{D})$. By Theorem 1.3 (Belyĭ's Theorem) we know that $M(\mathscr{D})$ and $M(X)$ are number fields. Then (4) and (5) follow from Galois theory. □

Next we modify the Belyĭ function by composing it with a Möbius transformation μ in such a way that the critical fibres of $B := \mu \circ \beta$ remain Galois-invariant.

Lemma 5.3 *Under the same hypotheses there is a Möbius transformation μ with the property that for all $g \in \mathbb{G}(X)$ we have*

$$\mu^g \circ m_g = \mu \ .$$

Proof By the Coombes-Harbater theorem we assume that β and X are both defined over $M(\mathscr{D})$, and we may regard g as an element of $\operatorname{Gal} M(\mathscr{D})/M(X)$. We need to replace $\{0, 1, \infty\}$ with three values on which the Galois group acts, in the opposite direction, via the antihomomorphism $g \mapsto m_g$. This can be done by a case-by-case analysis, as follows.

If $M(\mathscr{D}) = M(X)$, take $\mu = \mathrm{id}$.

In the case of a quadratic extension $M(\mathscr{D}) \ = \ M(X)(\sqrt{\alpha})$, with a nontrivial Galois conjugation $g : \sqrt{\alpha} \mapsto -\sqrt{\alpha}$, suppose without loss of generality that m_g fixes ∞ and transposes 0 and 1; then define $\mu(z) := \sqrt{\alpha}(2z - 1)$.

If $M(\mathscr{D})/M(X)$ is a cyclic cubic extension, choose a normal basis of this extension on which the Galois group acts by cyclic permutations. There is a Möbius transformation μ sending $\{0, 1, \infty\}$ to this basis, such that the claim of the lemma is satisfied.

The final possibility is that $M(\mathscr{D})/M(X)$ has a Galois group isomorphic to S_3, so that there are three Galois conjugate intermediate cubic extensions of $M(X)$. Here we can take one generator of such a cubic extension and its two conjugates, and choose μ so that $\{0, 1, \infty\}$ is sent to these three elements and the Galois action is opposite to the action $g \mapsto m_g$ on their pre-images $\{0, 1, \infty\}$ under μ. □

Now we can finish the proof of Theorem 5.4. For the modified Belyĭ function $B = \mu \circ \beta$ and its Galois conjugates under all $g \in \mathbb{G}(X)$ we have

$$B^g \circ f_g = \mu^g \circ \beta^g \circ f_g = \mu^g \circ m_g \circ \beta = \mu \circ \beta = B ,$$

so we can repeat the proof of Theorem 5.3 to prove that the pair (X, B) can be defined over $M(X)$. □

As mentioned earlier, if $M(\mathcal{D}) > M(X)$ then this extension has degree $2, 3$ or 6. In [52] one can find examples of degree 2 (see also Lemma 9.2 and Theorem 9.5) and of degree 3; examples of degree 6 are currently unknown.

5.5 Appendix: Linear and Projective Groups

Many of the groups appearing in this book are linear or projective groups, so we summarise their definitions and properties here. See [13], [27, Sect. II.6–II.8], [55, Sect. 4.5], or [57] for details.

If R is a commutative ring with identity, then for each integer $n \geq 1$ the *general linear group* $\mathrm{GL}_n(R)$ consists of the $n \times n$ matrices M over R which are invertible, that is, those for which the determinant $\det M$ is a unit in R. The function det is an epimorphism from $\mathrm{GL}_n(R)$ to the multiplicative group $U(R)$ of units in R, so its kernel, the *special linear group* $\mathrm{SL}_n(R)$ consisting of those M with $\det M = 1$, is a normal subgroup of $\mathrm{GL}_n(R)$ with $\mathrm{GL}_n(R)/\,\mathrm{SL}_n(R) \cong U(R)$. In particular, if R is a field K then $U(R) = K^* := K \setminus \{0\}$.

For any field K and $n \geq 2$, the natural action of $\mathrm{GL}_n(K)$ on the vector space $V = K^n$ induces an action of this group on the $(n-1)$-dimensional projective space $\mathbb{P}^{n-1}(K)$ formed by the 1-dimensional subspaces of V. The kernel of this action is a normal subgroup $Z \cong K^*$ consisting of the scalar matrices λI_n ($\lambda \in K^*$), and the group of transformations induced on $\mathbb{P}^{n-1}(K)$ is the *projective general linear group* $\mathrm{PGL}_n(K) \cong \mathrm{GL}_n(K)/Z$. In this action, the subgroup $\mathrm{SL}_n(K)$ induces the *projective special linear group* $\mathrm{PSL}_n(K) \cong \mathrm{SL}_n(K)/(\mathrm{SL}_n(K) \cap Z) \cong \mathrm{SL}_n(K)Z/Z$. This is a simple group for all $n \geq 2$ and all fields K, except when $n = 2$ and $|K| = 2$ or 3, in which case the group is isomorphic to S_3 or A_4. The kernel $\mathrm{SL}_n(K) \cap Z$ of the induced epimorphism $\mathrm{SL}_n(K) \to \mathrm{PSL}_n(K)$ is isomorphic to the group of nth roots of 1 in K, a cyclic group of order dividing n, while $\mathrm{PGL}_n(K)/\,\mathrm{PSL}_n(K)$ is isomorphic to the quotient of K^* by its subgroup of nth powers (so $\mathrm{PSL}_n(K) = \mathrm{PGL}_n(K)$ whenever K is algebraically closed).

The *affine group* $\mathrm{AGL}_n(K)$ consists of the affine transformations of V, those of the form $v \mapsto Av + b$ where $A \in \mathrm{GL}_n(K)$ and $b \in V$. This is a semidirect product of a normal subgroup, isomorphic to the additive group of V, consisting of the translations $v \mapsto v + b$, and a complement $\mathrm{GL}_n(K)$, consisting of the affine transformations with $b = 0$, that is, the linear transformations of V.

These groups $GL_n(K)$, $SL_n(K)$, $PGL_n(K)$, $PSL_n(K)$ and $AGL_n(K)$ can all be extended by the Galois group $\mathrm{Gal}\, K$ of the field K, acting on coordinates of vectors and points, to form groups

$$\Gamma L_n(K)\ ,\ \Sigma L_n(K)\ ,\ P\Gamma L_n(K)\ ,\ P\Sigma L_n(K)\quad \text{and}\quad A\Gamma L_n(K)$$

containing them as normal subgroups with quotients isomorphic to $\mathrm{Gal}\, K$.

If K is the finite field $\mathbb{F}_q$ of order q, then these groups such as $GL_n(K)$ are often denoted by $GL_n(q)$, etc., with the finite simple group $PSL_n(q)$ written as $L_n(q)$ in *ATLAS* notation [9]. In this case, $SL_n(q) \cap Z \cong PGL_n(q)/\, PSL_n(q) \cong C_d$, where $d = \gcd(n, q-1)$. We have $q = p^e$ for some prime p, so that $\mathrm{Gal}\, K$ is a cyclic group of order e, generated by the Frobenius automorphism $x \mapsto x^p$. Counting linearly independent row-vectors gives $|\, GL_n(q)| = \prod_{i=0}^{n-1}(q^n - q^i)$, and the orders of the other related groups are easily deduced from this.

The case $n = 2$ is particularly important. The projective line $\mathbb{P}^1(K)$ can be identified with $K \cup \{\infty\}$ by identifying the point with homogeneous coordinates $[x, y]$, that is the subspace of $V = K^2$ spanned by the non-zero vector (x, y), with the element x/y ($= \infty$ if $y = 0$). If we let matrices act on the left of column vectors, so that each $A \in GL_2(K)$ acts as

$$A = \begin{pmatrix} a & b \\ c & d \end{pmatrix} \ :\ \begin{pmatrix} x \\ y \end{pmatrix} \mapsto \begin{pmatrix} ax + by \\ cx + dy \end{pmatrix},$$

then the projective transformation of $\mathbb{P}^1(K)$ corresponding to A is given by

$$z = \frac{x}{y} \mapsto \frac{ax + by}{cx + dy} = \frac{az + b}{cz + d}.$$

This action of $PGL_2(K)$ on $\mathbb{P}^1(K)$ is sharply 3-transitive, meaning that the group acts regularly on ordered triples of distinct points, so a non-identity element has at most two fixed points. The subgroup fixing any one point is conjugate to the subgroup fixing ∞; this is induced by the matrices A with $c = 0$, and it is a semidirect product of an abelian normal subgroup isomorphic to the additive group of K, given by also taking $a = d = 1$, and a complement isomorphic to the multiplicative group K^*, given by taking $b = c = 0$. This complement is the subgroup of $PGL_2(K)$ fixing 0 and ∞, and the subgroup fixing any two points is conjugate to it. This describes all the elements with fixed points. There are also elements with no fixed points in $\mathbb{P}^1(K)$. When $K = \mathbb{F}_q$, so that $|\, PGL_2(K)| = q(q^2 - 1)$, these elements have orders dividing $q + 1$; they have two fixed points when regarded as elements of $PGL_2(q^2)$ acting on $\mathbb{P}^1(\mathbb{F}_{q^2})$ via the inclusion of $\mathbb{F}_q$ in $\mathbb{F}_{q^2}$. The elements of $PGL_2(q)$ fixing two points of $\mathbb{P}^1(\mathbb{F}_q)$ have orders dividing $q-1$, and those with one fixed point have order p, where $q = p^e$ for a prime p.

We are particularly interested in the groups $\mathrm{PSL}_2(q) = \mathrm{SL}_2(q)/\{\pm I\}$, often denoted by $L_2(q)$ in *ATLAS* notation [9]; see [13, Chap. XII] or [27, Sect. II.8] for full details and proofs of their following properties.

The group $\mathrm{PSL}_2(q)$ is a subgroup of $\mathrm{PGL}_2(q)$, of order $q(q^2-1)/d$ and of index $d = \gcd(2, q-1)$ equal to 1 or 2 as $p = 2$ or $p > 2$. If $p = 2$ then $\mathrm{GL}_2(q) = \mathrm{SL}_2(q) \times Z$ with $\mathrm{SL}_2(q) = \mathrm{PSL}_2(q) = \mathrm{PGL}_2(q)$, so in this case the above description of $\mathrm{PGL}_2(q)$ applies to $\mathrm{PSL}_2(q)$. For any q, a non-identity element of $\mathrm{PSL}_2(q)$ has order dividing $(q-1)/d$, equal to p, or dividing $(q+1)/d$, as it fixes two, one or no points in $\mathbb{P}^1(\mathbb{F}_q)$. Equivalently, if t is its trace (defined only up to multiplication by -1), then $t^2 - 4$ is respectively a non-zero square, equal to 0, or a non-square in $\mathbb{F}_q$. The automorphism group of $\mathrm{PSL}_2(q)$ can be identified with $\mathrm{P\Gamma L}_2(q)$, acting by conjugation on its normal subgroup $\mathrm{PSL}_2(q)$.

Dickson described the subgroups of $\mathrm{PSL}_2(q)$ in [13, Chap. XII], see also [27, Sect. II.8], and from this one can describe the maximal subgroups:

Proposition 5.3 *Any maximal subgroup of* $\mathrm{PSL}_2(q)$ *has one of the following forms, where* $d = (2, q-1)$*:*

1. *the stabiliser of a point in* $\mathbb{P}^1(\mathbb{F}_q)$*, isomorphic to the unique subgroup of order* $q(q-1)/d$ *in* $\mathrm{AGL}_1(q)$*;*
2. *a dihedral group of order* $2(q \pm 1)/d$*;*
3. *a group isomorphic to* $\mathrm{PSL}_2(r)$ *where* $\mathbb{F}_r$ *is a maximal subfield of* $\mathbb{F}_q$ *(that is* $r = p^f$ *with* e/f *prime);*
4. *a group isomorphic to* $\mathrm{PGL}_2(r)$ *where* $q = r^2$ *is a perfect square;*
5. *a group isomorphic to* A_4*,* S_4 *or* A_5*.* □

Maximal subgroups of types (1) and (2) exist for all q, and those of types (3) and (4) exist whenever r satisfies the stated conditions. Subgroups isomorphic to A_4 exist if and only if q is odd or $q = 2^e$ with e even; subgroups isomorphic to S_4 exist if and only if $q \equiv \pm 1 \bmod 8$; subgroups isomorphic to A_5 exist if and only if $p = 5$ or $q \equiv \pm 1 \bmod 5$; when they exist, these subgroups of type (5) are not always maximal.

We also need the character table for $\mathrm{PSL}_2(q)$. In fact we give three tables, for $q \equiv 1 \bmod 4$, $q \equiv -1 \bmod 4$ and $q = 2^e$.

In each table, the first row indicates the orders of the elements in the corresponding column: 1 for the identity element in the first column, p (where $q = p^e$) for the parabolic elements (two conjugacy classes when q is odd, one when $q = 2^e$), and then in the last two columns, representing the hyperbolic and elliptic classes, the divisors $n > 1$ of $(q-1)/2$ and $(q+1)/2$ (or of $q-1$ and $q+1$ when $q = 2^e$). The hyperbolic and elliptic classes are each represented by an element a^i or b^j, where a and b are elements of orders $(q \mp 1)/2$ respectively (or $q \mp 1$ when $q = 2^e$). In Table 5.5, where $q \equiv 1 \bmod 4$, we have $i, j = 1, \ldots, r := (q-1)/4$; for instance, the involutions are represented by the hyperbolic element a^r. In Table 5.6, where

Table 5.5 The character table of $\mathrm{PSL}_2(q)$ for $q \equiv 1 \bmod 4$

1	p	p	$1 < n \mid \frac{q-1}{2}$	$1 < n \mid \frac{q+1}{2}$
1	1	1	1	1
q	0	0	1	-1
$(q+1)/2$	$(1+\sqrt{q})/2$	$(1-\sqrt{q})/2$	$(-1)^i$	0
$(q+1)/2$	$(1-\sqrt{q})/2$	$(1+\sqrt{q})/2$	$(-1)^i$	0
$q+1$	1	1	$\zeta^{ik}+\zeta^{-ik}$	0
$q-1$	-1	-1	0	$-\xi^{jl}-\xi^{-jl}$

Table 5.6 The character table of $\mathrm{PSL}_2(q)$ for $q \equiv -1 \bmod 4$

1	p	p	$1 < n \mid \frac{q-1}{2}$	$1 < n \mid \frac{q+1}{2}$
1	1	1	1	1
q	0	0	1	-1
$(q-1)/2$	$(-1+\sqrt{-q})/2$	$(-1-\sqrt{-q})/2$	0	$(-1)^{j+1}$
$(q-1)/2$	$(-1-\sqrt{-q})/2$	$(-1+\sqrt{-q})/2$	0	$(-1)^{j+1}$
$q-1$	-1	-1	0	$-\xi^{jk}-\xi^{-jk}$
$q+1$	1	1	$\zeta^{il}+\zeta^{-il}$	0

Table 5.7 The character table of $\mathrm{PSL}_2(q)$ for $q = 2^e$

1	2	$1 < n \mid q-1$	$1 < n \mid q+1$
1	1	1	1
q	0	1	-1
$q+1$	1	$\zeta^{ik}+\zeta^{-ik}$	0
$q-1$	-1	0	$-\xi^{jl}-\xi^{-jl}$

$q \equiv -1 \bmod 4$, we have $i = 1, \ldots, r-1$ and $j = 1, \ldots, r := (q+1)/4$, with the involutions represented by the elliptic element b^r. In Table 5.7, where $q = 2^e$, we have $i = 1, \ldots, (q-2)/2$ and $j = 1, \ldots, q/2$; in this case the involutions are parabolic.

In each table, the entries in the first column are the degrees $\chi(1)$ of the corresponding irreducible characters χ. Each row represents a single character, except in the cases $\chi(1) = q \pm 1$. In Table 5.5, the last two rows represent $r-1$ and r characters, with $k = 1, \ldots, r-1$ and $l = 1, \ldots, r$ respectively. In Table 5.6, the last two rows each represent $r-1 = (q-3)/4$ characters, with $k, l = 1, \ldots, r-1$. In both tables, the preceding four rows each represent a single character, as do the first two rows of Table 5.7, where the last two rows represent $(q-2)/2$ and $q/2$ characters respectively. In Tables 5.5 and 5.6 we have $\zeta = \zeta_{(q-1)/2}$ and $\xi = \zeta_{(q+1)/2}$, but in Table 5.7 we have $\zeta = \zeta_{q-1}$ and $\xi = \zeta_{q+1}$, where $\zeta_n := \exp(2\pi i/n)$.

References

1. Accola, R.D.M.: On the number of automorphisms of a closed Riemann surface. Trans. Am. Math. Soc. **131**, 398–408 (1968)
2. Bolza, O.: On binary sextics with linear transformations onto themselves. Am. J. Math. **10**, 47–70 (1888)
3. Breda d'Azevedo, A.J., Jones, G.A.: Platonic hypermaps. Beiträge Algebra Geom. **42**, 1–37 (2001)
4. Burnside, W.: Note on the simple group of order 504. Math. Ann. **52**, 174–176 (1899)
5. Conder, M.D.E.: Regular maps and hypermaps of Euler characteristic -1 to -200. J. Comb. Theory Ser. B **99**, 455–459 (2009). Associated lists of computational data available at http://www.math.auckland.ac.nz/~conder/hypermaps.html
6. Conder, M.D.E.: An update on Hurwitz groups. Groups Complex. Cryptol. **2**, 35–49 (2010)
7. Conder, M.D.E., Jones, G.A., Streit, M., Wolfart, J.: Galois actions on regular dessins of small genera. Rev. Mat. Iberoam. **29**, 163–181 (2013)
8. Connor, T., Leemans, D.: An atlas of subgroup lattices of finite almost simple groups. Ars Math. Contemp. **8**(2), 259–266 (2015)
9. Conway, J.H., Curtis, R.T., Norton, S.P., Parker, R.A., Wilson, R.A.: ATLAS of Finite Groups. Clarendon Press, Oxford (1985)
10. Coombes, K., Harbater, D.: Hurwitz families and arithmetic Galois groups. Duke Math. J. **52**, 821–839 (1985)
11. Coxeter, H.S.M., Moser, W.O.J.: Generators and Relations for Discrete Groups. Springer, Berlin/Heidelberg (1980)
12. Dèbes, P., Emsalem, M.: On fields of moduli of curves. J. Algebra **211**, 42–56 (1999)
13. Dickson, L.E.: Linear Groups. Dover, New York (1958)
14. Downs, M.L.N.: The Möbius function of $PSL_2(q)$, with application to the maximal normal subgroups of the modular group. J. Lond. Math. Soc. **43**, 61–75 (1991)
15. Downs, M.L.N., Jones, G.A.: Enumerating regular objects with a given automorphism group. Discrete Math. **64**, 299–302 (1987)
16. Downs, M.L.N., Jones, G.A.: Möbius inversion in Suzuki groups and enumeration of regular objects. In: Proceedings in Mathematics and Statistics, SIGMAP 2014 Proceedings (to appear)
17. Džambić, A.: Macbeath's infinite series of Hurwitz groups. In: Holzapfel, R.-P., Uludağ, A.M., Yoshida, M. (eds.) Arithmetic and Geometry Around Hypergeometric Functions. Progress in Mathematics, vol. 260, pp. 101–108. Birkhäuser, Basel (2007)
18. Feierabend, F.: Galois-Operationen auf verallgemeinerten Macbeath-Hurwitz Kurven. Dissertation, Frankfurt (2008)
19. Fricke, R.: Ueber eine einfache Gruppe von 504 Operationen. Math. Ann. **52**, 321–339 (1899)
20. Fulton, W., Harris, J.: Representation Theory. Springer, Berlin (1991)
21. Garbe, D.: Über die regulären Zerlegungen geschlossener orientierbarer Flächen. J. Reine Angew. Math. **237**, 39–55 (1969)
22. Girondo, E., Wolfart, J.: Conjugators of Fuchsian groups and quasiplatonic surfaces. Q. J. Math. **56**, 525–540 (2005)
23. González-Diez, G., Jaikin-Zapirain, A.: The absolute Galois group acts faithfully on regular dessins and on Beauville surfaces. Proc. Lond. Math. Soc. (3) **111**(4), 775–796 (2015)
24. Hall, P.: The Eulerian functions of a group. Q. J. Math. **7**, 134–151 (1936)
25. Harvey, W.J.: Cyclic groups of automorphisms of a compact Riemann surface. Q. J. Math. **17**, 86–97 (1966)
26. Hidalgo, R.: Edmonds maps on the Fricke-Macbeath curve. Ars Math. Contemp. **8**, 275–289 (2015)

27. Huppert, B.: Endliche Gruppen I . Springer, Berlin (1967)
28. Jones, G.A.: Ree groups and Riemann surfaces. J. Algebra **165**, 41–62 (1994)
29. Jones, G.A.: Hypermaps and multiply quasiplatonic Riemann surfaces. Eur. J. Comb. **33**, 1588–1605 (2012)
30. Jones, G.A., Silver, S.A.: Suzuki groups and surfaces. J. Lond. Math. Soc. (2) **48**, 117–125 (1993)
31. Jones, G.A., Streit, M., Wolfart, J.: Wilson's map operations on regular dessins and cyclotomic fields of definition. Proc. Lond. Math. Soc. **100**, 510–532 (2010)
32. Kucharczyk, R.: On arithmetic properties of Fuchsian groups and Riemann surfaces. Dissertation, Bonn (2014)
33. Kulkarni, R.: A note on Wiman and Accola-Maclachlan surfaces. Ann. Acad. Sci. Fenn. Math. **16**, 83–94 (1991)
34. Kuribayashi, I., Kuribayashi, A.: Automorphism groups of compact Riemann surfaces of genera three and four. J. Pure Appl. Algebra **65**, 277–292 (1990)
35. Larsen, M.: How often is $84(g-1)$ achieved? Isr. J. Math. **126**, 1–16 (2001)
36. Macbeath, A.M.: On a theorem of Hurwitz. Proc. Glasg. Math. Assoc. **5**, 90–96 (1961)
37. Macbeath, A.M.: Generators of the linear fractional groups. In: Number Theory (Proc. Sympos. Pure Math., Vol. XII, Houston, Tex., 1967), pp. 14–32. American Mathematical Society, Providence (1969)
38. Maclachlan, C.: A bound for the number of automorphisms of a compact Riemann surface. J. Lond. Math. Soc. **44**, 265–272 (1969)
39. Magnus, W.: Noneuclidean Tesselations and Their Groups. Academic, New York (1974)
40. Pierro, E.: The Möbius function of the small Ree groups. arXiv:1410.8702v2 [math.GR] (2014). Accessed 20 Jan 2015
41. Popp, H.: On a conjecture of H. Rauch on theta constants and Riemann surfaces with many automorphisms. J. Reine Angew. Math. **253**, 66–77 (1972)
42. Potočnik, P.: Census of rotary maps. http://www.fmf.uni-lj.si/~potocnik/work.htm. Accessed 3 Feb 2015
43. Rauch, H.E.: Theta constants on a Riemann surface with many automorphisms. In: Symposia Mathematica III, pp. 305–322. Academic, Cambridge, Ma., (1970)
44. Schlage-Puchta, J.-C., Wolfart, J.: How many quasiplatonic surfaces?. Arch. Math. **86**, 129–132 (2006)
45. Serre, J.-P.: Topics in Galois Theory. Jones and Bartlett, Boston (1992)
46. Sherk, F.A., The regular maps on a surface of genus 3. Can. J. Math. **11**, 452–480 (1959)
47. Siegel, C.L.: Topics in Complex Function Theory, vol. II. Wiley, New York (1971)
48. Singerman, D.: Subgroups of Fuchsian groups and finite permutation groups. Bull. Lond. Math. Soc. **2**, 319–323 (1970)
49. Singerman, D.: Finitely maximal Fuchsian groups. J. Lond. Math. Soc. (2) **6**, 29–38 (1972)
50. Singerman, D.: Riemann surfaces, Belyi functions and hypermaps. In: Bujalance, E., Costa, A.F., Martinez, E. (eds.) Topics in Riemann Surfaces and Fuchsian Groups. London Mathematical Society Lecture Note Series, vol. 287, pp. 43–68. Cambridge University Press, Cambridge (2001)
51. Streit, M.: Field of definition and Galois orbits for the Macbeath-Hurwitz curves. Arch. Math. **74**, 342–349 (2000)
52. Streit, M., Wolfart, J.: Characters and Galois invariants of regular dessins. Rev. Mat. Complut. **13**, 49–81 (2000)
53. Takeuchi, K.: Arithmetic triangle groups. J. Math. Soc. Jpn. **29**, 29–38 (1977)
54. Takeuchi, K.: Commensurability classes of arithmetic triangle groups. J. Fac. Sci. Tokyo Sect. IA Math. **24**, 201–212 (1977)

55. Tsuzuku, T.: Finite Groups and Finite Geometries. Cambridge Tracts in Mathematics, vol. 78. Cambridge University Press, Cambridge (1982)
56. Turbek, P.: The full automorphism group of the Kulkarni surface. Rev. Mat. Complut. **10**, 265–276 (1997)
57. Wilson, R.A.: The Finite Simple Groups. Springer, London (2009)
58. Wiman, A.: Über die hyperelliptischen Curven und diejenigen von Geschlecht $p = 3$ welche eindeutige Transformationen in sich zulassen. Bihang till K. Svenska Vet.-Akad. Handlingar **21**, 1–23 (1895–1896)
59. Wolfart, J.: ABC for polynomials, dessins d'enfants, and uniformization – a survey. In: Schwarz, W., Steuding, J. (eds.) Elementare und Analytische Zahlentheorie (Tagungsband), Proceedings ELAZ-Conference May 24–28, 2004, pp. 313–345. Steiner, Stuttgart (2006)

Chapter 6
Regular Maps

Abstract Regular maps can be considered as special types of regular dessins. They include some of the oldest geometric objects known to mankind, in the form of the Platonic solids. These, together with the dihedra and their duals, the hosohedra (meaning 'with many faces'), are the regular maps on the sphere. In this chapter we study their generalisations to maps on compact Riemann surfaces of arbitrary genus. We show how to classify regular maps in terms of their genus, paying particular attention to the cases of genus 0—as above—and of genus 1, where there are infinitely many regular maps, associated with ideals in the rings of Gaussian and Eisenstein integers. For each genus $g \geq 2$ the Hurwitz bound implies that there are only finitely many regular maps, and we briefly consider the classification for $g = 2$. We also outline how one can classify regular maps in terms of their automorphism group, and we consider which groups can arise in this context.

Keywords Automorphism group • Hurwitz group • Quasiplatonic surface • Regular map

6.1 Regularity

A map $\mathscr{M}$ is simply a dessin $\mathscr{D}$ of type $(l, 2, n)$, where the vertices, edges and faces of $\mathscr{M}$ are regarded as the hypervertices, hyperedges and hyperfaces of a hypermap, as in Sect. 2.1.3. Conversely, any dessin $\mathscr{D}$ of type $(l, 2, n)$ can be regarded as a map $\mathscr{M}$ on the same surface, simply by representing it as a bipartite map $\mathscr{B}$ and ignoring the black vertices, which all have valency 2 or 1. (Any black vertex of valency 1 will give rise to a half-edge in $\mathscr{M}$, incident with only one white vertex; see Example 6.1, for instance.) The edges of $\mathscr{B}$ then correspond to the directed edges of $\mathscr{M}$, and the automorphisms of $\mathscr{D}$ are the (orientation-preserving) automorphisms of $\mathscr{M}$. We say that a map $\mathscr{M}$ is *regular* (as an oriented map) if and only if the group of all such automorphisms acts transitively on the directed edges of $\mathscr{M}$. Thus $\mathscr{D}$ is a regular dessin if and only if $\mathscr{M}$ is a regular map, in which case the automorphism groups $\mathrm{Aut}\,\mathscr{D}$ and $\mathrm{Aut}\,\mathscr{M}$ of $\mathscr{D}$ and $\mathscr{M}$ are the same.

Before proceeding, we must give a warning about terminology. Our use of the word 'regular' for maps is consistent with our terminology for dessins, and also with that used for maps by Coxeter and Moser in [3, Chap. 8]. However, in

Gareth A. Jones, J. Wolfart, *Dessins d'Enfants on Riemann Surfaces*, Springer Monographs in Mathematics, DOI 10.1007/978-3-319-24711-3_6

Topological Graph Theory, where non-orientable maps and orientation-reversing automorphisms are also studied, this word is often reserved for maps satisfying the stronger condition that the group of *all* automorphisms, preserving or reversing orientation, acts transitively on the set of flags (incident vertex-edge-face triples), so that the map is isomorphic to its mirror-image; orientable maps which are regular in our sense are then called ‘orientably regular’ or ‘rotary’. We will always use the term ‘regular’ in the sense defined above.

(There is an alternative method of converting a dessin $\mathscr{D}$ into a map, by retaining all the vertices of the bipartite map $\mathscr{B}$ but ignoring their colours. However, although this is simple and applies to dessins of *all* types, not just type $(l, 2, n)$, it is less convenient: the resulting map is regular if and only if $\mathscr{D}$ is regular and also self-dual, under the duality transposing the vertex colours. We will therefore generally avoid this method.)

A regular map $\mathscr{M}$, which as a dessin has type $(l, 2, n)$, is often referred to, for example by Coxeter and Moser in [3, Chap. 8], as being a map of type $\{n, l\}$: this means that it has n-gonal faces and an l-valent embedded graph. The automorphism group $G = \operatorname{Aut}\mathscr{M}$, also isomorphic to the monodromy group of the dessin, then has generators x, y and z of orders $l, 2$ and n, representing rotations around an incident vertex, edge and face respectively, satisfying $xyz = 1$. Conversely, if a finite group G has generators x, y and z of these orders, satisfying $xyz = 1$, then it is the automorphism group of a regular map of this type. Two such ordered generating triples for G correspond to isomorphic maps if and only if they are equivalent under $\operatorname{Aut} G$, that is, if and only if the obvious epimorphisms $\Delta(l, 2, n) \to G$ have the same kernel. Thus the regular maps of type $\{n, l\}$ with automorphism group G correspond to the torsion-free normal subgroups K of the triangle group $\Delta = \Delta(l, 2, n)$ with $\Delta/K \cong G$.

Most maps are not regular, in the sense that most subgroups of triangle groups are not normal. Nevertheless, regular maps play an important role within the theory of maps, just as regular dessins do in the more general theory of dessins: for instance, by Theorem 2.3 (adapted in the obvious way for maps), every map $\mathscr{M}$ is a quotient of a regular map $\hat{\mathscr{M}}$ of the same type by some group of automorphisms of $\hat{\mathscr{M}}$. This means that by studying regular maps and their automorphism groups one can, at least in principle, understand all maps. In practice, it can be difficult to relate $\mathscr{M}$ and $\hat{\mathscr{M}}$ since the genus $\hat{g}$ of $\hat{\mathscr{M}}$ may be considerably greater than the genus g of $\mathscr{M}$.

Example 6.1 Figure 6.1 shows a non-regular dessin $\mathscr{D}$ of type $(n, 2, n-1)$ and genus $g = 0$. If we remove the black vertices and retain the white vertex we obtain a non-regular map $\mathscr{M}$ of genus $g = 0$. This has a single (white) vertex, of valency $n \geq 2$; its edges consist of a single loop and $n - 2$ half-edges, and it has two faces.

By numbering the directed edges of $\mathscr{M}$ (equivalently, the edges of $\mathscr{D}$) $1, 2, \ldots, n$ we see that the cartographic group of $\mathscr{M}$ (that is, the monodromy group of $\mathscr{D}$) is the permutation group generated by

$$x = (1, 2, \ldots, n)\,, \quad y = (1, 2) \quad \text{and} \quad z = (xy)^{-1} = (n, n-1, \ldots, 2)\,,$$

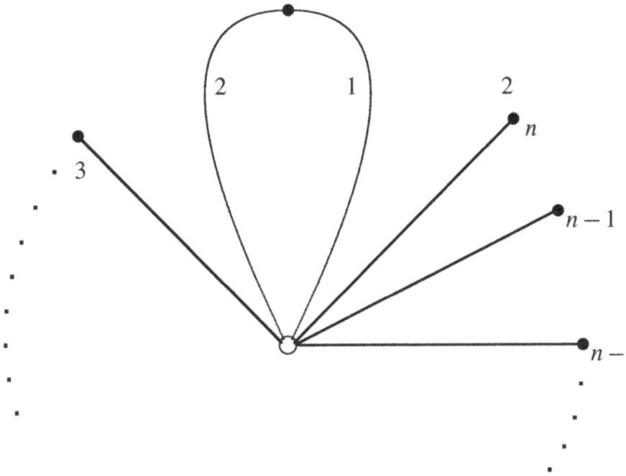

Fig. 6.1 A dessin with monodromy group S_n

where we use the convention that these permutations act from the right on $\{1, \ldots, n\}$. It is well known that x and y generate the symmetric group S_n: note that $y_i := x^{1-i}yx^{i-1} = (i, i+1)$, and the transpositions $(i,j) = y_{j-1} \ldots y_{i+1}y_iy_{i+1} \ldots y_{j-1}$ for $i < j$ generate S_n. It follows that the minimal regular cover $\hat{\mathcal{M}}$ of $\mathcal{M}$ is a regular map of type $\{n-1, n\}$, with automorphism group S_n, so it has genus

$$\hat{g} = 1 + \frac{|S_n|}{2}\left(1 - \frac{1}{n} - \frac{1}{2} - \frac{1}{n-1}\right) = 1 + \frac{(n-2)!}{4}(n^2 - 5n + 2).$$

Thus for $n = 2, 3, 4, 5, 6, 7, 8, \ldots$ we have $\hat{g} = 0, 0, 0, 4, 49, 481, 4681, \ldots$, increasing super-exponentially. For instance if $n = 4$ then $\hat{\mathcal{M}}$ is the octahedron, and if $n = 5$ it is a map on Bring's curve (see Example 5.7 in Sect. 5.1.4, also Sect. 5.2.3, entry 5.22 in Table 5.4). For $n = 6$ we obtain the map R49.33 of genus 49 in Conder's computer-generated list of regular maps [2]. When $n = 7$, however, the resulting genus, 481, is approaching the limits of computer-aided classifications. Nevertheless, the group S_7 is sufficiently small and well-understood that one can answer any reasonable questions about this particular map $\hat{\mathcal{M}}$, its automorphisms and its quotients, in many cases by inspection of its quotient $\mathcal{M}$.

A regular map $\mathcal{M}$ has three basic ingredients: a compact surface X, which may be assumed to be a Riemann surface, a graph $\mathcal{G}$ embedded in X, and an automorphism group G. It is reasonable to start to classify regular maps in terms of any one of these three ingredients, and we will describe how this has been done, at least in relatively simple cases. In the rest of this chapter we will consider classification by the genus of the surface, and then by the automorphism group of the map, while Chaps. 7 and 9 will be devoted to classification in terms of the embedded graph.

6.2 Classification by Genus

Although there are infinitely many regular maps of genus 0 and of genus 1, they lie in a small number of easily described families. For each genus $g \geq 2$ the Hurwitz bound (Sect. 5.1.2) gives $|G| \leq 84(g-1)$, so there are only finitely many possible automorphism groups G and hence only finitely many regular maps for each such g. For small values of g one can determine these by hand. However, as g increases this task becomes increasingly laborious, and it is more suitable in this case to use computers.

We start with genus 0, where the only compact Riemann surface X is the Riemann sphere $\hat{\mathbb{C}}$. If $\mathcal{D}$ is a regular dessin of genus g and type (l, m, n), with $\mathrm{Aut}\,\mathcal{D} = G$, then

$$2g - 2 = \left(1 - \frac{1}{l} - \frac{1}{m} - \frac{1}{n}\right)|G|. \tag{6.1}$$

If $g = 0$ then this implies that

$$\frac{1}{l} + \frac{1}{m} + \frac{1}{n} > 1.$$

To avoid trivial cases we will assume that $l, m, n \geq 2$. The inequality then shows that in each case at least one of l, m and n must be equal to 2, and since we are looking for maps we may assume that $m = 2$. (In fact this argument shows that every regular dessin on the sphere can be regarded as a regular map.) Simple arithmetic then shows that the only solutions for the type $(l, m, n) = (l, 2, n)$ are the following:

1. $(3, 2, 3)$ with $|G| = 12$;
2. $(3, 2, 4)$ and $(4, 2, 3)$ with $|G| = 24$;
3. $(3, 2, 5)$ and $(5, 2, 3)$ with $|G| = 60$;
4. $(2, 2, n)$ and $(n, 2, 2)$ with $|G| = 2n$.

This shows that each group G has the same order as the corresponding triangle group $\Delta = \Delta(l, 2, n)$; since G is an epimorphic image Δ/K of Δ we must have $G \cong \Delta$, so $K = 1$ and $\mathcal{M}$ is unique. In case (1) we obtain a regular map $\mathcal{M}$ of type $\{3, 3\}$, namely the tetrahedron, with $G \cong A_4$ acting naturally on the four vertices. In case (2) we have the cube, of type $\{4, 3\}$, and its dual, the octahedron of type $\{3, 4\}$, with $G \cong S_4$ permuting the four antipodal pairs of vertices of the cube (or faces of the octahedron). The dual maps in case (3) are the dodecahedron and the icosahedron, of types $\{5, 3\}$ and $\{3, 5\}$, with $G \cong A_5$ acting on the five tetrahedra inscribed in the icosahedron. In case (4), a regular dessin of type $(2, 2, n)$ is a regular map of type $\{n, 2\}$; this is the dihedron, consisting of a circuit of n vertices and n edges separating the sphere into two n-gonal faces. Its dual is the hosohedron, a map of type $\{2, n\}$ with two vertices joined by n edges, both maps having the dihedral group D_n of order $2n$ as their automorphism group. (Note that in this action of D_n, the elements of $D_n \setminus C_n$ all induce half-turns of the sphere, not reflections as in the action of D_n on an n-gon.)

Putting $g = 1$ in Eq. (6.1) gives

$$\frac{1}{l} + \frac{1}{m} + \frac{1}{n} = 1.$$

In this case the only solutions with $m = 2$ and $l, n \geq 2$ are $(4, 2, 4)$, $(3, 2, 6)$ and $(6, 2, 3)$. (Here we ignore the regular dessins of type $(3, 3, 3)$, which also have genus 1, since they do not correspond to maps.) In these cases there is no restriction on $|G|$, and in fact we will see that there are infinitely many groups which can arise.

The triangle group $\Delta = \Delta(4, 2, 4)$ is the automorphism group of the universal map of type $\{4, 4\}$ in $\mathbb{C}$. This is the tessellation of $\mathbb{C}$ by unit squares, with the ring $\mathbb{Z}[i] = \{a + bi \mid a, b \in \mathbb{Z}\}$ of Gaussian integers as vertex-set. There is a normal subgroup T of Δ consisting of the translations by elements of $\mathbb{Z}[i]$, and this is complemented by a cyclic group of order 4 generated by the rotation $z \mapsto iz$ around the vertex 0. If $0 \neq t = a + bi \in \mathbb{Z}[i]$ then the translations by t and by $it = -b + ai$ generate a subgroup K of index

$$\det \begin{pmatrix} a & b \\ -b & a \end{pmatrix} = a^2 + b^2$$

in T, which is normal in Δ. This subgroup K is torsion-free, and conversely, every torsion-free normal subgroup of finite index in Δ has this form for some t. (Check that these subgroups K are the non-zero ideals of $\mathbb{Z}[i]$.) The regular maps of genus 1 and type $\{4, 4\}$ correspond bijectively to these subgroups. Each such map $\mathcal{M}$, denoted in [3] by $\{4, 4\}_{a,b}$, has $a^2 + b^2$ vertices and faces, and $2(a^2 + b^2)$ edges; its automorphism group $G \cong \Delta/K$ has an abelian normal subgroup of index 4, isomorphic to T/K, complemented by the stabiliser C_4 of a vertex (or a face). Examples of fundamental regions for two such subgroups K are given in Fig. 7.1 in the next chapter. See [3, Sect. 8.3] for more details concerning these maps.

The situation is similar for $\Delta = \Delta(3, 2, 6) \cong \Delta(6, 2, 3)$, corresponding to the tessellations of $\mathbb{C}$ by hexagons and by triangles, with the ring of Eisenstein integers $\mathbb{Z}[\zeta_3] = \mathbb{Z}[\zeta_6]$ ($\zeta_m := e^{2\pi i/m}$) as the set of face-centres and vertices respectively. In this case Δ is a semidirect product of a normal translation group T by a complement C_6 fixing a face-centre or a vertex. The torsion-free normal subgroups K of finite index (again, the non-zero ideals of the ring) are those generated by translations through $t = a + b\zeta_6$ and $\zeta_6 t$, with index $a^2 + ab + b^2$ in T. These correspond to regular maps $\{6, 3\}_{a,b}$ of type $\{6, 3\}$ with $2(a^2 + ab + b^2)$ vertices and $a^2 + ab + b^2$ faces, or to the duals of these, denoted by $\{3, 6\}_{a,b}$. For examples see Fig. 7.2 in the next chapter. See also [3, Sect. 8.4] for more about these maps.

We make no attempt to be comprehensive in describing the regular maps of genus $g \geq 2$, even in the case $g = 2$ where there are, up to isomorphism, ten regular maps (see the table of quasiplatonic surfaces of genus 2 in Sect. 5.2.1 and the list of regular maps of this genus in [3, Table 9]): there are two self-dual maps of types $\{6, 6\}$ and

$\{8, 8\}$ with automorphism groups $C_3 \times C_2 \times C_2 \cong C_2 \times C_6$ and C_8, four maps of types $\{3, 8\}$, $\{4, 6\}$, $\{4, 8\}$ and $\{5, 10\}$ with automorphism groups of orders 48, 24, 16 and 10, and the duals of these last four. Instead, we will simply illustrate the methods by outlining one positive result, and one negative one.

Example 6.2 The negative result is the non-existence of a regular map $\mathcal{M}$ of genus 2 and type $\{3, 7\}$, or equivalently of a Hurwitz group $G = \operatorname{Aut}\mathcal{M}$ of genus 2. If it exists, such a group G achieves the maximum order $84(g-1) = 84$ for a group of automorphisms of a Riemann surface of genus 2. By Sylow's theorems, the Sylow 7-subgroups of G have order 7, and the number n_7 of them divides $|G|$ and satisfies $n_7 \equiv 1 \bmod 7$. The only possibility is that $n_7 = 1$, in which case the unique Sylow 7-subgroup S is normal in G. As a Hurwitz group, G has generators x, y and z satisfying $x^7 = y^2 = z^3 = xyz = 1$. Since S contains all the elements of order 7 in G, it contains x, so G/S is generated by the images $\overline{y}$ and $\overline{z}$ of y and z, satisfying $\overline{y}^2 = \overline{z}^3 = \overline{yz} = 1$. These relations imply that $\overline{y} = \overline{z} = 1$, so $|G/S| = 1$ and hence $G = S$, whereas G and S have orders 84 and 7. Thus no such group G or map $\mathcal{M}$ can exist. (This is the solution to Exercise 5.6 !)

Example 6.3 The positive result is the existence of a regular map $\mathcal{M}$ of genus 2 and type $\{3, 8\}$. By Eq. (6.1), $G := \operatorname{Aut}\mathcal{M}$ must have order 48, so we look for a group of this order with generators x, y and z satisfying the relations $x^8 = y^2 = z^3 = xyz = 1$ of $\Delta = \Delta(8, 2, 3)$. The relations $x^4 = y^2 = z^3 = xyz = 1$ define the triangle group $\Delta(4, 2, 3) \cong S_4$ of order 24, so we could look for a group G which has a normal subgroup $N = \langle x^4 \rangle$ of order 2 with $G/N \cong S_4$. Now $\mathrm{PGL}_2(3)$ acts on the projective line $\mathbb{P}^1(\mathbb{F}_3) = \{0, 1, 2, \infty\}$ (see Sect. 5.5), inducing all 4! permutations of its four points, so $\mathrm{PGL}_2(3) \cong S_4$. The group $G := \mathrm{GL}_2(3)$ has a normal subgroup $N = \{\pm I\}$ with $G/N \cong \mathrm{PGL}_2(3)$, so we could look for suitable generators of this group G. The elements

$$x = \begin{pmatrix} 0 & 1 \\ 1 & 2 \end{pmatrix}, \quad y = \begin{pmatrix} 0 & 1 \\ 1 & 0 \end{pmatrix}, \quad z = \begin{pmatrix} 1 & 0 \\ 1 & 1 \end{pmatrix}$$

of G have orders 8, 2 and 3, with $xyz = 1$. Since $x^4 = -I$, the subgroup $H \leq G$ they generate contains $N = \{\pm I\}$. Since x and y induce a 4-cycle $t \mapsto 1/(t-1)$ and a transposition $t \mapsto 1/t$ on $\mathbb{P}^1(\mathbb{F}_3)$, H acts as S_4 on $\mathbb{P}^1(\mathbb{F}_3)$ and thus maps onto $\mathrm{PGL}_2(3)$, so $H = G$. Thus x, y and z generate G, as required. The corresponding regular map $\mathcal{M}$, denoted by $\{3, 4+4\}$ in [3, Sect. 8.8], is a double covering of the octahedron (a regular map of type $\{3, 4\}$), branched over its six vertices. If we locate these on the sphere $\mathbb{P}^1(\mathbb{C})$ at $\pm 1, \pm i, 0$ and ∞ then the underlying surface of $\mathcal{M}$ is the hyperelliptic curve $y^2 = x^5 - x$, with the covering of the sphere given by $(x, y) \mapsto x$ (see Eq. (5.3) in Sect. 5.2.1).

Exercise 6.1 Construct the remaining regular maps of genus 2, using the information about them given earlier in this section.

In principle, arguments like these can be applied to classify the regular maps of low genus, since the Hurwitz bound implies that for each genus $g \geq 2$ there are only finitely many cases to consider. In fact, such a classification was achieved for $g = 2$ by Brahana [1] in 1927 and Threlfall [10] in 1932 (see [3, Table 9]), for $g = 3$ by Sherk [9] in 1959, and for $g = 4, 5$ and 6 by Garbe [5] in 1969. As g increases, the number of cases to consider increases steadily—for asymptotic estimates see [8]—and the task soon becomes too laborious for calculation by hand. However, the development of very efficient algorithms for group-theoretic computation has allowed significant advances, culminating in the publication by Conder [2] in 2009 of a computer-generated list of all regular dessins (including regular maps) of genera $g = 2, \ldots, 101$. (At the time of writing, this list, available on-line at Marston Conder's home page, has been extended up to genus 301. In addition, Primož Potočnik's home page [7] now contains a census of orientable rotary maps—regular maps in our terminology—with at most 3000 edges, that is, with $|G| \leq 6000$.)

6.3 Classification by Group

In order to classify the regular maps $\mathcal{M}$ with $\mathrm{Aut}\,\mathcal{M}$ isomorphic to a given group G, one can use the character-theoretic method of Sect. 5.1.5, or one can use direct arguments as in Sect. 5.1.4. There we classified the 5-valent regular maps with automorphism group A_5 (see also Exercise 5.9 in Sect. 5.1.5). We can generalise this as follows:

Example 6.4 Let us count the p-valent regular maps $\mathcal{M}$ with $\mathrm{Aut}\,\mathcal{M} \cong G = \mathrm{PSL}_2(p) := \mathrm{PSL}_2(\mathbb{F}_p)$, where p is an odd prime. (This generalises the classification mentioned above, since $\mathrm{PSL}_2(5) \cong A_5$; see Sect. 5.5 for properties of G used in what follows.)

There are $p^2 - 1$ elements $x \in G$ of order p, and $p(p \pm 1)/2$ elements y of order 2, as $p \equiv \pm 1 \bmod 4$. This gives $p(p^2 - 1)(p \pm 1)/2$ such pairs x, y respectively. Dickson's description [4, Chap. XII] of the subgroups of $\mathrm{PSL}_2(q)$, for $q = p^e$, shows that any such pair generates G or a subgroup $H \cong D_p$. If $p \equiv 1 \bmod 4$ there are $p + 1$ such subgroups H, each generated by $p(p - 1)$ such pairs x, y, so there are $p(p^2 - 1)$ pairs x, y generating subgroups $H \cong D_p$; the number of pairs generating G is therefore

$$\frac{p(p^2-1)(p+1)}{2} - p(p^2-1) = \frac{p(p^2-1)(p-1)}{2}.$$

If $p \equiv -1 \bmod 4$ there are no such subgroups H, so the number of generating pairs for G is again given by this formula. Now Aut G is isomorphic to $\mathrm{PGL}_2(p)$, acting by conjugation on its normal subgroup G. Since $\mathrm{PGL}_2(p)$ has order $p(p^2 - 1)$, and acts semiregularly on generating pairs for G, it follows that the number of orbits on these pairs, and hence the number of maps, is $(p - 1)/2$. One of these maps is the level p modular map of type $\{3, p\}$, arising from the reduction $\bmod p : \mathrm{PSL}_2(\mathbb{Z}) \to$

$\mathrm{PSL}_2(p) = G$; this map has genus $(p+2)(p-3)(p-5)/24$. For $p = 3, 5$ and 7 we obtain the tetrahedron, the icosahedron and Klein's map of genus 3, for example (see Fig. 3.4 in Sect. 3.1.6).

In fact, one can compute the type and genus of each of these $(p-1)/2$ maps $\mathcal{M}$, as follows. Each of the above generating pairs x, y is equivalent under automorphisms (conjugacy in $\mathrm{PGL}_2(p)$) to a pair with

$$x = \pm\begin{pmatrix} 1 & 1 \\ 0 & 1 \end{pmatrix}, \quad y = \pm\begin{pmatrix} 0 & -1/c \\ c & 0 \end{pmatrix} \quad \text{and hence} \quad z = \pm\begin{pmatrix} 0 & 1/c \\ -c & c \end{pmatrix}$$

for some $c \neq 0$ (see Exercise 6.2). Moreover, for a given pair x and y, the corresponding value of c is unique, up to multiplication by -1 (again, see Exercise 6.2), so we have $(p-1)/2$ equivalence classes of generating pairs, and hence this number of non-isomorphic maps, as shown above. Each map $\mathcal{M}$ has type $\{n, p\}$ where n is the order of z, and this can be computed from the eigenvalues $\lambda, \mu \in \mathbb{F}_{p^2}$ of the matrix representing z: since this has trace c and determinant 1 we have

$$\lambda, \mu = \frac{1}{2}(c \pm \sqrt{c^2 - 4}).$$

(Replacing c with $-c$ here is equivalent to multiplying the matrix by -1, with no effect on its image $z \in G$.) The order n of z is the least integer $i > 0$ such that $\lambda^i = \mu^i = \pm 1$, that is, the order of the images of λ and its inverse μ in $\mathbb{F}_{p2}^*/\{\pm 1\}$. Once n is known, the genus of each map $\mathcal{M}$ can be computed from the Riemann-Hurwitz formula (6.1), using the fact that $|G| = p(p^2-1)/2$.

The $(p-1)/2$ trace pairs $\pm c \neq 0$ in $\mathbb{F}_p$ correspond bijectively to the orbits of $\mathrm{PGL}_2(\mathbb{F}_p)$ on elements $z \in G$ of orders $n > 2$, so as $\mathcal{M}$ ranges over these $(p-1)/2$ maps, the corresponding face-valency n ranges over all such orders of elements of G. Specifically, we obtain one map with $n = p$, corresponding to $c = \pm 2$, and $\phi(n)/2$ maps for each divisor $n > 2$ of $(p-1)/2$ or of $(p+1)/2$, where ϕ is Euler's function. For instance the modular map, with $n = 3$, corresponds to $c = \pm 1$.

Exercise 6.2 Verify the above claim that each generating pair x, y of orders p and 2 is equivalent under automorphisms of G to a pair of the displayed form, with c uniquely determined up to multiplication by -1.

Exercise 6.3 Find the types and the genera of the regular p-valent maps $\mathcal{M}$ with $\mathrm{Aut}\,\mathcal{M} \cong \mathrm{PSL}_2(p)$ for $p = 7$ and $p = 11$.

At this point one might ask which finite groups G can arise as automorphism groups of regular maps. The simplest answer is that they are the finite groups with two generators, x of any order and y of order dividing 2, since these are the finite images of the triangle groups $\Delta(l, 2, n)$. These groups form a very rich and varied class: for instance, by a result of Malle, Saxl, and Weigel [6], it includes every finite simple group. It also includes the finite symmetric groups (see Example 6.1), so by Cayley's Theorem every finite group can be realised as a subgroup of the automorphism group of a regular map. In this sense, the automorphism groups of regular maps can be as complicated as any finite groups.

References

1. Brahana, H.R.: Regular maps and their groups. Am. J. Math. **49**, 268–284 (1927)
2. Conder, M.D.E.: Regular maps and hypermaps of Euler characteristic −1 to −200. J. Comb. Theory Ser. B **99**, 455–459 (2009). Associated lists of computational data available at http://www.math.auckland.ac.nz/~conder/hypermaps.html
3. Coxeter, H.S.M., Moser, W.O.J.: Generators and Relations for Discrete Groups. Springer, Berlin/Heidelberg (1980)
4. Dickson, L.E.: Linear Groups. Dover, New York (1958)
5. Garbe, D.: Über die regulären Zerlegungen geschlossener orientierbarer Flächen. J. Reine Angew. Math. **237**, 39–55 (1969)
6. Malle, G., Saxl, J., Weigel, Th.: Generation of classical groups. Geom. Dedicata **49**, 85–116 (1994)
7. Potočnik, P.: Census of rotary maps. http://www.fmf.uni-lj.si/~potocnik/work.htm. Accessed 3 Feb 2015
8. Schlage-Puchta, J.-C., Wolfart, J.: How many quasiplatonic surfaces?. Arch. Math. **86**, 129–132 (2006)
9. Sherk, F.A.: The regular maps on a surface of genus 3. Can. J. Math. **11**, 452–480 (1959)
10. Threlfall, W.: Gruppenbilder. Abh. Sächs. Akad. Wiss. Math. Phys. Kl. **41**, 1–59 (1932)

Chapter 7
Regular Embeddings of Complete Graphs

Abstract The methods described for classifying regular maps in terms of their genus or automorphism group can be extended, without much difficulty, to all regular dessins. However, the methods for classifying regular maps in terms of their embedded graphs are much more specific to maps, and do not extend so easily to other dessins. Gardiner, Nedela, Širáň and Škoviera developed a general strategy for this problem, involving the search for subgroups of the automorphism group of the graph which act transitively on the vertices, such that the stabiliser of a vertex is a cyclic group acting regularly on its neighbours. To illustrate the general methods available, we concentrate on a rather simple class of graphs, namely the complete graphs K_n. Using results about a class of permutation groups called Frobenius groups, we show that their only regular embeddings are those constructed by Biggs, using Cayley graphs for the additive groups of finite fields. We enumerate such maps, and show that their automorphism groups are 1-dimensional affine groups. We also show how these maps are related to cyclotomic polynomials, and to primitive polynomials over finite fields.

Keywords Affine group • Arc-transitive graph • Automorphism group • Biggs map • Cayley graph • Cayley map • Complete graph • Cyclotomic polynomial • Finite field • Frobenius group • Regular map

7.1 Examples of Regular Embeddings

If $\mathcal{M}$ is a regular map then its embedded graph $\mathcal{G}$ must be arc-transitive, meaning that its automorphism group $\mathrm{Aut}\,\mathcal{G}$ acts transitively on the directed edges, or arcs, of $\mathcal{G}$. In [8], Gardiner, Nedela, Širáň and Škoviera showed that a connected finite graph $\mathcal{G}$ is embedded in a regular map if and only if $\mathrm{Aut}\,\mathcal{G}$ has a subgroup G $(= \mathrm{Aut}\,\mathcal{M})$ acting transitively on the vertices, such that the stabiliser in G of a vertex v is a cyclic group acting regularly on the neighbours of v. One can therefore try to classify regular maps in terms of their embedded graphs by searching among the known families of arc-transitive graphs $\mathcal{G}$ for such subgroups G of $\mathrm{Aut}\,\mathcal{G}$. The first class of graphs to be treated in this way were the complete graphs: in 1971 Biggs [2] used Cayley maps over finite fields to show that a complete graph has a regular embedding if and only if the number of vertices is a prime power, and in 1985

Gareth A. Jones, J. Wolfart, *Dessins d'Enfants on Riemann Surfaces*, Springer Monographs in Mathematics, DOI 10.1007/978-3-319-24711-3_7

James and Jones [10] used the theory of Frobenius groups to enumerate and classify all such maps. Since then similar results have been obtained for a number of other classes of arc-transitive graphs, such as Johnson graphs [11], Hamming graphs [12] and n-dimensional cubes [4]—see Širáň's survey [16], for instance. In this chapter we will give a detailed description of the classification of regular embeddings of complete graphs, while two other classes, the generalised Paley graphs and the complete bipartite graphs, will be considered in Chap. 9.

7.1.1 Examples of Genus 0 and 1

The *complete graph* K_q has q vertices, each distinct pair joined by a single edge. Thus there are $q(q-1)/2$ edges, and hence $q(q-1)$ directed edges, so if $\mathcal{M}$ is any regular embedding of K_q then $|\mathrm{Aut}\,\mathcal{M}| = q(q-1)$. The subgroup fixing a vertex v is a cyclic group of order $q-1$, generated by a rotation fixing v and permuting its $q-1$ neighbours in a cycle of length $q-1$.

There are simple examples of regular embeddings $\mathcal{M}$ of K_q on the sphere for $q \leq 4$.

Example 7.1 When $q = 4$ we can take $\mathcal{M}$ to be the tetrahedral map on the sphere, with $\mathrm{Aut}\,\mathcal{M} \cong A_4$, the rotation group of the tetrahedron, inducing the even permutations of the four vertices.

Exercise 7.1 Show that for each $q \leq 4$ there is, up to isomorphism, a unique regular embedding of K_q, and that it has genus 0.

Example 7.2 The situation is rather different for $q = 5$: a simple argument (exercise!) using the Euler characteristic shows that any orientable map (regular or not) which embeds K_5 must have genus $g \geq 1$. The lower diagrams in Fig. 7.1 illustrate two regular embeddings of K_5 on a torus. These are the regular maps $\{4,4\}_{1,2}$ (on the left) and $\{4,4\}_{2,1}$ (on the right) of genus 1, described in Sect. 6.2. In each case, the identification of opposite sides of the tilted square means that its corners are identified to give a single vertex, in addition to the four in the interior of the square. These two maps form a *chiral pair*, that is, they are not isomorphic, but they are mirror-images of each other. The upper parallelogram—compare [1]—and the square below result from each other by a cut and paste procedure. In each case, the black vertices can be ignored; they are needed only if we consider the drawing as a hypermap or a dessin, and not as a map (see Sect. 6.1).

In each case, the automorphism group $G = \mathrm{Aut}\,\mathcal{M}$ is isomorphic to the affine group $\mathrm{AGL}_1(5)$, consisting of the transformations $t \mapsto at + b$ of the field $\mathbb{F}_5$ with $a, b \in \mathbb{F}_5$ and $a \neq 0$. The invariance of the maps under the translations $t \mapsto t + b$ is easier to see in the upper drawings, while the invariance under the rotations $t \mapsto at$ is best seen in the square drawings below. These two maps correspond to two normal subgroups K of the triangle group $\Delta = \Delta(4,2,4)$, of index 5 in the translation subgroup, with $\Delta/K \cong G$.

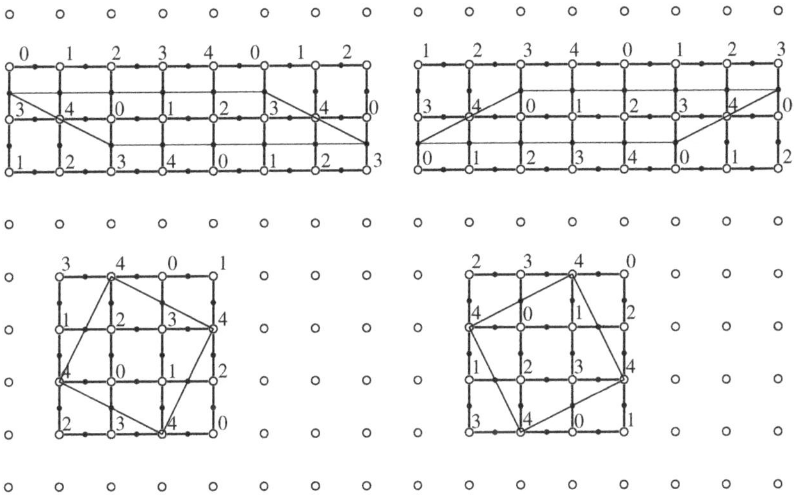

Fig. 7.1 Two versions of the two regular embeddings of K_5 (drawn in $\mathbb{C}$)

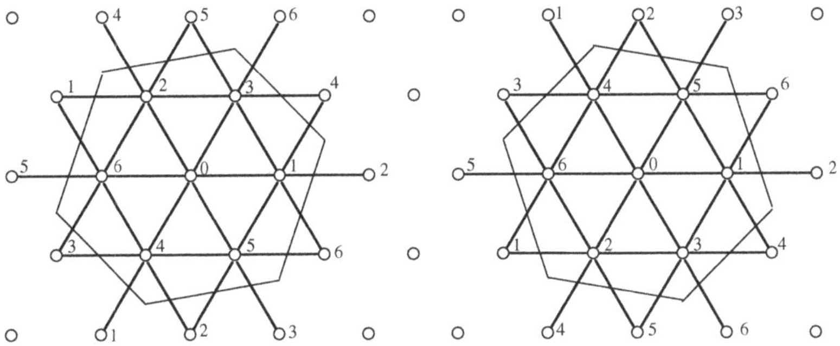

Fig. 7.2 The two regular embeddings of K_7 (drawn in $\mathbb{C}$)

Example 7.3 The situation for $q = 7$ is similar to that for $q = 5$: there is a chiral pair of regular embeddings $\mathcal{M} = \{3, 6\}_{2,1}$ and $\{3, 6\}_{1,2}$ of K_7 on a torus. These correspond to a pair of normal subgroups K of the triangle group $\Delta = \Delta(6, 2, 3)$, with $\mathrm{Aut}\,\mathcal{M} \cong \Delta/K \cong \mathrm{AGL}_1(7)$. In Fig. 7.2 we give only one drawing for each map, in which the fundamental region of the covering group of the underlying torus is a regular hexagon in each case. A parallelogram illustrating the invariance of the map under translations can be given in a similar way as for $q = 5$, see also [1].

Why did we omit the case $q = 6$? In 1971, Biggs [2] proved that there is a regular embedding of K_q if and only if q is a prime power, so 6 is the smallest value of q for which K_q has no regular embedding. The examples he constructed have the finite fields $\mathbb{F}_q$ as their vertex-sets, and such fields exist if and only if q is a prime power. They are examples of a more general class of maps, called *Cayley maps*, which we now describe.

7.1.2 Cayley Maps

Let V be a group, and S a set of non-identity elements of V with

$$S = S^{-1} := \{s^{-1} \mid s \in S\} .$$

The *Cayley graph* of V with respect to S is an undirected graph $C = C(V; S)$ which has vertex-set V, with an edge between v and vs for each $v \in V$ and $s \in S$. (In some contexts it is useful to regard such an edge as directed from v to vs, and as labelled with the generator s, but we do not need these extra features here.) Each vertex of C has valency $|S|$, and C is a connected graph if and only if S generates V, as we will always assume from now on.

Example 7.4 Here are some simple examples:

1. If V is a cyclic group of order $q > 2$, with $S = \{c, c^{-1}\}$ for some generator c of V, then C is a cycle of length q.
2. If V is a dihedral group of order $2q > 2$, with S consisting of a pair of involutions generating V, then C is a cycle of length $2q$.
3. If V is any group of order q, and $S = V \setminus \{1\}$, then C is a complete graph K_q.

Exercise 7.2 Show that the groups $V = A_4$ and $V = S_4$ can each be generated by a pair of elements of orders 2 and 3. Draw the resulting Cayley graphs (if possible, in the plane without crossings). How does this idea extend to the group $V = A_5$?

Remark 7.1 As epimorphic images of triangle groups, automorphism groups V of regular dessins are always generated by two elements. If both generators have orders greater than 2, the Cayley graph $C = C(V; S)$ for S consisting of these generators and their inverses has valency 4 at every vertex, and this seems to be very different from the graphs underlying most dessins. However, if we replace the usual Walsh representation with the Cori representation—see Fig. 2.4 in Sect. 2.1.3—it turns out that the underlying graph is precisely the Cayley graph $C(V; S)$, see [15]. When there is a generator of order 2, as in the case of maps, this idea has to be modified but still works similarly.

In what follows, we will not suppose S to be such a minimal generating set for V, and V will be only a subgroup of the full automorphism group of the resulting dessin. The group V acts as a group of automorphisms of $C = C(V; S)$, each $g \in V$ acting as the automorphism $\lambda_g : v \mapsto gv$. This is a permutation of the vertices, and it preserves the edges of C since if $v \in V$ and $s \in S$, that is, there is an edge between v and vs, then $(\lambda_g(v))s = (gv)s = g(vs) = \lambda_g(vs)$, so that there is an edge between $\lambda_g(v)$ and $\lambda_g(vs)$. The mapping $g \mapsto \lambda_g$ is an action of V on C: if we compose from right to left, then $\lambda_{g_1} \circ \lambda_{g_2}$ sends each v to $g_1(g_2v)$, while $\lambda_{g_1g_2}$ sends v to $(g_1g_2)v$, so $\lambda_{g_1} \circ \lambda_{g_2} = \lambda_{g_1g_2}$ for all $g_1, g_2 \in V$. (If we decide to compose from left to right, we need to define $\lambda_g : v \mapsto g^{-1}v$ in order to satisfy this equation.) The automorphism group Aut C of C thus acts transitively on the vertices of C: each vertex v is the image of the vertex 1 (the identity element of G) under the automorphism λ_g for

$g = v$, so the automorphisms λ_g $(g \in V)$ form a subgroup of Aut C, isomorphic to V, acting regularly on the vertices.

If we choose a cyclic permutation $\pi = (s_1, s_2, \ldots, s_{|S|})$ of the generators $s_i \in S$, then this determines a cyclic permutation of the neighbours vs_i of each vertex $v \in C$, and hence a cyclic permutation of the directed edges of C pointing to v. These are the disjoint cycles of a permutation x of all the directed edges of C, and this, together with the involution y which reverses every directed edge, generates the cartographic group of an oriented map with C as its embedded graph. We call this map the *Cayley map* $\mathcal{M} = \mathcal{M}(V; S, \pi)$ for V associated with S and π. It is easy to check that the graph automorphisms λ_g $(g \in V)$ act as automorphisms of $\mathcal{M}$, so Aut $\mathcal{M}$ acts transitively on the vertex-set V. Considerable research has been devoted to finding conditions under which $\mathcal{M}$ has extra automorphisms, and in particular, is a regular map (see [5, 14], for instance). Here we will restrict our attention to a simple example, used by Biggs [2] to construct regular embeddings of complete graphs (see also [3, Sect. 5.4]).

Exercise 7.3 Let $V = \langle c \mid c^q = 1\rangle$, a cyclic group of order $q > 2$. Describe the Cayley map $\mathcal{M}(v; S, \pi)$ where $S = \{c, c^{-1}\}$ and π is the transposition (c, c^{-1}). Is this map regular?

Exercise 7.4 Show that if π extends to an automorphism of V then the map $\mathcal{M}(v; S, \pi)$ is regular.

7.2 The Biggs Maps

We can now use the Cayley map construction, together with results from the theory of permutation groups to classify the regular embeddings of complete graphs.

7.2.1 Construction of the Biggs Maps

Our aim is to prove the following theorem, due to Biggs [2]:

Theorem 7.1 *The complete graph K_q has a regular embedding if and only if q is a prime power.*

In this section, we will show that this condition on q is sufficient by constructing a regular embedding for each prime power q; this will be a Cayley map, based on the additive group of a finite field. In the next section, we will introduce some ideas from permutation group theory, in order to show that the condition is necessary.

Proof of Sufficiency If q is a prime power p^e, let V be the additive group of the finite field $F = \mathbb{F}_q$ of order q, so V is an elementary abelian p-group $C_p \times \cdots \times C_p$ of rank e. The set $S = F^* := \mathbb{F}_q \setminus \{0\}$ of non-zero elements is closed under additive inverses, and does not contain the identity element 0 of V, so we can form the Cayley graph $C = C(V; S)$, isomorphic to K_q. Since F is a field, S is a group under multiplication. The multiplicative group of any finite field is cyclic, so let c be a generator for S. We have $S = \{c, c^2, c^3, \ldots, c^{q-1} = 1\}$, so let $\mathcal{M} = \mathcal{M}_q(c)$ denote the Cayley map $\mathcal{M}(V; S, \pi)$ determined by this cyclic ordering $\pi = (c, c^2, c^3, \ldots, 1)$ of S. Thus the vertices v of $\mathcal{M}$ are the elements of F, and the each vertex neighbours of v are $v + c, v + c^2, v + c^3, \ldots, v + 1$ in that cyclic order as we follow the orientation around v.

The 1-dimensional *affine group* $\mathrm{AGL}_1(q) := \mathrm{AGL}_1(\mathbb{F}_q)$ consists of the transformations of V of the form $v \mapsto av + b$ with $a, b \in F$ and $a \neq 0$. Such a transformation sends v to $v' = av + b$, and sends the neighbours $v + c^i$ of v to $a(v + c^i) + b = v' + ac^i$ for $i = 1, 2, \ldots, q - 1$. Since $a = c^j$ for some j, the cyclic order of these, as i increases, is

$$(v' + c^{j+1}, v' + c^{j+2}, \ldots, v' + c^{j+q-1}) = (v' + c, v' + c^2, \ldots, v' + c^{q-1}) .$$

Thus the elements of $\mathrm{AGL}_1(q)$ act as automorphisms of the graph C, preserving the cyclic order of neighbours around each vertex. It follows that they induce automorphisms of $\mathcal{M}$, that is, we can identify $\mathrm{AGL}_1(q)$ with a subgroup of $\mathrm{Aut}\,\mathcal{M}$. This subgroup acts doubly transitively on V, that is, it is transitive on ordered pairs of distinct vertices: if (v, w) and (v', w') are two such pairs, then we can find an affine transformation $v \mapsto av + b$ sending the first pair to the second by solving the simultaneous equations

$$av + b = v' \quad \text{and} \quad aw + b = w'$$

to give

$$a = \frac{v' - w'}{v - w} \quad \text{and} \quad b = \frac{vw' - v'w}{v - w}.$$

Since the directed edges of $\mathcal{M}$ can be identified with such ordered pairs, it follows that $\mathrm{AGL}_1(q)$ acts transitively on directed edges, so $\mathcal{M}$ is a regular map—see Exercise 7.4 for a generalisation of this. (In fact, since the automorphism group of any map acts semiregularly on its directed edges, we have $|\,\mathrm{Aut}\,\mathcal{M}| \leq q(q-1) = |\,\mathrm{AGL}_1(q)|$, so $\mathrm{Aut}\,\mathcal{M} = \mathrm{AGL}_1(q)$.)

Since there are $\phi(n-1)$ choices for a generator c of F^*, this construction yields that number of regular embeddings of K_q for each prime power q. However, as we will see later, there may be isomorphisms between some of these maps.

Exercise 7.5 Draw the Cayley maps $\mathcal{M}_q(c)$ for $q = 4, 5$ and 7, and all possible choices of c. Are there any isomorphisms between them?

We will prove the converse part of Theorem 7.1 in the next section.

7.2.2 *Frobenius Groups*

We have shown that if q is a prime power then K_q has a regular embedding. In order to prove the converse, we need to introduce some concepts from the theory of permutation groups.

If $\mathscr{M}$ is any regular map then the automorphism group $G = \mathrm{Aut}\,\mathscr{M}$ acts transitively on the vertex set V of $\mathscr{M}$, and the stabiliser G_v of each vertex $v \in V$ is a cyclic group, acting regularly on the set $N(v)$ of neighbours of v. If the embedded graph is a complete graph K_q then $N(v) = V \setminus \{v\}$, so G acts on V as a doubly transitive group. Moreover, since G acts semiregularly on directed edges, the stabiliser $G_{v,w}$ of any distinct ordered pair (v, w) is the identity, so G is *sharply 2-transitive* on V, that is, the element taking one pair to any other is unique. This shows that, provided $q > 2$, G acts on V as a *Frobenius group*: this is a transitive group in which the stabiliser of any two points is trivial, but the stabiliser of one point is not (this last condition simply excludes regular permutation groups). There is a very well-developed theory of Frobenius groups [7, 9, 13, 18], which we can apply to the regular embeddings of complete graphs (see [3, Sect. 5.4], for instance).

Exercise 7.6 For which $n \geq 3$ is the dihedral group D_n, acting on the n vertices of a regular n-gon, a Frobenius group? For which n is it doubly transitive?

Exercise 7.7 Show that if F is any field, finite or infinite, then $\mathrm{AGL}_1(F)$ acts on F as a doubly transitive Frobenius group.

If a group G acts on a set V as a Frobenius group, then the *Frobenius kernel* is the set

$$N = G \setminus \bigcup_{v \in V} (G_v \setminus \{1\})$$

consisting of the elements without fixed points, together with the identity element. This is a normal subset of G, in the sense that it is a union of conjugacy classes of G. If G is finite, with $|V| = q$ and $|G_v| = m$ for each $v \in V$, then since the stabilisers G_v have only trivial pairwise intersections we have

$$|N| = |G| - \sum_{v \in V} (|G_v| - 1) = qm - q(m-1) = q.$$

In fact, if G is finite then N is a normal *subgroup* of G. The proof of this is simplest for doubly transitive Frobenius groups, as in the case of regular embeddings of complete graphs:

Lemma 7.1 *If a finite group G acts on a set V as a doubly transitive Frobenius group, then the Frobenius kernel N is an elementary abelian normal subgroup of G. In particular, $|V|$ is a prime power.*

Proof Let $|V| = q$, so $|N| = q$ also by the argument above, and $|G| = q(q-1)$ by double transitivity. Let g be a non-identity element of N, and let $C = C_G(g)$, the subgroup of G centralising g. Then $C \leq N$, for if g commutes with some $h \in G_v$ then h fixes the distinct elements v and vg of V, so $h = 1$. The number of conjugates of g in G is

$$|G : C| = |G|/|C| \geq |G|/|N| = q(q-1)/q = q-1 \quad .$$

However, these conjugates are all non-identity elements of N, so there can be at most $|N|-1 = q-1$ of them. It follows that the conjugates of g are the non-identity elements of N, and that $|G : C| = q-1$, so $|C| = q = |N|$. Since N contains C we have $N = C$, which is a subgroup of G, and N is normal since it is a union of (two) conjugacy classes of G. This argument shows that each element g of N is in the centre of N, so N is abelian. Since all non-identity elements of N are conjugate in G they all have the same order; at least one of them must have prime order p, so they all do. Thus N is an elementary abelian p-group $C_p \times \cdots \times C_p$, and so $|V| = |N|$ is a power of p. □

(The result that N is a subgroup is also valid without the assumption of double transitivity, but the proof, due to Frobenius, is more complicated, requiring character theory, so we omit it. In this more general situation, N is not necessarily elementary abelian, but a deep theorem of Thompson [17] shows that it is nilpotent, and thus a direct product of p-groups for various primes p.)

Proof of Theorem 7.1, continued We showed earlier that if $\mathcal{M}$ is a regular embedding of K_q then Aut $\mathcal{M}$ acts on the vertex-set V as a doubly transitive Frobenius group. Lemma 7.1 therefore implies that q must be a prime power, completing the proof of Theorem 7.1. □

7.2.3 Classifying the Regular Embeddings

In 1985 James and Jones [10] proved that the *Biggs maps* $\mathcal{M}_q(c)$ constructed in the proof of Theorem 7.1 are, in fact, the only regular embeddings of complete graphs, and they gave a formula for the number of such maps for each q. In order to prove this, we first show that the automorphism groups of such regular maps are the same as those of the maps $\mathcal{M}_q(c)$.

Lemma 7.2 *If* $\mathcal{M}$ *is a regular embedding of a complete graph* K_q *then the automorphism group* $G = \mathrm{Aut}\,\mathcal{M}$ *is isomorphic to* $\mathrm{AGL}_1(q)$.

Proof We have seen that G acts as a doubly transitive Frobenius group on the vertex set V, with an elementary abelian regular normal subgroup N, the Frobenius kernel. We can identify the vertex-set V with N, so that V acts on itself by translations. The stabiliser G_0 of the vertex 0 is a cyclic group of order $q-1$; its action on V is identified with its action by conjugation on N. We can regard the elementary abelian group $V = N$ as an e-dimensional vector space over $\mathbb{F}_p$, where $q = p^e$. The stabiliser G_0 acts on V as a group of linear transformations, so that G, which is a semidirect product of V by G_0, is a subgroup of the e-dimensional affine group $\mathrm{AGL}_e(p)$ over $\mathbb{F}_p$.

Let g be a generator of G_0. This acts on V as a linear transformation over $\mathbb{F}_p$, so let $f(t)$ be its minimal polynomial in the polynomial ring $\mathbb{F}_p[t]$. Since G_0 acts transitively on $V \setminus \{0\}$, it acts irreducibly on V, so $f(t)$ is an irreducible polynomial. We can identify V with the quotient $\mathbb{F}_p[t]/(f(t))$ of $\mathbb{F}_p[t]$ by the ideal $(f(t))$ generated by $f(t)$, with multiplication by t corresponding to the action of g. Since V is irreducible this ideal is maximal, so $\mathbb{F}_p[t]/(f(t))$ is a field; having order $p^e = q$, it is isomorphic to the unique field $\mathbb{F}_q$ of order q, with g acting as multiplication $v \mapsto cv$ by a generator c of the multiplicative group $\mathbb{F}_q^*$. (Such elements c are called *primitive elements* of $\mathbb{F}_q$, and the linear transformations of V which they induce are called *Singer cycles*.)

We have shown that G is a semidirect product of a normal subgroup V, isomorphic to the additive group of the field $\mathbb{F}_q$, by a complement G_0, isomorphic to the multiplicative group $\mathbb{F}_q^*$ of $\mathbb{F}_q$, with the action of G_0 by conjugation on V corresponding to the action of $\mathbb{F}_q^*$ by multiplication on $\mathbb{F}_q$. This description shows that G is isomorphic to $\mathrm{AGL}_1(q)$, which has the same structure as a semidirect product. □

We aim to show that the Biggs maps $\mathcal{M}_q(c)$ are the only regular embeddings of complete graphs K_q. For later purposes, concerning Galois conjugacy of the associated dessins, we will in fact prove a more general result, classifying the regular maps $\mathcal{M}$ of valency $q-1$ with automorphism group $G = \mathrm{Aut}\,\mathcal{M} \cong \mathrm{AGL}_1(q)$. Specifically, we will prove the following result:

Theorem 7.2 *Let $q = p^e$ where p is prime.*

1. *Any regular map $\mathcal{M}$ of valency $q-1$ with automorphism group $G = \mathrm{Aut}\,\mathcal{M} \cong \mathrm{AGL}_1(q)$ is isomorphic to $\mathcal{M}_q(c)$ for some primitive element c of $\mathbb{F}_q$.*
2. *Two such maps $\mathcal{M}_q(c)$ and $\mathcal{M}_q(c')$ are isomorphic if and only if c and c' have the same minimal polynomial in $\mathbb{F}_p[t]$.*
3. *The number of such maps, up to isomorphism, is $\phi(q-1)/e$ where ϕ denotes Euler's function.*

Proof When $q = 2$ there is, up to isomorphism, only one regular embedding, with automorphism group $C_2 \cong \mathrm{AGL}_1(2)$, so the result is obvious. To avoid trivial cases we will therefore assume for the rest of this proof that $q > 2$.

The maps $\mathcal{M}$ satisfying the hypotheses of Theorem 7.2 correspond to the orbits of $\mathrm{Aut}\,G$ on pairs x, y of generators of $G = \mathrm{AGL}_1(q)$ of orders $q-1$ and 2. We shall first describe all such generating pairs, then determine $\mathrm{Aut}\,G$, and finally determine its orbits on these pairs.

As before, let V denote the translation subgroup of $\mathrm{AGL}_1(q)$, and G_0 the subgroup fixing 0; thus G_0 is a cyclic group of order $q-1$, consisting of the transformations $t \mapsto at$ of $\mathbb{F}_q$ with $a \neq 0$. As in the proof of Lemma 7.2, let us choose a generator g of G_0.

Lemma 7.3 *Let $q > 2$. The elements x of order $q-1$ in G are those in the cosets Vg^i with i coprime to $q-1$. If $p = 2$ the elements y of order 2 in G are the non-identity elements in V, and every such pair x, y generates G. If $p > 2$ the elements y of order 2 in G are those in the coset Vg^m with $m = (q-1)/2$, and each such pair generates G provided $y \neq x^m$. In either case there are $q(q-1)\phi(q-1)$ such ordered pairs x, y generating G.*

Proof First suppose that $p = 2$, so $q = 2^e > 2$. Then the elements x of order $o(x) = q-1$ in G are those in the cosets Vg^i of V where i is coprime to $q-1$, and the elements y of order $o(y) = 2$ are the non-identity elements of V. Any such pair generates G, since y is conjugate, under the powers of x, to every non-identity element of V, so that $\langle x, y\rangle$ contains V; now G/V is generated by the image of x, so $\langle x, y\rangle = G$. There are $q\phi(q-1)$ choices for x (namely q in each of the $\phi(q-1)$ cosets Vg^i with i coprime to $q-1$), and there are $q-1$ choices for y, so there are $q(q-1)\phi(q-1)$ such generating pairs (x, y).

Now suppose that p is odd. In this case the elements x of order $q-1$ are as described in the case $p = 2$, but now the elements of order 2 are those in the coset Vg^m where $m = (q-1)/2$. Each of x and y has a unique fixed point in V, say v and w respectively. If $v = w$ then x and y are both in the proper subgroup $G_v = G_w$ of G fixing this element, so they cannot generate G. If $v \neq w$, however, then $x^m y^{-1}$ is an element of V, nontrivial since x^m and y have different fixed points v and w, so an argument similar to that in the case $p = 2$ shows that G is generated by x and $x^m y^{-1}$, and hence by x and y. As before, it follows that we obtain $q(q-1)\phi(q-1)$ generating pairs (x, y). □

Continuing the proof of Theorem 7.2, let

$$\Sigma = \{(x, y) \in G \times G \mid o(x) = q-1, o(y) = 2, \langle x, y\rangle = G\}$$

denote the set of generating pairs (x, y) described in Lemma 7.3, so

$$|\Sigma| = q(q-1)\phi(q-1)\ .$$

We need to determine $\mathrm{Aut}\, G$ and its orbits on this set. The Galois group $\mathrm{Gal}\,\mathbb{F}_q$ of $\mathbb{F}_q$, the group of all field automorphisms of $\mathbb{F}_q$, is a cyclic group of order e, generated by the Frobenius automorphism $v \mapsto v^p$. We define $\mathrm{A\Gamma L}_1(q) := \mathrm{A\Gamma L}_1(\mathbb{F}_q)$ to be the group of all transformations of $\mathbb{F}_q$ of the form

$$v \mapsto av^{\gamma} + b \quad (a, b \in \mathbb{F}_q,\ a \neq 0,\ \gamma \in \mathrm{Gal}\,\mathbb{F}_q).$$

This has a normal subgroup $\mathrm{AGL}_1(q)$, consisting of the transformations for which γ is the identity, complemented by a subgroup isomorphic to $\mathrm{Gal}\,\mathbb{F}_q$, consisting of the transformations with $a = 1$ and $b = 0$. There are $q-1, q$ and e independent choices for a, b and γ, so $|\mathrm{A\Gamma L}_1(q)| = eq(q-1)$.

Lemma 7.4 *Let $q > 2$.*

1. *The automorphism group* $\mathrm{Aut}\, G$ *of the group* $G = \mathrm{AGL}_1(q)$ *is isomorphic to* $\mathrm{A\Gamma L}_1(q)$*, acting by conjugation on its normal subgroup* G.
2. $\mathrm{Aut}\, G$ *acts semiregularly on* Σ*, with* $\phi(q-1)/e$ *regular orbits.*
3. *Two pairs* $(x, y), (x', y') \in \Sigma$ *are in the same orbit of* $\mathrm{Aut}\, G$ *if and only if they have the form* $x : v \mapsto av + b$ *and* $x' : v \mapsto a'v + b'$ *with the elements* $a, a' \in \mathbb{F}_q$ *having the same minimal polynomial in* $\mathbb{F}_p[t]$.

(When $q = 2$ we have $G \cong C_2$, with only the identity automorphism.)

Proof Since the group $G := \mathrm{AGL}_1(q)$ is a normal subgroup of $A := \mathrm{A\Gamma L}_1(q)$, the action of A by conjugation on G induces a group of automorphisms of G. If $q > 2$ then the centraliser $C_A(G)$ of G in A is trivial, so this action embeds A in $\mathrm{Aut}\, G$.

This gives us at least $|A| = eq(q-1)$ automorphisms of G. We will now show that there are at most this number of automorphisms, so $\mathrm{Aut}\, G = A$.

Any automorphism of G must preserve Σ. Moreover, any automorphism fixing a pair $(x, y) \in \Sigma$ must be the identity, since x and y generate G, so $\mathrm{Aut}\, G$ acts semiregularly on Σ. In order to determine the orbits of $\mathrm{Aut}\, G$ on Σ, suppose that $(x, y) \in \Sigma$. The element x has the form $v \mapsto av + b$ for some $a, b \in \mathbb{F}_q$, and acts by conjugation on V as multiplication by a. This is a linear transformation of V, regarded as a vector space over $\mathbb{F}_p$, and it is a root of a polynomial $f(t) \in \mathbb{F}_p[t]$ if and only if $f(a) = 0$. In particular, the minimal polynomial of x, as a linear transformation of V, is the same as that of a, as an element of the extension field $\mathbb{F}_q$ of $\mathbb{F}_p$. If some automorphism α of G sends (x, y) to (x', y'), where $x' : v \mapsto a'v + b'$, then the polynomials $f(t) \in \mathbb{F}_p[t]$ satisfied by x are the same as those satisfied by x': to see this, note that x is a root of $f(t) = a_0 + a_1 t + \cdots + a_n t^n$ if and only if

$$a_0 v + a_1 v^x + \cdots + a_n v^{x^n} = 0$$

for all $v \in V$, or in multiplicative notation

$$v^{a_0} (v^x)^{a_1} \ldots (v^{x^n})^{a_n} = 1$$

for all $v \in V$; now α preserves V (as the subgroup generated by the elements of order p in G), and since α is a group automorphism, x satisfies this condition for all $v \in V$ if and only if x' does. In particular, x and x' have the same minimal polynomial in $\mathbb{F}_p[t]$, and this is also the minimal polynomial of a and a'. This polynomial has degree at most $\dim V = e$ (in fact, exactly e), so, for a given pair $(x, y) \in \Sigma$, there are at most e possibilities for the coefficient a' of x'; in addition, there are at most q possibilities for the coefficient b', and at most $q - 1$ for y', so there are at most $eq(q-1)$ possibilities for (x', y'). Thus each orbit of $\mathrm{Aut}\, G$ on Σ contains at most $eq(q-1)$ pairs. Since $\mathrm{Aut}\, G$ acts regularly on each orbit, it follows that $|\mathrm{Aut}\, G| \leq eq(q-1)$. Since we have shown that $\mathrm{Aut}\, G$ contains a subgroup A of order $eq(q-1)$, it follows that $\mathrm{Aut}\, G = A$, proving (1).

Since $\mathrm{Aut}\, G$ acts semiregularly on Σ, the number of orbits it has on Σ is

$$\frac{|\Sigma|}{|\mathrm{Aut}\, G|} = \frac{q(q-1)\phi(q-1)}{eq(q-1)} = \frac{\phi(q-1)}{e},$$

proving (2).

We showed earlier that if (x, y) and (x', y') are in the same orbit of $\mathrm{Aut}\, G$ then a and a' have the same minimal polynomial in $\mathbb{F}_p[t]$. We also showed that the number of pairs with a given minimal polynomial is at most equal to the size of each orbit, so these numbers are equal, and (3) is proved. □

We now return to the proof of Theorem 7.2. Since the $(q-1)$-valent regular maps with automorphism group G correspond bijectively to the orbits of $\mathrm{Aut}\, G$ on Σ, it follows from Lemma 7.4 that there are, up to isomorphism, $\phi(q-1)/e$ such maps. Each map corresponds to the orbit consisting of those $(x, y) \in \Sigma$ for which x has the form $v \mapsto av + b$ where a is a primitive element of $\mathbb{F}_q$ with a given minimal polynomial $f(t)$. By its construction, each Biggs map $\mathcal{M}_q(c)$ corresponds to such a pair (x, y), with $a = c$, so this completes the proof of Theorem 7.2. □

Any regular embedding of K_q has valency $q - 1$, and by Lemma 7.2 it has automorphism group $\mathrm{AGL}_1(q)$, so it follows from Theorem 7.2 that the only such maps are the Biggs maps $\mathcal{M}_q(c)$, with $\mathcal{M}_q(c) \cong \mathcal{M}_q(c')$ if and only if c and c' have the same minimal polynomial in $\mathbb{F}_p[t]$. This is equivalent to c and c' being in the same orbit of the Galois group $\mathrm{Gal}\, \mathbb{F}_q$ of $\mathbb{F}_q$. Since this is generated by the Frobenius automorphism $v \mapsto v^p$ of $\mathbb{F}_q$, we have the following:

Theorem 7.3 *Let $q = p^e$ for some prime p. The regular embeddings of K_q are the Biggs maps $\mathcal{M}_q(c)$ where c is a primitive element of $\mathbb{F}_q$. Such maps $\mathcal{M}_q(c)$ and $\mathcal{M}_q(c')$ are isomorphic if and only if $c' = c^{p^f}$ for some $f = 0, 1, \ldots, e-1$. Up to isomorphism there are $\phi(q-1)/e$ such maps.*

This shows that the examples we discussed at the start of this chapter, namely one embedding each for $q = 2, 3$ and 4, and two each for $q = 5$ and 7, are in fact the only regular embeddings for these values of q.

Exercise 7.8 Let $\mathcal{M}$ be a regular embedding of K_q, where $q \geq 4$. Show that $\mathcal{M}$ has type $\{q-1, q-1\}$ and genus $(q-1)(q-4)/4$, unless $q \equiv 3 \bmod 4$, in which case the type is $\{(q-1)/2, q-1\}$ and the genus is $(q^2 - 7q + 4)/4$. What happens when $q = 2$ or $q = 3$?

Following Coxeter and Moser [6, Chap. 8], we define a regular map $\mathcal{M}$ to be *reflexible* if it is isomorphic to its mirror image $\overline{\mathcal{M}}$; this is equivalent to $\mathcal{M}$ having an orientation-reversing automorphism, as happens for the regular embeddings of K_2, K_3 and K_4, but not for K_5 or K_7 (see Examples 7.2 and 7.3). In the case of the Biggs maps, reversing orientation corresponds to inverting the chosen generator c of $\mathbb{F}_q^*$, so $\overline{\mathcal{M}_q(c)} = \mathcal{M}_q(c^{-1})$. Theorem 7.3 implies that $\overline{\mathcal{M}_q(c)} \cong \mathcal{M}_q(c)$ if and only if $c^{-1} = c^{p^f}$, or equivalently $p^e - 1$ divides $p^f + 1$, for some $f = 0, 1, \ldots, e-1$. Simple arithmetic shows that this happens if and only if $q = 2, 3$ or 4, so the unique regular embeddings of K_q for these values of q are reflexible, whereas those for $q > 4$ occur in chiral (mirror-image) pairs: see Examples 7.2 and 7.3, for instance.

Exercise 7.9 Fill in the details for the statement in the last sentence.

7.2.4 Regular Embeddings and Cyclotomic Polynomials

The primitive elements of $\mathbb{F}_q$ are the primitive $(q-1)$th roots of unity in that field, that is, the elements of multiplicative order $q-1$. These, and their minimal polynomials, can be obtained by first considering the corresponding elements and polynomials for the field $\mathbb{C}$. For any integer $n \geq 1$, the primitive nth roots of unity in $\mathbb{C}$ are the roots of the nth *cyclotomic polynomial*

$$\Phi_n(t) = \prod_{d|n} (t^d - 1)^{\mu(n/d)},$$

where μ is the *Möbius function* on $\mathbb{N}$ (see Sect. 5.1.6), with values

$$\mu(n) = \begin{cases} (-1)^k & \text{if } n \text{ is a product of } k \text{ distinct primes,} \\ 0 & \text{otherwise.} \end{cases}$$

In the formula for $\Phi_n(t)$, the factor $t^n - 1$ (with $d = n$ and $\mu(n/d) = \mu(1) = 1$) has all the nth roots of unity as its roots; the remaining factors $(t^d-1)^{\pm 1}$ (with $d < n$) use the inclusion-exclusion principle to remove (once each) those roots of multiplicative order d properly dividing n, so that only the primitive nth roots remain.

Exercise 7.10 Check that for $n = 1, 2, 3, 4, 5, 6$ we have

$$\Phi_n(t) = t-1,\ t+1,\ t^2+t+1,\ t^2+1,\ t^4+t^3+t^2+t+1,\ t^2-t+1.$$

Exercise 7.11 Show that $\Phi_n(t)$ has degree

$$\deg \Phi_n(t) = \sum_{d|n} \mu\left(\frac{n}{d}\right) d = n \prod_{p|n} \left(1 - \frac{1}{p}\right) = \phi(n),$$

where the product is over the distinct primes p dividing n.

Each cyclotomic polynomial $\Phi_n(t)$ has integer coefficients, that is, it is an element of $\mathbb{Z}[t]$; it is the minimal polynomial over $\mathbb{Q}$ of the primitive nth roots of unity, so it is irreducible in $\mathbb{Q}[t]$.

For applications to regular embeddings of K_q, we take $n = q-1$, where $q = p^e$ for some prime p. By reducing its coefficients mod p, we can regard $\Phi_{q-1}(t)$ as an element of $(\mathbb{Z}/p\mathbb{Z})[t] = \mathbb{F}_p[t]$. As such, its roots are the primitive $(q-1)$th roots of unity in the extension field $\mathbb{F}_q \geq \mathbb{F}_p$, that is, the primitive elements of $\mathbb{F}_q$. In general, $\Phi_{q-1}(t)$ need not be irreducible as an element of $\mathbb{F}_p[t]$: it splits into $\phi(q-1)/e$ distinct irreducible factors $f(t)$ of degree e, the roots of each forming an orbit under the action of Gal $\mathbb{F}_q$. These are the *primitive polynomials* of degree e in $\mathbb{F}_p[t]$, the minimal polynomials of the primitive elements of $\mathbb{F}_q$. The regular embeddings of K_q correspond bijectively to these primitive polynomials. If c is a primitive element with minimal polynomial $f(t)$, we can denote the corresponding map $\mathcal{M}_q(c)$ by $\mathcal{M}_q(f)$ since, up to isomorphism, it depends only on f, and not on the choice of a particular root c of f.

Example 7.5 Let $q = 27$, so $p = e = 3$. Since $27 \equiv 3 \bmod 4$ the regular embeddings of K_{27} have type $\{13, 26\}$ and genus 136 (see Exercise 7.8), with automorphism group $\mathrm{AGL}_1(27)$ of order 702. Now $\phi(q-1) = \phi(26) = 12$, so there are, up to isomorphism, $\phi(q-1)/e = 4$ such embeddings. The cyclotomic polynomial

$$\Phi_{26}(t) = \frac{(t^{26}-1)(t-1)}{(t^{13}-1)(t^2-1)} = \frac{t^{13}+1}{t+1} = t^{12} - t^{11} + t^{10} - \cdots + t^2 - t + 1,$$

when regarded as an element of $\mathbb{F}_3[t]$, has irreducible factors

$$f_i(t) = t^3 - t + 1, \quad t^3 - t^2 + 1, \quad t^3 + t^2 - t + 1, \quad t^3 - t^2 + t + 1$$

for $i = 1, \ldots, 4$, namely the primitive polynomials of degree 3 over $\mathbb{F}_3$. The multiplicative group $\mathbb{F}_{27}^* \cong C_{26}$ has $\phi(26) = 12$ generators, forming four orbits under the Galois group Gal $\mathbb{F}_{27} \cong C_3$, and each polynomial $f_i(t)$ is the minimal polynomial over $\mathbb{F}_3$ of the generators in one orbit. These four polynomials thus correspond to the four mutually non-isomorphic regular embeddings $\mathcal{M}_{27}(f_i)$ of K_{27}.

Exercise 7.12 Show that the polynomials $f_i(t)$ $(i = 1, \ldots, 4)$ appearing in Example 7.5 are irreducible in $\mathbb{F}_3[t]$, and that their product is $\Phi_{26}(t)$.

If a primitive element c has minimal polynomial

$$f(t) = t^e + a_{e-1}t^{e-1} + \cdots + a_1 t + a_0,$$

then c^{-1} is also a primitive element, with minimal polynomial

$$\bar{f}(t) = \frac{t^e}{a_0} f\left(\frac{1}{t}\right) = t^e + \frac{a_1}{a_0}t^{e-1} + \cdots + \frac{a_{e-1}}{a_0}t + \frac{1}{a_0},$$

the monic polynomial whose roots are the inverses of those of $f(t)$. Thus $\overline{\mathcal{M}_q(f)} = \mathcal{M}_q(\bar{f})$. For instance, in the example above, the maps $\mathcal{M}_{27}(f_1)$ and $\mathcal{M}_{27}(f_2)$ form a chiral pair, as do $\mathcal{M}_{27}(f_3)$ and $\mathcal{M}_{27}(f_4)$.

References

1. Bauer, M., Itzykson, C.: Triangulations. In: Schneps, L. (ed.) The Grothendieck Theory of Dessins d'Enfants. London Mathematical Society Lecture Note Series, vol. 200, pp. 179–236. Cambridge University Press, Cambridge (1994)
2. Biggs, N.L.: Classification of complete maps on orientable surfaces. Rend. Math. (6) **4**, 645–655 (1971)
3. Biggs, N.L., White, A.T.: Permutation Groups and Combinatorial Structures. London Mathematical Society Lecture Note Series, vol. 33. Cambridge University Press, Cambridge/New York (1979)
4. Catalano, D.A., Conder, M.D.E., Du, S.F., Kwon, Y.S., Nedela, R., Wilson, S.: Classification of regular embeddings of n-dimensional cubes. J. Algebraic Combin. **33**, 215–238 (2011)
5. Conder, M.D.E., Jajcay, R., Tucker, T.: Regular Cayley maps for finite abelian groups. J. Algebraic Combin. **25**, 259–283 (2007)
6. Coxeter, H.S.M., Moser, W.O.J.: Generators and Relations for Discrete Groups. Springer, Berlin/Heidelberg/New York (1980)
7. Dixon, J.D., Mortimer, B.: Finite Permutation Groups. Graduate Texts in Mathematics, vol. 163. Springer, Berlin/Heidelberg/New York (1996)
8. Gardiner, A.D., Nedela, R., Širáň, J., Škoviera, M.: Characterization of graphs which underlie regular maps on closed surfaces. J. Lond. Math. Soc. (2) **59**, 100–108 (1999)
9. Huppert, B.: Endliche Gruppen I . Springer, Berlin/Heidelberg/New York (1967)
10. James, L.D., Jones, G.A.: Regular orientable imbeddings of complete graphs. J. Combin. Theory Ser. B **39**, 353–367 (1985)
11. Jones, G.A.: Automorphisms and regular embeddings of merged Johnson graphs. Eur. J. Combin. **26**, 417–435 (2005)
12. Jones, G.A.: Classification and Galois conjugacy of Hamming maps. Ars Math. Contemp. **4**, 313–328 (2011)
13. Passman, D.S.: Permutation Groups. Benjamin, New York (1968)
14. Richter, R.B., Širáň, J., Jajcay, R., Tucker, T.W., Watkins, M.E.: Cayley maps. J. Combin. Theory Ser. B **95**, 189–245 (2005)
15. Singerman, D., Wolfart, J.: Cayley graphs, Cori hypermaps, and dessins d'enfants. ARS Math. Contemp. **1**, 144–153 (2008)
16. Širáň, J.: How symmetric can maps on surfaces be? In: Blackburn, S.R., Gerke, S., Wildon, M. (eds.) Surveys in Combinatorics 2013. London Mathematical Society Lecture Note Series, vol. 409, pp. 161–238. Cambridge University Press, Cambridge (2013)
17. Thompson, J.G.: Finite groups with fixed-point-free automorphisms of prime order. Proc. Natl. Acad. Sci. USA **45**, 578–581 (1959)
18. Tsuzuku, T.: Finite Groups and Finite Geometries. Cambridge Tracts in Mathematics, vol. 78. Cambridge University Press, Cambridge (1982)

Chapter 8
Wilson Operations

Abstract As shown in Chap. 3, a dessin uniquely determines all the relevant properties of its underlying Belyĭ surface. A key invariant is the moduli field of the corresponding algebraic curve, and a major step in determining this is to understand the action of the absolute Galois group $\mathbb{G}$ on regular dessins. Under relatively mild conditions this action can be described combinatorially with some map and hypermap operations, the so-called Wilson (hole) operations, introduced around the same time as Belyĭ functions. However, their role in the understanding of Galois actions on dessins has only recently been discovered. We give several examples, based on the regular embeddings of complete graphs classified in Chap. 7. In the final section we consider the group of all operations on dessins, introduced by James, showing that it is isomorphic to the outer automorphism group of the free group of rank 2, and hence to $\mathrm{GL}_2(\mathbb{Z})$. As an example we consider the action of this group on the 19 regular dessins with automorphism group A_5.

Keywords Absolute Galois group • Cyclotomic field • Galois action • Galois orbit • Hypermap operation • Outer automorphism group • Regular dessin • Regular map • Wilson operation

8.1 A Class of Map Operations

In 1979, Wilson published a paper [14] describing certain operations which convert one regular map into another regular map with the same automorphism group. An obvious example is the classical vertex-face duality operation, which preserves the underlying surface of a map. Some of Wilson's operations change the surface, in certain cases converting an orientable surface into one which is non-orientable. This is not so useful for the study of dessins, which are always on orientable surfaces. However, one class of Wilson's operations preserves the orientability of surfaces, and consequently has applications to dessins.

A regular map $\mathcal{M}$ of valency l is simply a regular dessin of type $(l, 2, n)$ for some n. Its automorphism group $G = \mathrm{Aut}\,\mathcal{M}$ has standard generators x, y and z satisfying $x^l = y^2 = z^n = xyz = 1$. If j is any integer coprime to l, then since $\langle x^j \rangle = \langle x \rangle$ we have $G = \langle x^j, y \rangle$. This new pair of generators therefore realises G as the automorphism group of a regular dessin of type $(l, 2, n')$ for some n', namely the

Gareth A. Jones, J. Wolfart, *Dessins d'Enfants on Riemann Surfaces*, Springer Monographs in Mathematics, DOI 10.1007/978-3-319-24711-3_8

order of the element $z' = (x^j y)^{-1}$. This dessin is a regular map of valency l which we will denote by $H_j(\mathcal{M})$.

The graphs embedded by $\mathcal{M}$ and $H_j(\mathcal{M})$ are the same, but the rotation of edges around each vertex is changed: if the edges around a vertex v of $\mathcal{M}$ are in the cyclic order $e_1, e_2, e_3, \ldots, e_l$ as we follow the orientation around v, then in $H_j(\mathcal{M})$ they are in the cyclic order $e_1, e_{j+1}, e_{2j+1}, \ldots, e_{1-j}$, where we regard subscripts as elements of $\mathbb{Z}_l := \mathbb{Z}/l\mathbb{Z}$. The operation H_j depends only on the congruence class of j in $\mathbb{Z}_l$, and we have $H_j \circ H_{j'} = H_{jj'}$ for all j and j' coprime to l, so we can regard these operations as providing an action of the multiplicative group $U_l = \mathbb{Z}_l^*$ of units mod l on regular maps of valency l. The group $\mathrm{Exp}\,\mathcal{M} = \{j \in U_l \mid H_j(\mathcal{M}) \cong \mathcal{M}\}$ of *exponents* of $\mathcal{M}$, introduced by Nedela and Škoviera in [12], is the subgroup of U_l stabilising the isomorphism class of $\mathcal{M}$ in this action; since U_l is abelian, $\mathrm{Exp}\,\mathcal{M}$ is the kernel of the action of U_l on this orbit.

Apart from the identity operation H_1, the simplest of Wilson's operations is H_{-1}, which reverses the cyclic rotation of edges around each vertex, so that $H_{-1}(\mathcal{M})$ is the mirror-image $\overline{\mathcal{M}}$ of a regular map $\mathcal{M}$. This operation preserves the genus and the type $\{n, l\}$ of a map, but for other values of j these may be changed.

Example 8.1 Let $\mathcal{M}$ be the icosahedron, regarded as a regular map of type $\{3, 5\}$ on the sphere. Then G is isomorphic to the alternating group A_5, and we can identify its generators x, y and z with the permutations $(1, 2, 3, 4, 5)$, $(1, 2)(3, 4)$ and $(2, 5, 4)$, as in Sect. 5.1.4, Example 5.7. Here $l = 5$, so $U_5 = \mathbb{F}_5^* = \langle 2 \rangle \cong C_4$ can act. To apply H_2 we replace x with x^2, corresponding to $(1, 3, 5, 2, 4)$, so $z' = (x^2 y)^{-1}$ corresponds to $(1, 5, 3, 2, 4)$. Thus $H_2(\mathcal{M})$ is a regular map of type $\{5, 5\}$, with automorphism group G, so it has genus

$$1 + \frac{1}{2}|G|\left(1 - \frac{1}{2} - \frac{1}{5} - \frac{1}{5}\right) = 4.$$

This map is the great dodecahedron, described in Sect. 5.1.4, Example 5.7. The generating pair for A_5 used there, namely $(1, 2, 3, 4, 5)$ and $(1, 3)(2, 4)$, are conjugated to the pair $(1, 3, 5, 2, 4)$ and $(1, 2)(3, 4)$ used here by the permutation $(1, 2, 4, 3)$; thus they are equivalent under $\mathrm{Aut}\,A_5 = S_5$, so they determine isomorphic maps. Applying $H_4 = H_{-1}$ to $\mathcal{M}$ gives $\overline{\mathcal{M}} \cong \mathcal{M}$, so $\mathrm{Exp}\,\mathcal{M}$ is the subgroup $\{\pm 1\} \cong C_2$ of U_5.

As we saw in Chap. 7, the regular embeddings of the complete graph K_q are the Biggs maps $\mathcal{M}_q(c)$, all of valency $q - 1$, so in their case one can apply the Wilson operators H_j with j coprime to $m = q - 1$. Since these operations preserve the regularity and the embedded graph of a map, it follows from Theorem 7.3 that $H_j(\mathcal{M}_q(c))$ is another Biggs map $\mathcal{M}_q(c')$. Conversely, given any two Biggs maps $\mathcal{M}_q(c)$ and $\mathcal{M}_q(c')$, corresponding to primitive elements c and c' of $\mathbb{F}_q$, we have $c' = c^j$ for some j coprime to $q-1$; thus $\mathcal{M}_q(c')$ is obtained from $\mathcal{M}_q(c)$ by replacing the rotation x around vertices with its jth power, so $\mathcal{M}_q(c') = H_j(\mathcal{M}_q(c))$. This shows that for each prime power $q = p^e$ the Wilson operations H_j $(j \in U_{q-1})$ act transitively on the regular embeddings of K_q. Theorem 7.3 also shows that

$$c, c^3, c^9; \quad c^5, c^{15}, c^{19}; \quad c^{17}, c^{25}, c^{23}; \quad c^7, c^{21}, c^{11}.$$

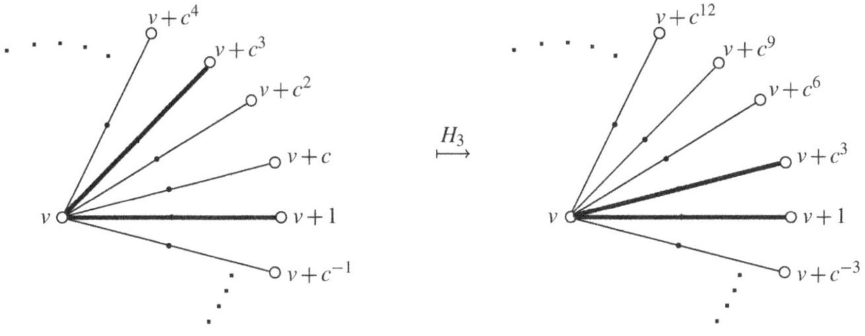

Fig. 8.1 The effect of the Wilson operation H_3 on $\mathcal{M}_q(c)$, locally around v

$H_j(\mathcal{M}_q(c)) \cong \mathcal{M}_q(c)$ if and only if $c^j = c^{p^f}$ for some $f = 0, 1, \ldots, e-1$, so the kernel of this action of U_{q-1} is the subgroup of order e generated by the unit p. For example, H_3 in Fig. 8.1 is an isomorphism if q is a power of 3:

Example 8.2 Let $q = 27$ again, as in Example 7.5. If c is any primitive element of $\mathbb{F}_{27}$, then the full set of primitive elements, grouped into their four orbits under $\mathrm{Gal}\,\mathbb{F}_{27}$, is:

$$c,\ c^3,\ c^9; \quad c^5,\ c^{15},\ c^{19}; \quad c^{17},\ c^{25},\ c^{23}; \quad c^7,\ c^{21},\ c^{11}.$$

By Theorem 7.3, these orbits correspond to the four isomorphism classes of regular embeddings of K_{27}. Since $c^{-1} = c^{25}$ and $c^{-5} = c^{21}$, the Wilson operation H_{-1} transposes the first and third of these embeddings, and also the second and fourth, so they form two chiral pairs. Similarly, since $c^{125} = c^{21}$, the operation H_5 permutes them in a 4-cycle. One can see this in terms of the primitive polynomials $f_i(t)$, as follows. Suppose, for instance, that we choose c to have $f_1(t) = t^3 - t + 1$ as its minimal polynomial. A simple calculation shows that c^5 is a root of $f_4(t) = t^3 - t^2 + t - 1$, so $H_5(\mathcal{M}_{27}(f_1)) = \mathcal{M}_{27}(f_4)$. Iterating, we have $H_5^2(\mathcal{M}_{27}(f_1)) = \mathcal{M}_{27}(f_2)$ and $H_5^3(\mathcal{M}_{27}(f_1)) = \mathcal{M}_{27}(f_3)$, so H_5 induces the 4-cycle $(1, 4, 2, 3)$ on the subscripts. In this particular example, we have $U_{q-1} = U_{26} = \langle 3 \rangle \times \langle 5 \rangle \cong C_3 \times C_4$, with the first factor the kernel $\mathrm{Exp}\,\mathcal{M}$ of the action on these regular embeddings $\mathcal{M}$, and the second factor acting regularly on them.

Exercise 8.1 Show that if $\mathcal{M}$ is a regular map and $\mathrm{Aut}\,\mathcal{M}$ is abelian, then $H_j(\mathcal{M}) \cong \mathcal{M}$ for all Wilson operations H_j.

8.2 Wilson Operations and Galois Actions

The action of the Wilson operations H_j on l-valent regular maps closely resembles that of the Galois group $\mathrm{Gal}\,\mathbb{Q}(\zeta_l)$ of the cyclotomic field $\mathbb{Q}(\zeta_l)$, which consists of the field-automorphisms sending the primitive lth root of unity $\zeta_l = e^{2\pi i/l}$ to ζ_l^j where j is coprime to l. This is particularly clear in the case of the Biggs maps, where a primitive root of unity in a finite field is also raised to its jth power. This suggests that there may be some more fundamental connections between the actions on regular maps of the Wilson operations and of the elements of the absolute Galois group. This question is explored in some detail by Jones, Streit and Wolfart in [8], both for the original Wilson operations H_j on maps and also for some generalisations of them applicable to all dessins. It is shown there that in certain cases there are strong connections between these two actions, whereas in others the actions seem to have very little in common. Here we will consider how this applies to the Biggs maps, where the connection is particularly strong.

We saw in the preceding section that, for any given prime power $q = p^e$, the regular embeddings $\mathcal{M}_q(c)$ of K_q form a single orbit under the action of the Wilson operations H_j $(j \in U_{q-1})$. From the point of view of dessins, we would like to know how the absolute Galois group $\mathbb{G} = \mathrm{Gal}\,\overline{\mathbb{Q}}$ acts on these maps (or, indeed, on any other class of maps). In general this is a very difficult question, but here we can exploit our knowledge of how the Wilson operations act and which properties of dessins are Galois-invariant.

Theorem 2 of [8] is a general result about the connections between Wilson operations and Galois conjugations for dessins of arbitrary types. Here we will state and use a slightly simpler version, which is applicable only to maps.

Theorem 8.1 *Suppose that a set S of regular maps $\mathcal{M}$ of valency l forms a single orbit under the Wilson operations H_j $(j \in U_l)$, and is invariant under the action of $\mathbb{G}$. Then*

1. *S forms a single orbit of $\mathbb{G}$;*
2. *the minimal field of definition of each map in S is the subfield K of $\mathbb{Q}(\zeta_l)$ fixed by those elements $\sigma_j : \zeta_l \mapsto \zeta_l^j$ of $\mathrm{Gal}\,\mathbb{Q}(\zeta_l)$ such that $H_j(\mathcal{M}) \cong \mathcal{M}$ for each $\mathcal{M}$ in S, that is, by those σ_j with $j \in \mathrm{Exp}\,\mathcal{M}$.*

Proof Since the maps in S form a single orbit under the Wilson operations H_j, we may denote one of them by $\mathcal{M}_1$ and then label the whole set by $\mathcal{M}_j := H_j(\mathcal{M}_1)$ for $j \in U_m$. It also follows that these maps share the same automorphism group G, though they correspond to different generators: if $\mathcal{M}_1$ corresponds to generators x and y, satisfying $x^l = y^2 = 1$, then the generators corresponding to $\mathcal{M}_j$ are x^j and y.

In particular, x fixes a vertex P of $\mathcal{M}_1$ (a white vertex in the language of dessins) and sends every edge incident with P to the next edge, following the orientation around P. In the underlying Riemann surface there is a local coordinate $z : U \to \mathbb{C}$ in a neighbourhood U of P such that P corresponds to $z = 0$, and composing z with

x gives

$$z \circ x = \zeta_l \cdot z + \text{higher order terms in } z\ .$$

By Belyĭ's theorem we may suppose that the surface is an algebraic curve C defined over $\overline{\mathbb{Q}}$, that x is an automorphism defined over $\overline{\mathbb{Q}}$ (otherwise field automorphisms acting on x would produce infinitely many automorphisms of C), and hence that P is a $\overline{\mathbb{Q}}$-rational point on C. Moreover, we may assume that z is a rational function on C defined over $\overline{\mathbb{Q}}$ with a simple zero at P.

Now consider the action of some $\sigma \in \mathbb{G}$ on curves, maps, their automorphisms, points and functions. Locally around the fixed point P^σ of x^σ on the curve C^σ we have

$$z^\sigma \circ x^\sigma \;=\; \sigma(\zeta_l) \cdot z^\sigma + \text{higher order terms in } z^\sigma \tag{8.1}$$

where z^σ is now a function with a simple zero at P^σ (since zero orders are Galois-invariant), so it serves there as a local coordinate.

By hypothesis, the regular map $\mathcal{M}_1^\sigma$ on C^σ belongs to S, so $\mathcal{M}_1^\sigma \cong \mathcal{M}_j$ for some $j \in U_l$. If we identify G with G^σ and x with x^σ, the new standard generator $(x^\sigma)^j$ has the fixed point P^σ. The local behaviour of x^j at P is $z \circ x^j = \zeta_l^j \cdot z + \ldots$, and at the corresponding vertex P^σ in $\mathcal{M}_j$ the standard generator acts as $z^\sigma \circ (x^\sigma)^j = \zeta_l \cdot z^\sigma + \ldots$. Comparing this with Eq. (8.1) shows that

$$\sigma(\zeta_l)^j = \zeta_l\ .$$

It follows that if σ fixes the cyclotomic field $\mathbb{Q}(\zeta_l)$ element-wise, then $j = 1$. In other words, the moduli field of $\mathcal{M}_1$—hence by Theorem 5.3 its minimal field of definition—is contained in $\mathbb{Q}(\zeta_l)$. If σ does not fix $\mathbb{Q}(\zeta_l)$, its action on this field of definition is completely determined by its effect $\sigma : \zeta_l \mapsto \zeta_l^k$ $(k \in U_l)$ on ζ_l. Comparison of the standard generators on $\mathcal{M}_1^\sigma$ and $H_j(\mathcal{M}_1)$ shows that $kj \equiv 1 \bmod l$. Identifying U_l on the one hand with the group of Wilson operations acting on S, and on the other hand with $\operatorname{Gal} \mathbb{Q}(\zeta_l)/\mathbb{Q}$, we see that the two actions on S are inverse to each other. □

In general, in order to apply this result we need to verify three hypotheses: that the set S is invariant under the Wilson operations, that it forms a single orbit under them, and that it is invariant under all Galois conjugations. The first two can be verified by studying the effect of the transformations $x \mapsto x^j$ on the appropriate generating pairs x, y of the common automorphism group G of these maps: if G is well understood, as in the case of the Biggs maps, then this can be a fairly routine process (see Chap. 7). In the case of the third hypothesis, we know from Theorem 4.11 that certain properties of dessins are Galois invariants, so if S consists of all the dessins having a particular set of Galois-invariant properties, then it must be a Galois-invariant set. It is therefore a union of orbits of $\mathbb{G}$, and part (1) of the theorem asserts that it is, in fact, a single orbit. Proving such a result directly, without the help of the Wilson operations, is generally much more difficult: one would need to find explicit models of the dessins, and explicit Galois conjugations between them, something which has proved to be feasible in only a few simple cases. As an example of what can be done without the explicit knowledge of algebraic equations and Belyĭ functions, we have:

Theorem 8.2 *For each prime power $q = p^e$, the regular embeddings $\mathcal{M}_q(c)$ of K_q form a single orbit under the action of $\mathbb{G}$. Each of these dessins is defined over the splitting subfield K of the prime p in the cyclotomic field $\mathbb{Q}(\zeta_{q-1})$, that is, the subfield fixed by the automorphism $\sigma_p : \zeta_{q-1} \mapsto \zeta_{q-1}^p$, an extension of $\mathbb{Q}$ of degree $\phi(q-1)/e$.*

(Algebraic number theory provides alternative descriptions of these splitting fields: K is the largest subfield of $\mathbb{Q}(\zeta_{q-1})$ in which p splits into $|K : \mathbb{Q}|$ prime ideals of residue degree 1, and it is the smallest subfield in which p splits into as many prime ideals as in $\mathbb{Q}(\zeta_{q-1})$.)

Proof We have already verified the first two hypotheses of Theorem 8.1: the Wilson operations permute the maps $\mathcal{M}_q(c)$, and do so in a single orbit. It is therefore sufficient to show that these maps form a Galois-invariant set. Now Theorem 4.11 shows that regularity, automorphism group and valency (of black or white vertices) are all Galois-invariant properties of dessins, and Theorem 7.2(1) characterises these maps, up to isomorphism, as the regular dessins of type $(q-1, 2, r)$ for any r, with automorphism group $\mathrm{AGL}_1(q)$. (This was why we needed to prove Theorem 7.2, a stronger form of Theorem 7.3.) It follows that these Biggs maps form a Galois-invariant set, so by Theorem 8.1 they form an orbit of $\mathbb{G}$.

We have also seen in Theorem 7.3 that $H_j(\mathcal{M}_q(c)) \cong \mathcal{M}_q(c)$ if and only if j is in the subgroup $\langle p \rangle$ of U_{q-1}. It therefore follows from Theorem 8.1(2) that the minimal field of definition of each $\mathcal{M}_q(c)$ is the subfield K of $\mathbb{Q}(\zeta_{q-1})$ fixed by the automorphism $\sigma_p : \zeta_{q-1} \mapsto \zeta_{q-1}^p$. The Galois correspondence gives

$$|\mathbb{Q}(\zeta_{q-1}) : K| = |\langle \sigma_p \rangle| = |\langle p \rangle| = e,$$

so $|K : \mathbb{Q}| = \phi(q-1)/e$.

Since $\mathrm{Gal}\,\mathbb{Q}(\zeta_{q-1})$ is an abelian group (isomorphic to U_{q-1}), $\langle \sigma_p \rangle$ is a normal subgroup and so $K/\mathbb{Q}$ is a Galois extension, with $\mathrm{Gal}\,K \cong \mathrm{Gal}\,\mathbb{Q}(\zeta_{q-1})/\langle \sigma_p \rangle$. This is a group of order $\phi(q-1)/e$, permuting the maps $\mathcal{M}_q(c)$ regularly. The absolute Galois group $\mathbb{G}$ leaves K invariant; restriction of automorphisms maps it onto $\mathrm{Gal}\,K$, inducing a transitive action of $\mathbb{G}$ on the maps $\mathcal{M}_q(c)$. □

Example 8.3 Once again, let $q = 27$, so there are four regular maps $\mathcal{M}_{27}(c)$ forming a single Galois orbit. If we define $\zeta = \zeta_{26}$, then the minimal field of definition of each of these maps is the subfield K of the cyclotomic field $\mathbb{Q}(\zeta)$ fixed by the automorphism $\sigma_3 : \zeta \mapsto \zeta^3$. This has the form $K = \mathbb{Q}(\eta)$ where $\eta = \zeta + \zeta^3 + \zeta^9$, with $|K : \mathbb{Q}| = 4$. The Galois group $\mathrm{Gal}\,K$ is a cyclic group of order 4, isomorphic to $U_{26}/\langle 3 \rangle$, generated by the restriction to K of the automorphism $\sigma_5 : \zeta \mapsto \zeta^5$ of $\mathbb{Q}(\zeta)$. (Recall that $U_{26} = \langle 3 \rangle \times \langle 5 \rangle \cong C_3 \times C_4$.) The involution

$\sigma_5^2 : \zeta \mapsto \zeta^{25} = \zeta^{-1} = \overline{\zeta}$ is simply complex conjugation, sending each $\mathcal{M}_{27}(c)$ to its mirror-image $\mathcal{M}_{27}(c^{-1})$.

Exercise 8.2 Determine the degrees and generators for the minimal fields of definition K of $\mathcal{M}_{16}(c)$ and $\mathcal{M}_{32}(c)$.

A second look at the proof of Theorem 8.1 shows that the fact that the second generator of G has order 2 is completely irrelevant. This observation is the source of many possible generalisations of the Wilson operations and of their role for a better understanding of Galois actions on dessins. For later use, we mention just one possible statement where we replace the order 2 with l and interchange the role of the two generators. This result can be proved in the same way or derived from [8, Theorem 2].

Theorem 8.3 *Let l and m be coprime integers, both greater than* 1. *Suppose that a set S of regular bipartite maps $\mathcal{M}$ of type (l, m, n) forms a single orbit under the Wilson operations H_j $(j \in U_m)$, acting on the pair of standard generators of the automorphism group by*

$$(x, y) \mapsto (x, y^j) .$$

Suppose moreover that S is invariant under the action of $\mathbb{G}$. Then

1. *S forms a single orbit of $\mathbb{G}$;*
2. *the minimal field of definition of each map in S is the subfield K of $\mathbb{Q}(\zeta_m)$ fixed by those elements $\sigma_j : \zeta_m \mapsto \zeta_m^j$ of* Gal $\mathbb{Q}(\zeta_m)$ *for which $H_j(\mathcal{M}) \cong \mathcal{M}$ for each $\mathcal{M}$ in S.*

For further generalisations of these results—useful for many examples—see also Theorem 2 of [1].

8.3 The Group of Operations on Dessins

The Wilson operations H_j discussed in Sect. 8.1 can be interpreted algebraically as outer automorphisms of certain triangle groups, as follows. The maps of valency dividing l can be identified with the conjugacy classes of subgroups of the triangle group $\Delta = \Delta(l, 2, \infty)$. The automorphism group Aut Δ permutes these conjugacy classes, with the inner automorphism group Inn Δ preserving each class, so there is an induced action of the outer automorphism group Out Δ = Aut Δ/ Inn Δ on these conjugacy classes, and hence on the corresponding maps.

Now $\Delta(l, 2, \infty)$ is the free product $C_l * C_2$ of cyclic groups $\langle X \rangle \cong C_l$ and $\langle Y \rangle \cong C_2$ of orders l and 2. A theorem of Schreier shows that if $l > 2$ then Out Δ is isomorphic to the group $U_l = (\mathbb{Z}/l\mathbb{Z})^*$ of units mod l. Indeed, Aut Δ is a semidirect product of Inn Δ and a complement, isomorphic to U_l, consisting of the automorphisms $X \mapsto X^j$, $Y \mapsto Y$ where $j \in U_l$. This action of U_l then corresponds to its action on maps of valency dividing l as the group of Wilson operations H_j.

The operations which apply to *all* maps are rather more restricted. In this case we take Δ to be the triangle group $\Delta(\infty, 2, \infty) \cong C_\infty * C_2$, so that Out Δ is now a Klein four-group $V_4 = C_2 \times C_2$, induced by the automorphisms $X \mapsto X^{-1}$, $Y \mapsto Y$, sending each map to its mirror image, and $X \mapsto Z$ $(= (XY)^{-1})$, $Y \mapsto Y$, sending each map to its dual.

For dessins, however, we have a much richer supply of operations, first described by James in [6] in the context of hypermaps. Here we use the triangle group

$$\Delta = \Delta(\infty, \infty, \infty) = \langle X, Y, Z \mid XYZ = 1 \rangle,$$

a free group F_2 of rank 2, freely generated by any two of X, Y and Z. In this case it follows from a theorem of Nielsen (which we shall prove later in this section) that the group Ω of operations on dessins induced by Out Δ is isomorphic to $\mathrm{GL}_2(\mathbb{Z})$.

Each automorphism α of Δ leaves the commutator subgroup Δ' invariant, so it induces an automorphism $\overline{\alpha}$ of the abelianisation $\Delta^{\mathrm{ab}} = \Delta/\Delta'$. This is a free abelian group of rank 2, and we can take the images of X and Y as an ordered basis, so that $\overline{\alpha}$ corresponds to a matrix $A \in \mathrm{GL}_2(\mathbb{Z})$ acting on the left of column vectors. This gives us a homomorphism

$$\theta : \mathrm{Aut}\,\Delta \to \mathrm{GL}_2(\mathbb{Z}), \quad \alpha \mapsto A.$$

We will denote the corresponding operation on dessins by H_α.

For instance, the automorphism $\beta : X \mapsto Y$, $Y \mapsto X$ is mapped by θ to the matrix

$$B = \begin{pmatrix} 0 & 1 \\ 1 & 0 \end{pmatrix};$$

the corresponding operation H_β transposes the colours of the black and white vertices of a dessin, changing its type from (l, m, n) to (m, l, n) while preserving the underlying surface. The automorphism $\gamma : X \mapsto Y \mapsto Z \mapsto X$ of Δ is mapped to the matrix

$$C = \begin{pmatrix} 0 & -1 \\ 1 & -1 \end{pmatrix};$$

the corresponding operation H_γ permutes the colours white, black and red of the vertices and face-centres in a 3-cycle, again preserving the surface of a dessin. These two operations generate a subgroup

$$\Sigma = \langle H_\beta, H_\gamma \rangle \cong S_3$$

of Ω, inducing all 3! permutations of the colours and preserving the surface; it was described as a group of hypermap operations by Machì in [11].

The automorphism δ of Δ inverting X and Y is mapped to the matrix

$$D = -I = \begin{pmatrix} -1 & 0 \\ 0 & -1 \end{pmatrix};$$

the corresponding operation H_δ sends each dessin to its mirror image. This operation commutes with Σ, giving a subgroup

$$\Omega_1 = \langle H_\beta, H_\gamma, H_\delta \rangle = \Sigma \times \langle H_\delta \rangle \cong S_3 \times C_2 \cong D_6$$

of Ω, preserving the genus of each dessin.

The automorphism $\varepsilon : X \mapsto X^{-1}$, $Y \mapsto Y$ of Δ is mapped to the matrix

$$E = \begin{pmatrix} -1 & 0 \\ 0 & 1 \end{pmatrix};$$

the corresponding operation H_ε respectively reverses and preserves the rotation of edges around the white and black vertices. This can (and usually does) change the type and the genus of a dessin: for instance, it transposes faces and Petrie polygons, while preserving the embedded bipartite graph. (This is an instance $H_{-1,1}$ of the generalised Wilson operations $H_{i,j}$ introduced in [8], raising x and y to powers x^i and y^j where i and j are coprime to the valencies of the white and black vertices.)

The operations H_β and H_ε generate a subgroup

$$\Omega_2 = \langle H_\beta, H_\varepsilon \rangle \cong D_4$$

of Ω, containing $H_\delta = (H_\beta H_\varepsilon)^2$. We then have

$$\Omega_0 := \Omega_1 \cap \Omega_2 = \langle H_\beta, H_\delta \rangle \cong V_4,$$

a Klein four-group.

Proposition 8.1 $\mathrm{SL}_2(\mathbb{Z})$ *is generated by the matrices*

$$R := CD = -C = \begin{pmatrix} 0 & -1 \\ 1 & 1 \end{pmatrix} \quad \textit{and} \quad S := EB = \begin{pmatrix} 0 & -1 \\ 1 & 0 \end{pmatrix},$$

with defining relations

$$R^6 = S^4 = 1, \quad R^3 = S^2.$$

This is an easy consequence of the facts that $R^3 = S^2 = -I$, and that the modular group $\mathrm{PSL}_2(\mathbb{Z}) = \mathrm{SL}_2(\mathbb{Z})/\{\pm I\}$ is the free product of the cyclic groups of orders 3 and 2 generated by the images of R and S (see Sect. 1.2.4, where a conjugate pair of generators are used). In fact, Proposition 8.1 shows that $\mathrm{SL}_2(\mathbb{Z})$ is a free product with amalgamation $C_6 *_{C_2} C_4$ of the cyclic groups generated by R and S, amalgamating their isomorphic subgroups $\langle R^3\rangle$ and $\langle S^2\rangle$ of order 2.

Exercise 8.3 Fill in the details of the proof of Proposition 8.1.

Corollary 8.1 $\mathrm{GL}_2(\mathbb{Z})$ *is generated by the matrices* R, S *and* B, *with defining relations*

$$R^6 = S^4 = R^3S^{-2} = B^2 = (BR)^2 = (BS)^2 = 1.$$

This follows easily from Proposition 8.1, using the fact that $\mathrm{GL}_2(\mathbb{Z})$ is the semidirect product of a normal subgroup $\mathrm{SL}_2(\mathbb{Z})$ and a complement $\langle B\rangle \cong C_2$, with B acting by conjugation by inverting the generators R and S of $\mathrm{SL}_2(\mathbb{Z})$. One can also deduce this presentation from that for $\mathrm{GL}_2(\mathbb{Z})$ given in [2, Sect. 7.2]. Corollary 8.1 shows that $\mathrm{GL}_2(\mathbb{Z})$ is a free product with amalgamation $D_6 *_{D_2} D_4$ of the dihedral groups $\langle R, B\rangle \cong D_6$ and $\langle S, B\rangle \cong D_4$, amalgamating their subgroups $\langle R^3, B\rangle$ and $\langle S^2, B\rangle$, both of which are Klein four-groups $V_4 \cong D_2$.

We can now prove the main theorem of this section, a classical result due to Nielsen (see [10, Ch. I, Prop. 4.5]):

Theorem 8.4 *The action of* $\mathrm{Aut}\,\Delta$ *on* Δ^{ab} *induces an isomorphism*

$$\mathrm{Out}\,\Delta \cong \mathrm{GL}_2(\mathbb{Z})\,.$$

Proof By Corollary 8.1, $\mathrm{GL}_2(\mathbb{Z})$ is generated by the matrices B, R and S, and hence by the matrices B, C, D and E. Since these are the images under θ of automorphisms β, γ, δ and ε of Δ, the homomorphism $\theta : \mathrm{Aut}\,\Delta \to \mathrm{GL}_2(\mathbb{Z})$ is an epimorphism.

Now $\mathrm{Inn}\,\Delta$ is contained in the kernel K of θ: if α is an inner automorphism of Δ, then $g\Delta' = \alpha(g)\Delta'$ for each $g \in \Delta$ since $g^{-1}\alpha(g)$ is a commutator. In fact, we also have the reverse inclusion $K \le \mathrm{Inn}\,\Delta$. To see this, it is sufficient to check that, for each of the defining relators for $\mathrm{GL}_2(\mathbb{Z})$ in Corollary 8.1, the corresponding word in the automorphisms β, γ, δ and ε represents an inner automorphism of Δ. For instance, the first relator $R^6 = (CD)^6$ corresponds to the word $(\gamma\delta)^6$; composing from right to left, we have

$$\gamma\delta : X \mapsto Y^{-1},\ Y \mapsto XY,$$

so

$$(\gamma\delta)^3 : X \mapsto Y^{-1}X^{-1}Y = (X^{-1})^{XY},\ Y \mapsto (Y^{-1})^{XY},$$

and hence

$$(\gamma\delta)^6 : X \mapsto X^{(XY)^2},\ Y \mapsto Y^{(XY)^2}.$$

Thus R^6 corresponds to the inner automorphism induced by $(XY)^2$. All the other relators yield the identity automorphism, except R^3S^{-2} which gives conjugation by XY. This shows that $K = \mathrm{Inn}\,\Delta$, so the result follows from the first isomorphism theorem. □

Exercise 8.4 Verify the final remark about the other relators of $\mathrm{GL}_2(\mathbb{Z})$.

Remark 8.1 One can apply similar arguments to the free group F_n of any finite rank n, not just $n = 2$. As above, the action of $\mathrm{Aut}\,F_n$ on $F_n^{\mathrm{ab}} \cong \mathbb{Z}^n$ induces an epimorphism $\mathrm{Aut}\,F_n \to \mathrm{GL}_n(\mathbb{Z})$, but if $n > 2$ then $\mathrm{Inn}\,F_n$ is a proper subgroup of the kernel, so although $\mathrm{Out}\,F_n$ maps onto $\mathrm{GL}_n(\mathbb{Z})$ these groups are not isomorphic.

Exercise 8.5 Find an automorphism of the free group F_n $(n \geq 3)$ which acts trivially on F_n^{ab} but is not inner.

The following result was given, with a slightly different proof, by James in [6]:

Theorem 8.5 $\mathrm{Out}\,\Delta$ *acts faithfully on regular dessins.*

Proof We need to show that if α is a non-inner automorphism of Δ then $H_\alpha(\mathscr{D}) \not\cong \mathscr{D}$ for some regular dessin $\mathscr{D}$.

Theorem 8.4 shows that α is mapped by θ to a non-identity matrix $A \in \mathrm{GL}_2(\mathbb{Z})$. If $A = -I$ $(= D)$ then H_α is the operation H_8 sending each dessin to its mirror image, so we could take $\mathscr{D}$ to be one of the regular torus embeddings of the complete graph K_5 (see Example 7.2). Otherwise, there is some prime p such that A is not congruent to a scalar matrix mod p, so on reduction mod p it is mapped to a non-trivial element of $\mathrm{PGL}_2(\mathbb{F}_p)$. This means that α acts non-trivially on the projective line $\mathbb{P}^1(\mathbb{F}_p)$ formed by the $p + 1$ normal subgroups of index p in Δ, so H_α acts non-trivially on the $p + 1$ corresponding regular dessins (see Example 5.6 for a description of these). □

In particular, this result shows that $\mathrm{Out}\,\Delta$ acts faithfully on all dessins, so it can be identified with the group Ω of operations on dessins which it induces. Thus $\Omega \cong \mathrm{GL}_2(\mathbb{Z})$, so Ω is a free product with amalgamation

$$\Omega = \Omega_1 *_{\Omega_0} \Omega_2 \cong D_6 *_{D_2} D_4 \,.$$

The automorphisms of Δ send normal subgroups of Δ to normal subgroups with isomorphic quotient groups, so the operations in Ω preserve regularity and automorphism groups. It is useful to study how Ω acts on the set of all regular dessins with a given automorphism group.

For any finite group G, let $\mathscr{N}(G)$ denote the set of normal subgroups N of Δ with $\Delta/N \cong G$. The orbits of $\mathrm{Aut}\,\Delta$ on $\mathscr{N}(G)$ correspond to what are known in combinatorial group theory as the T_2-systems in G, that is, the orbits of $\mathrm{Aut}\,\Delta \times \mathrm{Aut}\,G$ acting by composition on epimorphisms $\Delta \to G$ and thus on generating pairs (x, y) for G. These orbits correspond to those of Ω on the corresponding set $\mathscr{R}(G)$ of regular dessins with automorphism group G. Let $\nu(G)$ denote the number of orbits in each of these actions.

Example 8.4 It is known [3, 13] that if G is abelian then $\nu(G) = 1$, so the dessins in $\mathscr{R}(G)$ form a single orbit under Ω. For example, if p is prime then Ω acts on the $p+1$ dessins in $\mathscr{R}(C_p)$ as the image of $\mathrm{GL}_2(\mathbb{Z})$ in $\mathrm{PGL}_2(\mathbb{F}_p)$ (see the proof of Theorem 8.5); this image is $\mathrm{PSL}_2(\mathbb{F}_p)$ unless $p \equiv -1 \bmod 4$, in which case -1 is a non-square in $\mathbb{F}_p$ and the image is $\mathrm{PGL}_2(\mathbb{F}_p)$.

Nielsen proved that every automorphism of the free group $\Delta \cong F_2$ sends the commutator $[X, Y]$ of the free generators X, Y to a conjugate of $[X, Y]^{\pm 1}$. This implies that the order of the commutator $[x, y] \in G$ is invariant under Ω, and hence so is the Petrie length of a regular dessin in $\mathscr{R}(G)$. This can often be used to give lower bounds for $\nu(G)$.

Example 8.5 Example 5.14 showed that there are 19 regular dessins in $\mathscr{R}(A_5)$. In [13], Neumann and Neumann showed that the 19 elements of $\mathscr{N}(A_5)$ form two orbits, of lengths 9 and 10, under $\mathrm{Aut}\,\Delta$, so the same applies to the corresponding dessins in $\mathscr{R}(A_5)$ under the action of Ω.

To see this, here is an outline of the argument in [13], rewritten in the language of dessins. Let $\mathscr{D}_1$ and $\mathscr{D}_2$ be the two regular dessins, of types $(5, 2, 3)$ and $(5, 2, 5)$, constructed in Example 5.7 (see also Exercise 5.9): these are the icosahedron and the great dodecahedron, regarded as dessins, and both are elements of $\mathscr{R}(A_5)$. By using the corresponding generating pairs x, y for A_5 given there, one can check that they have different Petrie lengths, so by thc invariance of this parameter they lie in different orbits of Ω. By starting with $\mathscr{D}_1$ and successively applying various generating operations of Ω, one can obtain dessins of ten different types, so there are at least ten non-isomorphic dessins in the orbit of Ω containing $\mathscr{D}_1$. Similarly, one can find at least nine in the orbit containing $\mathscr{D}_2$. Since $|\mathscr{R}(A_5)| = 19$ by Example 5.14, Ω has just two orbits on $\mathscr{R}(A_5)$, consisting of these two sets of dessins. (This does not contradict what we showed in Example 8.1, that the Wilson operation H_2 transposes $\mathscr{D}_1$ and $\mathscr{D}_2$: since H_2 is an operation on maps of odd valency, and not on all dessins, it is not an element of Ω.)

Exercise 8.6 Fill in the details of this argument, and find the types, Petrie lengths and genera of the dessins in these two orbits.

Garion and Shalev [4] have shown that if G is a non-abelian finite simple group then $\nu(G) \to \infty$ as $|G| \to \infty$,

Exercise 8.7 Verify this for the groups $G = \mathrm{PSL}_2(p)$ (prime $p \geq 5$) by using generating pairs

$$x = \pm \begin{pmatrix} 1 & a \\ 0 & 1 \end{pmatrix} \quad \text{and} \quad y = \pm \begin{pmatrix} 1 & 0 \\ 1 & 1 \end{pmatrix}$$

where $a \neq 0$.

We now have two infinite groups, the absolute Galois group $\mathbb{G}$ and the group Ω of operations, which act on dessins, preserving regularity and automorphism

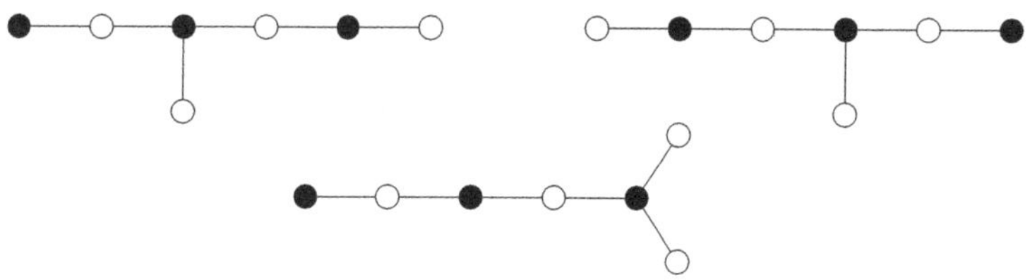

Fig. 8.2 Three dessins of genus 0

groups (though not type or genus, in the case of Ω). How do they interact with each other? The following example shows that they do not commute, even though their restrictions to certain specific sets $\mathscr{R}(G)$ often do so (see [7, Sect. 7] for examples and for further details of these actions). It would be interesting to know more about the relationship between these two groups.

Example 8.6 Figure 8.2 shows three non-regular dessins of genus 0 and type $(6, 2, 6)$: one is symmetric and the other two are mirror images of each other. It is known (see [5, Example 4.58] or [9, §2.2.2.3], for instance) that they are defined over the splitting field of the polynomial $25t^3 - 12t^2 - 24t - 16$, and that they form an orbit of $\mathbb{G}$, which acts on them as the Galois group S_3 of this polynomial. One can show (exercise!) that these dessins each have monodromy group $G \cong S_6$, so their minimal regular covers, which also form an orbit of $\mathbb{G}$, consist of three dessins of type $(6, 2, 6)$ and genus 61 in $\mathscr{R}(S_6)$; again, one is symmetric and the other two are mirror images of each other. Since $\mathbb{G}$ induces S_3 on them it has an element permuting them in a 3-cycle. The orientation-reversing operation $H_\delta \in \Omega$ fixes one of these dessins and transposes the other two. These two permutations do not commute, so the actions of $\mathbb{G}$ and Ω on $\mathscr{R}(S_6)$ do not commute.

References

1. Conder, M.D.E., Jones, G.A., Streit, M., Wolfart, J.: Galois actions on regular dessins of small genera. Rev. Mat. Iberoam. **29**, 163–181 (2013)
2. Coxeter, H.S.M., Moser, W.O.J.: Generators and Relations for Discrete Groups. Springer, Berlin/Heidelberg/New York (1980)
3. Dunwoody, M.J.: On T-systems of groups. J. Aust. Math. Soc. **3**, 172–179 (1963)
4. Garion, S., Shalev, A.: Commutator maps, measure preservation, and T-systems. Trans. Am. Math. Soc. **361**, 4631–4651 (2009)
5. Girondo, E., González–Diez, G.: Introduction to Compact Riemann Surfaces and Dessins d'Enfants. London Mathematical Society Student Texts, vol. 79. Cambridge University Press, Cambridge (2012)
6. James, L.D.: Operations on hypermaps and outer automorphisms. Eur. J. Combin. **9**, 551–560 (1988)
7. Jones, G.A.: Regular dessins with a given automorphism group. Contemp. Math. **629**, 245–260 (2014)

8. Jones, G.A., Streit, M., Wolfart, J.: Wilson's map operations on regular dessins and cyclotomic fields of definition. Proc. Lond. Math. Soc. **100**, 510–532 (2010)
9. Lando, S.K., Zvonkin, A.K.: Graphs on Surfaces and Their Applications. Springer, Berlin/Heidelberg/New York(2004)
10. Lyndon, R.C., Schupp, P.E.: Combinatorial Group Theory. Springer, Berlin/Heidelberg/New York (1977)
11. Machì, A.: On the complexity of a hypermap. Discrete Math. **42**, 221–226 (1982)
12. Nedela, R., Škoviera, M.: Exponents of orientable maps. Proc. Lond. Math. Soc. (3) **75**, 1–31 (1997)
13. Neumann, B.H., Neumann, H.: Zwei Klassen charakteristischer Untergruppen und ihre Faktorgruppen. Math. Nachr. **4**, 106–125 (1951)
14. Wilson, S.E.: Operators over regular maps. Pac. J. Math. **81**, 559–568 (1979)

Chapter 9
Further Examples

Abstract In the first part of this chapter we outline the classification of the regular embeddings of two families of regular graphs, namely certain generalised Paley graphs (using Cayley maps for finite fields), and the complete bipartite graphs $K_{n,n}$. We describe their automorphism groups, and characterise the generalised Paley maps as those maps for which the automorphism group acts primitively and faithfully on the vertices. In the case of the complete bipartite graphs, results of Huppert, Itô and Wielandt on factorisations of groups, and of Hall on solvable groups, are used in the classification. We show how Wilson operations act on these two families of maps, and we use this to investigate their Galois orbits and fields of definition.

In the second part of this chapter we extend the action of Wilson operations from regular maps to regular dessins, concentrating on those dessins which embed a complete bipartite graph $K_{p,q}$ where p and q are distinct primes. Under suitably favourable conditions on p and q, we can classify these dessins and use Wilson operations to determine their Galois orbits and fields of definition. Finally, we determine explicit equations for the associated quasiplatonic curves, a problem which is completely intractable in most other cases.

Keywords Automorphism group • Complete bipartite graph • Galois orbit • Generalised Paley graph • Group factorisation • Metacyclic group • Paley graph • Regular dessin • Regular map • Wilson operation

9.1 Galois Orbits

The methods used in Chaps. 7 and 8 to classify the regular embeddings of complete graphs, and to study their Galois orbits, will now be applied to two other classes of arc-transitive graphs, namely certain generalised Paley graphs and complete bipartite graphs.

Gareth A. Jones, J. Wolfart, *Dessins d'Enfants on Riemann Surfaces*, Springer Monographs in Mathematics, DOI 10.1007/978-3-319-24711-3_9

9.1.1 Generalised Paley Maps

One can generalise the construction of the Biggs maps $\mathcal{M}_q(c)$ in Theorem 7.1. Again let V be the additive group of a finite field $\mathbb{F}_q$, where q is a prime power p^e. Instead of taking $S = \mathbb{F}_q \setminus \{0\}$ as a generating set for V, let S be any subgroup of the multiplicative group $\mathbb{F}_q^*$ containing the element -1, so that $S = -S$; since $\mathbb{F}_q^*$ is cyclic, it has one subgroup of order n for each n dividing $q-1$, and this contains -1 if and only if $p = 2$ or n is even. As before, one can form the Cayley graph $C(V; S)$, with vertex set V and an edge between v and vs whenever $v \in V$ and $s \in S$. This graph, which has valency $n = |S|$, is known as the *generalised Paley graph* $P = P_q^{(n)}$; the classical Paley graph [24] corresponds to the case where $q \equiv 1 \bmod 4$ and S is the group of squares in $\mathbb{F}_q^*$, so that $n = (q-1)/2$, while taking $n = q-1$ gives the complete graph K_q. The following simple lemma tells us when P is connected, as we shall assume from now on:

Lemma 9.1 *The following are equivalent:*

1. *the graph P is connected;*
2. *S generates the additive group V;*
3. *S acts irreducibly on V, regarded as a vector space over $\mathbb{F}_p$;*
4. *p has multiplicative order e in the group of units $U_n = \mathbb{Z}_n^*$.*

Exercise 9.1 Prove Lemma 9.1.

Example 9.1 Let $q = 16$, so $p = 2$ and $e = 4$. Then $\mathbb{F}_q^* \cong C_{15}$, with unique subgroups $S \cong C_n$ for $n = 3$ and 5, both containing $-1 = 1$. Now 2 has multiplicative order $2 < e$ in U_3, so condition (4) of Lemma 9.1 fails for $n = 3$, and in fact $P_{16}^{(3)}$ is the disjoint union of four copies of K_4. However, 2 has multiplicative order 4 in U_5, so condition (4) is satisfied for $n = 5$, and $P_{16}^{(5)}$ is connected.

It is straightforward to verify that $\operatorname{Aut} P$ contains the subgroup $\mathrm{A\Gamma L}_1^{(n)}(q)$ of order nqe in $\mathrm{A\Gamma L}_1(q)$ consisting of the transformations

$$t \mapsto at^{\gamma} + b \quad \text{with} \quad a \in S,\ b \in \mathbb{F}_q,\ \gamma \in \operatorname{Gal}\mathbb{F}_q\ .$$

Lim and Praeger [20] have shown that if S is 'large', but not too large, these are the only automorphisms of P:

Theorem 9.1 *Suppose that $n < q-1$. If $|\mathbb{F}_q^* : S|$ divides $p-1$, or equivalently if $(q-1)/(p-1)$ divides n, then* $\operatorname{Aut} P_q^{(n)} = \mathrm{A\Gamma L}_1^{(n)}(q)$.

Their proof relies on the classification of finite simple groups (see Sect. 2.2). Of course, if $n = q-1$ then $P = K_q$ and $\operatorname{Aut} P$ is the symmetric group S_q. Lim and Praeger also give examples, such as $P_{81}^{(20)}$, where $n < q-1$ and $|\mathbb{F}_q^* : S|$ does not divide $p-1$, and the automorphism group properly contains $\mathrm{A\Gamma L}_1^{(n)}(q)$. Determining $\operatorname{Aut} P$ in general seems to be a difficult problem.

Suppose now that the hypotheses of Lemma 9.1 are satisfied, so that P is connected. As a subgroup of the cyclic group $\mathbb{F}_q^*$, S is cyclic, so let c be a generator for S. The successive powers of c define a cyclic order π on S, and we can use this, as in the case of the Biggs maps (see Chap. 7), to define a Cayley map $C(V; S, \pi)$. This is an embedding of $P_q^{(n)}$ in an oriented surface, called a *generalised Paley map* $\mathcal{M}_q(c)$, in which the neighbours of each vertex $v \in V$ are $v+c, v+c^2, \ldots, v+c^n = v+1$ in this cyclic order around v. It generalises both the Paley maps described by Biggs and White in [1, 32] and the Biggs maps described in Chap. 7.

The methods of the preceding chapter have been extended in [15] to show that provided q and n satisfy the hypotheses of Lemma 9.1 and Theorem 9.1 we have the following results, which are direct generalisations of those for the Biggs maps:

- the maps $\mathcal{M}_q(c)$ are the only regular embeddings of the graphs $P_q^{(n)}$;
- $\mathrm{Aut}\,\mathcal{M}_q(c)$ is the subgroup $\mathrm{AGL}_1^{(n)}(q)$ of $\mathrm{AGL}_1(q)$ consisting of the transformations $t \mapsto at + b$ with $a \in S$ and $b \in \mathbb{F}_q$;
- $\mathcal{M}_q(c) \cong \mathcal{M}_q(c')$ if and only if c and c' are equivalent under $\mathrm{Gal}\,\mathbb{F}_q$;
- there are, up to isomorphism, $\phi(n)/e$ maps $\mathcal{M}_q(c)$;
- the maps $\mathcal{M}_q(c)$ form a single orbit under the Wilson operations H_j for $j \in U_n$;
- the dessins $\mathcal{M}_q(c)$ form a single orbit under the absolute Galois group $\mathbb{G} = \mathrm{Gal}\,\overline{\mathbb{Q}}$, and are defined over the subfield of $\mathbb{Q}(\zeta_n)$ fixed by the automorphism $\zeta_n \mapsto \zeta_n^p$.

The arguments in Sect. 7.2.3 show that if $\mathcal{M}$ is a regular map such that the group $G = \mathrm{Aut}\,\mathcal{M}$ acts faithfully and doubly transitively on the vertex-set V, then $\mathcal{M}$ is isomorphic to a Biggs map, a regular embedding of a complete graph K_q. The reason is that G must act on V as a doubly transitive Frobenius group with a cyclic point stabiliser, so that G can be identified with $\mathrm{AGL}_1(q)$ acting naturally on the field $V = \mathbb{F}_q$, and hence Theorem 7.2(1) applies.

In fact, there is a similar result under the weaker hypothesis that G acts faithfully and primitively on V. Recall that a group acts *primitively* if the only equivalence relations it preserves are the universal relation (with a single equivalence class) and the trivial relation (with all classes singletons). This is equivalent to the group acting transitively, with the point-stabilisers maximal subgroups. Every doubly transitive group action is primitive: given an equivalence relation other than the universal or trivial relation, take two points which are related and two which are not; by double transitivity there is a group element sending the first pair to the second, so this relation is not preserved by the group. On the other hand, there are many groups which act primitively but not doubly transitively.

Example 9.2 Let G be a group of affine transformations of a vector space V over $\mathbb{F}_p$, containing the translation group. Then G is a semidirect product of a normal subgroup V, acting on itself by translations, by the subgroup G_0 fixing 0, acting as a subgroup of the general linear group $\mathrm{GL}(V)$. In this action of G, the V-invariant equivalence relations on V are those given by congruence modulo some subgroup (equivalently subspace) U of V, and such a relation is G_0-invariant if and only if U is G_0-invariant. It follows that G acts primitively on V if and only if G_0 acts irreducibly on V. This is certainly the case if G_0 acts transitively on $V \setminus \{0\}$, so that G is doubly transitive on V, but there are many examples where G_0 is irreducible but intransitive on $V \setminus \{0\}$: for instance, this happens if the conditions in Lemma 9.1 are satisfied, with $G_0 = S \cong C_n$ and $n < q - 1$.

Theorem 9.2 *Let* $\mathcal{M}$ *be a regular map on a compact surface. Then* $\mathrm{Aut}\,\mathcal{M}$ *acts primitively and faithfully on the vertices of* $\mathcal{M}$ *if and only if* $\mathcal{M}$ *is isomorphic to a generalised Paley map* $\mathcal{M}_q(c)$.

Proof Each map $\mathcal{M}_q(c)$ has these properties since its automorphism group G acts faithfully on $V = \mathbb{F}_q$, with the stabiliser $G_0 = S$ of 0 acting as an irreducible subgroup of $\mathrm{GL}(V)$ by Lemma 9.1.

Conversely, if the group $G = \mathrm{Aut}\,\mathcal{M}$ acts primitively on the vertex set V, then the stabiliser G_v of each vertex v is a maximal subgroup of G. If $G_v = 1$ then G is cyclic of prime order; as $\mathcal{M}$ is a regular map, G contains an involution, so we have $G \cong C_2$, and hence $\mathcal{M} \cong \mathcal{M}_2(1)$, the spherical embedding of K_2. We may therefore assume that $G_v \neq 1$. If $v \neq w$ in V, then $G_{vw} = 1$: for if $g \in G_{vw} = G_v \cap G_w$ then since G_v and G_w are abelian, distinct and maximal, the centraliser $C_G(g)$ of g contains $\langle G_v, G_w \rangle = G$, so g is in the centre of G; but g fixes at least one vertex, and the vertex stabilisers are all conjugate, so g fixes every vertex, giving $g = 1$.

Thus G acts on V as a Frobenius group, so it has a Frobenius kernel N, a normal subgroup acting regularly on V (see Sect. 7.2.2). We may identify V with N, acting by multiplication on itself. The stabiliser in G of the identity vertex is a complement for N in G, acting on V as it acts by conjugation on N. Since G acts primitively on V, no proper subgroup of N can be normal in G, so N is characteristically simple, that is, a direct product of isomorphic simple groups. By a theorem of Thompson [31], finite Frobenius kernels are nilpotent, so N is an elementary abelian p-group for some prime p.

We can therefore regard V as a vector space of some dimension e over $\mathbb{F}_p$. By primitivity, G_0 acts on V as an irreducible subgroup of $GL(V)$. Since G_0 is cyclic we can therefore identify V with the field $\mathbb{F}_q$ of order $q = p^e$ so that G_0 acts by multiplication as a subgroup S of $\mathbb{F}_q^*$ (see [12, Satz II.3.10], for instance). Since G contains an involution, $-1 \in S$. Thus G, a semidirect product of $V = \mathbb{F}_q$ by $G_0 = S \leq \mathbb{F}_q^*$, is isomorphic to $\mathrm{AGL}_1^{(n)}(q)$ where $n = |S|$ is the valency of $\mathcal{M}$. An argument similar to the proof of Theorem 7.2(1) now shows that $\mathcal{M} \cong \mathcal{M}_q(c)$ for some generator c of S. For full details of this proof, see [15]. □

The condition in Theorem 9.2 that $G = \mathrm{Aut}\,\mathcal{M}$ should act faithfully on the vertex set V is not particularly restrictive: if we merely assume that G acts primitively on V, then the kernel K of the action on V is cyclic, since it is a subgroup of each vertex-stabiliser G_v, which is itself cyclic. The map $\mathcal{M}/K$, with automorphism group G/K, satisfies the hypotheses of Theorem 9.2, so $\mathcal{M}$ is a cyclic covering of a generalised Paley map $\mathcal{M}_q(c)$, where $q = |V|$, branched over the vertices (and possibly the face-centres). For instance, if $|V| = 2$ we obtain the regular embeddings of dipoles, classified by Nedela and Škoviera in [22], as cyclic coverings of $\mathcal{M}_2(1)$; if $\mathcal{M}$ has k edges then

$$G = \langle x, y \mid x^k = y^2 = 1\ ,\ y^{-1}xy = x^u \rangle$$

is of order $2k$, where $u^2 = 1$ in U_k and $K = \langle x \rangle \cong C_k$.

9.1.2 *Complete Bipartite Maps*

We saw in Theorem 7.1 that a complete graph has a regular embedding if and only if the number of vertices is a prime power. Thus 'most' complete graphs have no regular embeddings, in the sense that the set of all prime powers has asymptotic density

$$\lim_{x\to\infty} \frac{1}{x} |\{n \in \mathbb{N} \mid n \leq x \text{ and } n \text{ is a prime power}\}| = 0$$

in $\mathbb{N}$. (This follows easily from the Prime Number Theorem, which states that the number $\pi(x)$ of primes $p \leq x$ satisfies $\pi(x) \sim x/\log x$ as $x \to \infty$.)

There is, however, an apparently similar class of graphs, *all* of whose members have at least one regular embedding. These are the complete bipartite graphs $K_{n,n}$, which have n black and n white vertices, each pair of black and white vertices joined by a single edge. We saw in Example 1.10 that the nth degree Fermat curve F_n, given as a projective algebraic curve by $x^n + y^n = z^n$, has a Belyĭ function

$$\beta : F_n \to \hat{\mathbb{C}}, \quad [x, y, z] \mapsto \frac{x^n}{z^n}.$$

The corresponding dessin $\mathcal{D}$ has n white vertices $v_j = [0, \zeta_n^j z, z]$ over 0, and n black vertices $w_k = [\zeta_n^k z, 0, z]$ over 1, with $\zeta_n = e^{2\pi i/n}$ and $j, k = 0, 1, \ldots, n-1$. For each j and k there is an edge e_{jk} from v_j to w_k, consisting of the points

$$[t\zeta_n^k z, (1-t^n)^{1/n}\zeta_n^j z, z] \quad \text{for} \quad 0 \leq t \leq 1,$$

so the embedded bipartite graph is isomorphic to $K_{n,n}$. These dessins are drawn in Fig. 2.3 in Example 2.3 for the cases $n = 3$ and 4, with the latter only topologically and combinatorially correct. Geometrically correct representations are shown in Fig. 1.2 in Sect. 1.4.3 for $n = 4$, and in Fig. 9.1 for $n = 5$; in each case the map is shown as a quotient of a tessellation of the unit disc, invariant under the triangle group $\Delta = \Delta(n, n, n)$, with numbers indicating the identification of sides (not part of the dessin) of a fundamental region for the surface group $K = \Delta'$ (see Examples 3.10 and 3.11).

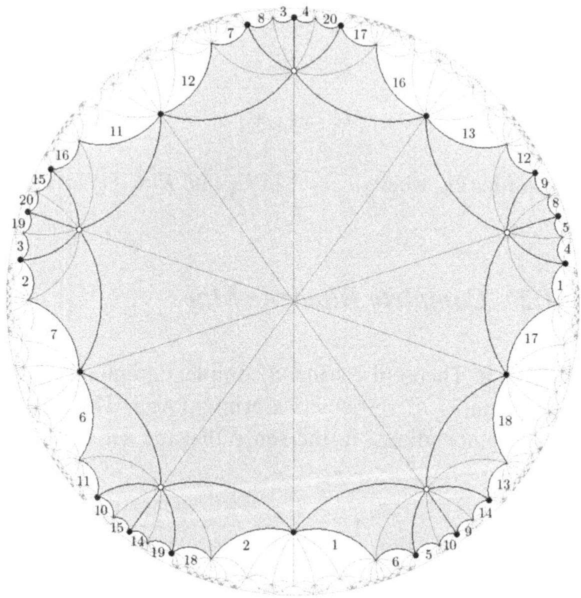

Fig. 9.1 $K_{5,5}$ regularly embedded in the Fermat curve F_5 of genus 6

The automorphisms

$$\alpha_{jk} : [x, y, z] \mapsto [\zeta_n^k x, \zeta_n^j y, z] \quad (j, k = 0, 1, \ldots, n-1)$$

of the curve F_n form a group isomorphic to $C_n \times C_n$; they preserve the dessin $\mathscr{D}$, and act transitively on its edges, so $\mathscr{D}$ is a regular dessin with

$$\mathrm{Aut}\,\mathscr{D} = \{\alpha_{jk} \mid j, k = 0, 1, \ldots, n-1\} \cong C_n \times C_n \cong \Delta/\Delta'.$$

Since $\mathscr{D}$ has type (n, n, n) with $|\,\mathrm{Aut}\,\mathscr{D}\,| = n^2$, the Riemann-Hurwitz formula shows that it has genus

$$g = 1 + \frac{n^2}{2}\left(1 - \frac{3}{n}\right) = \frac{(n-1)(n-2)}{2}.$$

By ignoring the vertex-colours we can regard $\mathscr{D}$ as a map $\mathscr{M}$. We have seen that $\mathrm{Aut}(\mathscr{M})$ is edge-transitive; since the automorphism

$$\alpha : [x, y, z] \mapsto [y, x, z]$$

of F_n preserves $\mathscr{M}$ and reverses the edge e_{00}, $\mathscr{M}$ is in fact a regular map. Its automorphism group $\mathrm{Aut}\,\mathscr{M}$ is a semidirect product of a normal subgroup $\mathrm{Aut}\,\mathscr{D}$ by a complement $\langle\alpha\rangle \cong C_2$. Since α, acting by conjugation on $\mathrm{Aut}\,\mathscr{D}$, transposes the two direct factors C_n, it follows that $\mathrm{Aut}\,\mathscr{M}$ is isomorphic to the wreath product

$C_n \wr C_2$ of C_n by C_2. The fact that $\mathcal{M}$ is regular corresponds to the surface group Δ' being normal, not just in $\Delta = \Delta(n,n,n)$, but also in the triangle group $\Delta^\dagger = \Delta(n,2,2n)$ which contains Δ as a subgroup of index 2 (see inclusion (a) in Theorem 3.12), with $\Delta^\dagger/\Delta' \cong C_n \wr C_2 \cong \mathrm{Aut}\,\mathcal{M}$ (see Sect. 3.3.3).

This regular embedding of $K_{n,n}$, known as the *standard embedding* $\mathcal{S}_n$, can also be constructed as a Cayley map $\mathcal{M}(V;S,\pi)$ (see Sect. 7.1.2), where the vertex set V is the additive group $\mathbb{Z}_{2n} = \mathbb{Z}/2n\mathbb{Z}$ with generating set

$$S = \{1,3,5,\ldots,2n-1\}$$

in that cyclic order π (see [1, §5.6.7]). The partition of $K_{n,n}$ into two monochrome sets of n vertices corresponds to the partition of $\mathbb{Z}_{2n}$ into even and odd elements.

Recent years have seen a prolonged effort to classify the regular embeddings of $K_{n,n}$ for all n, and to understand the corresponding dessins. The first result was that of Nedela, Škoviera and Zlatoš [23], who showed that if n is prime there is only the standard embedding $\mathcal{S}_n$. This was extended by Jones, Nedela, and Škoviera [18], who showed that the integers n with this uniqueness property are the products $n = p_1 \ldots p_k$ of distinct primes p_i such that $p_i \not\equiv 1 \bmod p_j$ for all i and j. (Burnside showed that these integers are also those for which all groups of order n are cyclic, see [26, Exercise 575] or [25, §10.1, Exercise 12], for example. Erdős [6] showed that the proportion of integers $n \leq N$ with this property is asymptotic to

$$\frac{e^{-\gamma}}{\log\log\log N}$$

as $N \to \infty$, where γ is Euler's constant

$$\lim_{n\to\infty}\left(\sum_{k=1}^{n}\frac{1}{k} - \log n\right) = 0.57721\ldots .$$

This proportion approaches 0 as $N \to \infty$, but does so very slowly.)

The same authors classified the regular embeddings for which n is an odd prime power in [17], and, together with Du and Kwak, dealt with the rather harder case $n = 2^e$ in [4, 5]. The complete classification, building on these earlier results and using induction on the number of primes dividing n, was published in [14].

The main technique for studying these embeddings is first to determine the subgroup $G = \mathrm{Aut}\,\mathcal{D}$ of index 2 in $\mathrm{Aut}\,\mathcal{M}$ which preserves the vertex colours. Then $\mathrm{Aut}\,\mathcal{M}$ is a semidirect product of G by a subgroup C_2 which reverses an edge and hence transposes the colours. A group G arises in this way from a regular embedding of $K_{n,n}$ if and only if

1. G has subgroups $X, Y \cong C_n$ such that $G = XY$ and $X \cap Y = 1$;
2. G has an automorphism transposing generators x and y of X and Y.

Exercise 9.2 Prove the above assertion.

When the above conditions hold we say that G is *n-isobicyclic*, or simply *isobicyclic*, and we call (G, x, y) an isobicyclic triple. Condition (1) implies that every element of G has a unique expression as $x^i y^j$ with $i, j \in \mathbb{Z}_n$, so $|G| = n^2$. For instance, if $\mathcal{M}$ is the standard embedding $\mathcal{S}_n$ then $G = X \times Y$, but in general G need not be a direct, or even a semidirect product, as neither X nor Y need be normal in G.

By good fortune, in the 1950s a number of group-theorists such as Huppert, Itô and Wielandt had studied groups which factorise as a product XY of two cyclic groups, and it was possible to exploit their results in this situation.

For example, if $n = p^e$ for some prime p then $|G| = p^{2e}$, so G is a p-group. Huppert [11] showed that if a p-group is a product of two cyclic groups, where $p > 2$, then it is metacyclic, that is, it has a cyclic normal subgroup with a cyclic quotient. Using this, one can show that the n-isobicyclic groups, for odd $n = p^e$, are the groups

$$G_f := \langle g, h \mid g^{p^e} = h^{p^e} = 1,\ h^g = h^{1+p^f} \rangle$$

where $f = 1, 2, \ldots, e$, with X and Y generated by $x = g^u$ and $y = g^u h$ for some $u = 1, \ldots, p^{e-f}$ coprime to p. Different choices of u give $\phi(p^{e-f})$ non-isomorphic regular embeddings $\mathcal{M}_{f,u}$ for each f, and hence a total of

$$\sum_{f=1}^{e} \phi(p^{e-f}) = (p^{e-1} - p^{e-2}) + (p^{e-2} - p^{e-3}) + \cdots + (p-1) + 1 = p^{e-1}$$

regular embeddings for each $n = p^e$. (Note that $G_e \cong C_n \times C_n$, with $u = 1$ giving the standard embedding $\mathcal{S}_n$.) These dessins all have type (n, n, n); as maps they have type $\{2n, n\}$, and their surfaces have genus $(n-1)(n-2)/2$.

Unfortunately, Huppert's result does not extend to the prime $p = 2$. If $n = 2^e$ then, in addition to the embeddings $\mathcal{M}_{f,u}$ for $f = 2, \ldots, e$ (but not $f = 1$) and odd $u = 1, \ldots, 2^{e-f} - 1$, associated with metacyclic groups $G = G_f$ and described in [4], there are embeddings for which G is not metacyclic: there is one when $e = 2$, namely the torus map $\{4, 4\}_{2,2}$ (see Sect. 6.2) shown in Fig. 9.2 (as usual, identify opposite sides of the outer square to get a map on a torus), and there are four for each $e > 2$. These maps, which all arise as branched coverings of $\{4, 4\}_{2,2}$, are classified by induction on e in [5].

The problem now is to extend these results for prime powers to all values of n. The objective is to show that the regular embeddings of $K_{n,n}$ can all be obtained from the prime-power embeddings described above by using two basic constructions: a cartesian product and a semidirect product.

Exercise 9.3 Show that if (G_i, x_i, y_i) is an n_i-isobicyclic triple for each $i = 1, \ldots, k$, with $n_1, \ldots, n_k$ mutually coprime, then (G, x, y) is an n-isobicyclic triple, where $G = G_1 \times \cdots \times G_k$, $x = (x_i)$, $y = (y_i)$, and $n = n_1 \ldots n_k$.

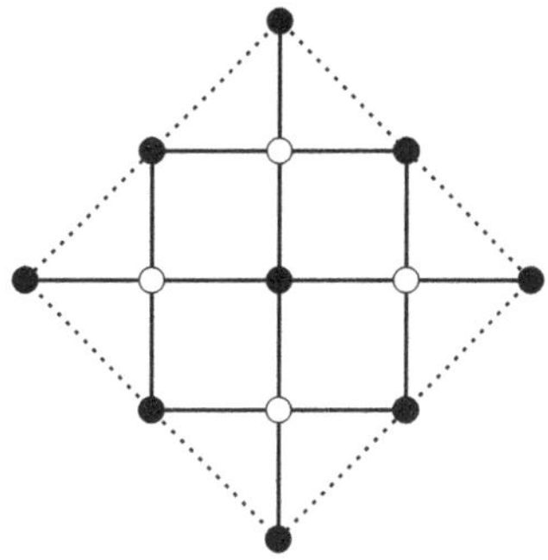

Fig. 9.2 A regular embedding $\{4,4\}_{2,2}$ of $K_{4,4}$ on a torus

In this situation, if $\mathscr{M}_i$ is the regular embedding of K_{n_i,n_i} corresponding to (G_i, x_i, y_i) then the regular embedding of $K_{n,n}$ corresponding to (G, x, y) is called the *cartesian product* $\mathscr{M}_1 \times \cdots \times \mathscr{M}_k$ of the embeddings $\mathscr{M}_i$.

Now let (S, x_S, y_S) be an s-isobicyclic triple and let (T, x_T, y_T) be the standard t-isobicyclic triple (so that $T \cong C_t \times C_t$), where s and t are mutually coprime. Let G be the semidirect product of T by S, where the generators x_S and y_S of the complement S act by conjugation on the normal subgroup T by

$$x_T^{x_S} = x_T, \quad y_T^{x_S} = y_T^{\lambda}, \quad x_T^{y_S} = x_T^{\lambda} \quad \text{and} \quad y_T^{y_S} = y_T$$

for some $\lambda \in U_t$. If we regard T as $(\mathbb{Z}/t\mathbb{Z})^2$, then the actions of x_S and y_S on T can be represented with respect to the basis x_T, y_T of T by the matrices

$$\begin{pmatrix} 1 & 0 \\ 0 & \lambda \end{pmatrix} \quad \text{and} \quad \begin{pmatrix} \lambda & 0 \\ 0 & 1 \end{pmatrix}.$$

For instance, if we take $\lambda = 1$ then $G = S \times T$.

Exercise 9.4 Show that (G, x, y) is an n-isobicyclic triple, where $x = x_S x_T$, $y = y_S y_T$ and $n = st$.

Exercise 9.5 Show that if (S, x_S, y_S) is the standard s-isobicyclic triple, then G is a direct product of two isomorphic subgroups, each a semidirect product of C_t by C_s. What are these subgroups if we take $s = 2$ and $\lambda = -1$?

This triple is called a *semidirect product* of (T, x_T, y_T) by (S, x_S, y_S). If $\mathscr{S}$ and $\mathscr{T}$ $(= \mathscr{S}_t)$ are the regular embeddings of $K_{s,s}$ and $K_{t,t}$ corresponding to these two triples, then the regular embedding of $K_{n,n}$ corresponding to (G, x, y) is called a *semidirect product* of $\mathscr{T}$ by $\mathscr{S}$. (For instance, if we take $\lambda = 1$ it is just the cartesian product defined earlier.)

In [13], Itô proved that any group which is a product of two abelian groups is metabelian, that is, an extension of one abelian group by another. It follows that the group $G = \operatorname{Aut} \mathscr{D} = XY$ is solvable (of derived length at most 2). Now Sylow's theorems, which apply to individual primes dividing the order of a finite group, have been generalised by Hall [8], in the case of finite solvable groups, to apply to arbitrary sets of primes dividing the group order. Hall's theorems can therefore be applied to G, together with more specific results of Wielandt [33] relevant to a product of two cyclic groups. One of these implies that if $p_1, \ldots, p_k$ are the distinct primes dividing n, with $p_1 > \cdots > p_k$, then G has a series of normal subgroups

$$1 = N_0 < N_1 = P_1 < N_2 = P_1P_2 < \cdots < N_k = P_1 \ldots P_k = G,$$

where each P_i is a Sylow p_i-subgroup of G. This means that we will know G completely if we can identify its Sylow subgroups P_i and how they act by conjugation on each other, or more precisely, how each P_i acts by conjugation on each quotient $N_j/N_{j-1} \cong P_j$. It is shown in [14] that for each prime p the possible Sylow p-subgroups are precisely the groups G which arise when n is a power of p; these are known by the prime power classifications mentioned earlier. It is also shown that P_i acts trivially on P_j unless the latter is abelian, in which case only the semidirect product action $P_i \to \operatorname{Aut} P_j$ defined above is possible.

Using these ingredients, and arguing by induction on the number of primes dividing n, one can prove the following (see [14] for details):

Theorem 9.3 *G is a semidirect product of a normal subgroup $T \cong C_t \times C_t$ by a complement S of order s^2, where*

1. *$n = st$ with s and t coprime,*
2. *S is a cartesian product of $p_i^{e_i}$-isobicyclic groups for $i = 1, \ldots, k$, where $s = p_1^{e_1} \ldots p_k^{e_k}$ for distinct primes p_i.*

Corollary 9.1 *A map $\mathcal{M}$ is a regular embedding of $K_{n,n}$ if and only if it is a semidirect product of the standard embedding of $K_{t,t}$ by the cartesian product of regular embeddings of K_{n_i,n_i}, where $n_i = p_i^{e_i}$ for distinct primes $p_i \ldots, p_k$, and $n = st$ with $s = p_1^{e_1} \ldots p_k^{e_k}$ and t mutually coprime.*

Exercise 9.6 Find a nonstandard regular embedding of $K_{6,6}$, and calculate its type and genus.

These methods also yield a formula for the number of regular embeddings of $K_{n,n}$, up to isomorphism. Unfortunately it is too complicated to state here, involving a summation over certain subgraphs of a directed graph formed from the prime factors of n.

Exercise 9.7 Show that if $n = 2p$ for some odd prime p, then there just two regular embeddings of $K_{n,n}$.

For some values of n one can determine the Galois orbits and fields of definition of the dessins corresponding to these maps. For instance, Jones, Streit, and Wolfart [16] have shown that if n is an odd prime power p^e then the $\phi(p^{e-f})$ maps $\mathcal{M}_{f,u}$ with a given colour-preserving automorphism group G_f form an orbit under Wilson's operations H_j $(j \in U_{p^{e-f}})$ and a Galois-invariant set; it therefore follows from Theorem 8.1 that they form an orbit of $\mathbb{G} = \operatorname{Gal} \overline{\mathbb{Q}}$, defined over $\mathbb{Q}(\zeta_{p^{e-f}})$.

Indeed, explicit equations for the corresponding algebraic curves are obtained in the cases $f \geq e/2$, where G_f is 'close to abelian'. The same authors, together with Coste [3], have obtained similar but more complicated results in the more general situation where $K_{n,n}$ is embedded as a regular dessin (rather than a regular map) for an odd prime power $n = p^e$; here the same automorphism groups $G = G_f$ arise, but extra generating pairs x, y must be considered, since we no longer assume that x and y are transposed by an automorphism of G.

9.2 Regular Dessins and Equations

9.2.1 $K_{p,q}$-Dessins, Classification

In this section, we again consider regular dessins $\mathcal{D}$ coming from embeddings of complete bipartite graphs $K_{p,q}$, but now—instead of the hypothesis $p = q = n$ we discussed in Sect. 9.1—we take p and q to be distinct primes, say $p > q$. Such graphs are edge-transitive (with Aut $K_{p,q} \cong S_p \times S_q$), but no longer arc-transitive, so they yield regular dessins rather than regular maps. With p and q as the valencies of the white and black vertices, the number of edges is pq. By regularity, this is also the order of the automorphism group $G = \text{Aut}\,\mathcal{D}$. By Sylow's theorems, G is generated by the unique normal cyclic subgroup C_p with generator a and a cyclic subgroup C_q with generator b. We may assume that a and b are the standard generators fixing neighbouring white and black vertices respectively and acting there (in suitable local coordinates, see the proof of Theorem 8.1) as

$$a : z \mapsto \zeta_p z + \dots \quad , \qquad b : w \mapsto \zeta_q w + \dots \quad .$$

Group theory and number theory predict that

$$b^{-1}ab = a^u \quad \text{for some} \quad u \in (\mathbb{Z}/p\mathbb{Z})^*,$$

where either $u = 1$ or u has order q. In the first case we have $G \cong C_{pq}$; the dessin $\mathcal{D}$ then has type (p, q, pq) and is a quotient of the regular dessin on the Fermat curve F_{pq} we met earlier (see Exercise 5.8). We will consider this case next, while the second case will be considered in Sect. 9.2.3.

9.2.2 $K_{p,q}$-Dessins, Abelian Case

In the first case we can weaken the primality condition, and prove the following more general result:

Theorem 9.4 *Let p and q be coprime integers, both greater than* 1, *let* $\mathcal{D}$ *be a regular dessin with white and black vertices of valencies p and q, and suppose that*

Aut $\mathcal{D}$ is abelian. Then Aut $\mathcal{D} \cong C_p \times C_q \cong C_{pq}$, *and* $\mathcal{D}$ *is of type* (p, q, pq), *with embedded graph* $K_{p,q}$. *Its quasiplatonic surface X is of genus* $(p-1)(q-1)/2$. *The surface X and its dessin* $\mathcal{D}$ *are quotients of the Fermat curve* F_{pq} *and its regular dessin of type* (pq, pq, pq) *by a subgroup* $C_q \times C_p$ *of their automorphism group* $C_{pq} \times C_{pq}$. *Up to isomorphism, X and* $\mathcal{D}$ *are uniquely determined by p and q. An affine (singular) model of X is given by the equation*

$$y^{pq} = x^q (x-1)^p ,$$

and in this model, the Belyĭ function for $\mathcal{D}$ *is* $\beta : (x, y) \mapsto x$.

Proof As $G = \operatorname{Aut}\mathcal{D}$ is abelian and p and q are coprime, the canonical generators satisfy $a^p = b^q = 1$, with ab of order pq. This implies all the stated claims about the type, automorphism group, and graph. Up to composition with automorphisms of G there is only one epimorphism from the triangle group $\Delta(p, q, pq)$ onto G, so its kernel, the surface group of X, is uniquely determined by p and q. If we compose this epimorphism with the natural epimorphism $\Delta := \Delta(pq, pq, pq) \to \Delta(p, q, pq)$, then since G is abelian the kernel contains the commutator subgroup Δ', so $\mathcal{D}$ is a quotient of the regular dessin of type (pq, pq, pq) on the Fermat curve F_{pq} corresponding to Δ'. Moreover, the mapping $(x, y) \mapsto x$ has the correct ramifications: one has to take into account the facts that the point $(x, y) = (0, 0)$ corresponds to q points on a smooth model of the curve, and similarly the point $(1, 0)$ on the singular model splits into p points, whereas (∞, ∞) corresponds to the unique pole of order pq of the Belyĭ function (see Exercises 9.8 and 9.9). The automorphism group acts on the singular model by

$$(x, y) \mapsto (x, \zeta_{pq}^k y) , \quad k \in \mathbb{Z}/pq\mathbb{Z} ,$$

and the genus of X can be found from the Riemann-Hurwitz formula. □

Remark 9.1 Theorem 9.4 is a special case of a much more general result [3, Proposition 3] that each regular dessin with abelian automorphism group is a quotient of the regular dessin of type (n, n, n) on the Fermat curve F_n for some n, and that the underlying quasiplatonic curve can be defined over the rationals (an idea due to Benjamin Mühlbauer).

Exercise 9.8 Prove that a non-singular affine model for the curve $y^{pq} = x^q(x-1)^p$ is given by the equations

$$xz = y^p , \quad (x-1)w = y^q , \quad z^q = (x-1)^p , \quad w^p = x^q$$

in $\mathbb{C}^4$, and determine the points on this model with $x = 0$ and $x = 1$.

Exercise 9.9 Prove that $K_{p,q}$ is the graph underlying $\mathcal{D}$, and that $\mathcal{D}$ has only one face. Deduce that β has only one pole, and that the affine model has a smooth projective compactification given by adding a single point.

Exercise 9.10 Draw the dessin of Theorem 9.4 for $K_{3,2}$ with automorphism group C_6. Is there a regular embedding for $K_{3,2}$ with a non-abelian automorphism group?

9.2.3 $K_{p,q}$-Dessins, Semidirect Product Case

We now return to the second case of the situation considered in Sect. 9.2.1, where $\mathscr{D}$ is a regular dessin which embeds the graph $K_{p,q}$ for distinct primes p and q, and $G := \operatorname{Aut}\mathscr{D}$ is a semidirect product $C_p \rtimes C_q$, with standard generators a and b of order p and q satisfying

$$a^b := b^{-1}ab = a^u \quad \text{for some} \quad u \in (\mathbb{Z}/p\mathbb{Z})^* \quad \text{with} \quad u \neq 1\,. \tag{9.1}$$

Since b has prime order q, so has u; it then follows that $p \equiv 1 \bmod q$ and that all elements of G not in the normal subgroup C_p have order q. The resulting regular dessins are now of type (p, q, q); there is no problem if $q = 2$ (why?), but if q is odd (as we will assume from now on) then we have to consider many more epimorphisms from the corresponding triangle group $\Delta(p, q, q)$ onto G than in the abelian case. The classification of the regular dessins therefore becomes more delicate, and finding explicit equations for the quasiplatonic curves can no longer be done by making and then verifying a good guess. Nevertheless we will see that—in contrast with many other families of quasiplatonic curves—we can find explicit models without too much effort. The following is essentially taken from [30].

Lemma 9.2 *Let p and q be odd primes with $p \equiv 1 \bmod q$, and let $G = C_p \rtimes C_q$, generated by a of order p and b of order q, satisfying condition (9.1). Then*

1. *different choices of u lead to isomorphic groups, so we can fix one u of order q;*
2. *the automorphisms of G are given by*

$$a \mapsto a^k\,,\ k \in (\mathbb{Z}/p\mathbb{Z})^* \quad \text{and} \quad b \mapsto a^m b\,,\ m \in \mathbb{Z}/p\mathbb{Z}\,;$$

3. *there are $q - 1$ different normal subgroups N_s of $\Delta = \Delta(p, q, q)$ with quotient $\Delta/N_s \cong G$; they are the kernels of the epimorphisms $h := h_s$ given by*

$$h(\gamma_0) = a \quad \text{and} \quad h(\gamma_1) = b^s\,,\ s \in (\mathbb{Z}/q\mathbb{Z})^*\,,$$

   where γ_0 and γ_1 are the canonical generators of Δ as in Sect. 3.1.2;
4. *if $(p, q, q) \neq (7, 3, 3)$, the normaliser of N_s in $\mathrm{PSL}_2(\mathbb{R})$ is Δ;*
5. *in any case, N_s is conjugate in $\mathrm{PSL}_2(\mathbb{R})$ to N_t if and only if $t \equiv \pm s \bmod q$, so there are precisely $(q-1)/2$ non-isomorphic quasiplatonic surfaces $X_s := N_s\backslash\mathbb{H}$, all of genus $(p-1)(q-2)/2$;*
6. *in the case $\Delta = \Delta(7, 3, 3)$, we have $N_s \lhd \Delta(7, 7, 7) \lhd \Delta$, the normaliser of N_s is a triangle group $\Delta(2, 3, 7)$, and X_s is isomorphic to Klein's quartic curve.*

Proof

(1) Let v be another element of order q in $(\mathbb{Z}/p\mathbb{Z})^*$. Together with 1 the elements of order q form a cyclic subgroup of $(\mathbb{Z}/p\mathbb{Z})^*$, so there is some $k \not\equiv 0 \bmod q$ such that $u^k = v$, and hence

$$b^{-k}ab^k = b^{-k+1}a^u b^{k-1} = \ldots = a^{u^k} = a^v \ .$$

Replacing u with v is therefore equivalent to replacing b with b^k.

(2) This follows from direct computations using generators and defining relations (exercise!).

(3) This follows from counting the possible epimorphisms $\Delta \to G$ modulo composition with automorphisms of G (see Proposition 5.1).

(4) If we set aside the special case $p = 7$, $q = 3$, then by Singerman's list of inclusions between triangle groups (see [28] or Theorem 3.12), the only other possible candidate for the normaliser of N_s could be the triangle group $\Delta(2, 2p, q)$. If this were the normaliser, then G would have an automorphism of order 2 transposing

$$b^s \quad \text{and} \quad (ab^s)^{-1} = b^{-s}a^{-1} = a^{-u^s}b^{-s} \ .$$

This would imply—see point (2) of this lemma—that $s \equiv -s \bmod q$, which is a contradiction.

(5) All isomorphisms between the surfaces X_s are induced by conjugations of their surface groups in $\mathrm{PSL}_2(\mathbb{R})$, and by a deep theorem of Margulis [21], the conjugating elements lie in maximal Fuchsian groups containing Δ provided Δ is not an arithmetic group: this exception occurs precisely for $p = 7$, $q = 3$, and in all other cases one can argue exactly as in point (4). The two dessins corresponding to $\pm s$ are in fact dual to each other by the transposition of black vertices and face centres.

(6) This is more or less classical; for a more modern approach see [7].

□

Passing from the pair of generators a, b to a, b^s corresponds to applying the generalised Wilson operator H_s introduced in Theorem 8.3. Since regularity, type and automorphism group (up to isomorphism) are Galois-invariant properties of a dessin by Theorem 4.11, we have the following:

Lemma 9.3 *Let S be the family of dessins $\mathcal{D}_s$, $s \in (\mathbb{Z}/q\mathbb{Z})^*$, on the surfaces X_s induced by the inclusions of their surface groups $N_s \lhd \Delta(p, q, q)$, as in Lemma 9.2. Then S is a single orbit of the Wilson operators H_s and also a Galois-invariant family.*

Thus the hypotheses of Theorem 8.3 are satisfied, so we have:

Theorem 9.5 *Let p and q be odd primes with $p \equiv 1 \bmod q$ and let G be the non-abelian group $C_p \rtimes C_q$. The family S of regular dessins of type (p, q, q) with automorphism group G forms a single Galois orbit of dessins with minimal field of definition $\mathbb{Q}(\zeta_q)$. The underlying quasiplatonic curves have its maximal real subfield $\mathbb{Q}(\cos(2\pi/q))$ as minimal field of definition.*

Proof The last claim follows from the fact that Wilson's operation H_{-1} transposes the two dessins $\mathscr{D}_{\pm s}$, forming chiral pairs for all s. The underlying surfaces $X_s \cong X_{-s}$ are therefore complex conjugate, and are therefore defined over the reals. □

9.2.4 Equations

It is sufficient to find equations for just one of the curves X_s, since they are all Galois conjugate by Theorem 9.5. We therefore try to find equations for $X := X_1$. Since C_p is a normal subgroup of G, the complement C_q has a well-defined action as a group of automorphisms of the quotient curve $Y := C_p \backslash X$, and the quotient map $X \to Y$ must ramify at the q white vertices of the dessin, with order p at each of them. By the Riemann-Hurwitz formula, Y has genus 0 and the quotient dessin on Y has one black vertex of valency q joined to q white vertices of valency 1. It is also regular, and we may assume that $Y = \hat{\mathbb{C}}$, the black vertex is at $z = 0$, the white vertices are at the q-th roots of unity, and the corresponding Belyĭ function $\hat{\mathbb{C}} \to \hat{\mathbb{C}}$ is

$$z \mapsto 1 - z^q .$$

From Sect. 4.1.3 we know that the function field $\mathbb{C}(X)$ is a cyclic extension of $\mathbb{C}(Y)$ of degree p, totally ramified above the q points $z = \zeta_q^k$ where $k \in \mathbb{Z}/q\mathbb{Z}$, and hence of the form

$$\mathbb{C}(X) = \mathbb{C}(z, y) \quad \text{with defining equation} \quad y^p = \prod_{k \bmod q} (z - \zeta_q^k)^{c_k}$$

with exponents $c_k \in \mathbb{Z}$ to be calculated. According to classical results about cyclic extensions or Kummer extensions (see [19]), the right-hand side is unique up to p-th power factors and up to taking powers with exponents $\not\equiv 0 \bmod p$. So if we assume that $c_1 = 1$, the exponents c_k are unique mod p, and we can assume that they satisfy $c_k \equiv 1 \bmod q$. To determine them we may regard a and b as function field automorphisms acting on the generators by

$$a(y) := y \circ a^{-1} = \zeta_p^{-1} y\,, \quad a(z) := z \circ a^{-1} = z \quad \text{and} \quad b(z) := z \circ b^{-1} = \zeta_q^{-1} z\,.$$

To determine the action of b on y note that $b(y)$ is also a p-th root generating the field extension $\mathbb{C}(z,y)/\mathbb{C}(z)$, so $b(y)^p$ has to be a power of y^p; on the other hand

$$b(y)^p = \prod_{k \bmod q} (b(z) - \zeta_q^k)^{c_k} = \prod_{k \bmod q} (\zeta_q^{-1} z - \zeta_q^k)^{c_k} =$$
$$\zeta_q^{-\sum c_k} \prod_{k \bmod q} (z - \zeta_q^{k+1})^{c_k} = \zeta_q^{-\sum c_k} \prod_{k \bmod q} (z - \zeta_q^k)^{c_{k-1}} \quad .$$

By our assumption that $c_k \equiv 1 \bmod q$ we can ignore the power of ζ_q in front of the product, and see that there is a factor c (also $\equiv 1 \bmod q$ and of order q in $(\mathbb{Z}/p\mathbb{Z})^*$) such that

$$c_k \equiv c \cdot c_{k-1} \bmod p \quad \text{for all} \quad k \bmod q \,.$$

We may therefore assume that $(c_1, c_2, \ldots, c_q) = (1, c, c^2, \ldots, c^{q-1})$. To find c, write the function y locally around $z = \zeta_q^k$ as a Puiseux series

$$y = (z - \zeta_q^k)^{c^k/p} + \ldots$$

and obtain

$$b(y) = (\zeta_q^{-1} z - \zeta_q^k)^{c^k/p} + \ldots = \zeta_q^{-c^k/p} (z - \zeta_q^{k+1})^{c^k/p} + \ldots \quad ,$$

a Puiseux series locally around $z = \zeta_q^{k+1}$, hence with

$$b(y^c) = \zeta_q^{-c^{k+1}/p} y + \ldots \quad \text{and} \quad ab(y^c) = \zeta_p^{-1} \zeta_q^{-c^{k+1}/p} y + \ldots \,.$$

A similar calculation gives

$$ba^u(y^c) = \zeta_p^{-cu} \zeta_q^{-c^{k+1}/p} y + \ldots \,,$$

so by the defining relation $ab = ba^u$ we therefore have $cu \equiv 1 \bmod p$. This gives the following:

Theorem 9.6 *The curves X_s described in Lemma 9.2 have an affine (singular) model described by the equation*

$$y^p = \prod_{k \bmod q} (z - \zeta_q^{\bar{s}k})^{c^k}$$

where $\bar{s}$ and c are integers with

$$\bar{s} \cdot s \equiv 1 \bmod q \,, \quad c \equiv 1 \bmod q \quad \textit{and} \quad c \cdot u \equiv 1 \bmod p \,.$$

Proof The only missing point of the proof is the extension of the result for X_1 to all $X_s,\ s \in (\mathbb{Z}/q\mathbb{Z})^*$. This is easy, considering the effect of the Wilson operators H_j in Theorem 8.3 and recalling that they act as 'inverse' Galois conjugations $\zeta_q^j \mapsto \zeta_q$, as in the proof of Theorem 8.1. □

Remark 9.2 In this model, the Belyĭ function for the dessin is given as $(y,z) \mapsto 1 - z^q$. A more direct way to see the isomorphism $X_{-s} \cong X_s$ can now be derived from the equation

$$y^p = \prod_{k \bmod q} (z - \zeta_q^{-\bar{s}k})^{c^k} = z^{pN} \prod (\zeta_q^{\bar{s}k} - \frac{1}{z})^{c^k}$$

for X_{-s}. Observe that $\sum c^k = pN$ for some integer N because $c^q \equiv u^{-q} \equiv 1 \bmod p$; with new variables $y' := \zeta_{2p} y/z^N$ and $z' := 1/z$ we get the equation for X_s, corresponding to a duality of dessins transposing face centres (where $\beta = \infty$, $z = \infty$) and black vertices (where $\beta = 1$, $z = 0$).

Remark 9.3 The key advantage of this class of examples is the fact that the function field of X_s is a cyclic extension of a rational function field. Similar ideas are used in [9] for more general abelian extensions, and in Sects. 5.2.1–5.2.3 for almost all low genus examples of quasiplatonic surfaces, see Eqs. (5.1)–(5.20). A more systematic approach to the determination of such equations, using canonical models and representation theory, has been given by Streit [29]. Finally, for the explicit determination of curves and Belyĭ functions by using computational methods one should consult the articles by Schneps [27], Birch [2], or recent work of Hidalgo [10] based on a computational version of invariant theory.

Exercise 9.11 Let $p \equiv 1 \bmod 3$ be prime and let X be a quasiplatonic surface with a regular dessin $\mathcal{D}$ of type $(3,3,p)$ and automorphism group G of order $3p$. Show that $G \cong C_p \rtimes C_3$, that $\mathcal{D}$ has three faces, p white and p black vertices, and that every vertex lies on the boundary of every face. Show moreover that X is defined over $\mathbb{Q}$, but that $\mathcal{D}$ is not: complex conjugation exchanges the colours of the vertices. The minimal field of definition of $\mathcal{D}$ is in fact $\mathbb{Q}(\sqrt{-3})$.

References

1. Biggs, N.L., White, A.T.: Permutation Groups and Combinatorial Structures. London Mathematical Society Lecture Note Series, vol. 33. Cambridge University Press, Cambridge/New York (1979)
2. Birch, B.: Noncongruence subgroups, coverings and drawings. In: Schneps, L. (ed.) The Grothendieck Theory of Dessins d'Enfants. London Mathematical Society Lecture Note Series, vol. 200, pp. 25–46. Cambridge University Press, Cambridge (1994)
3. Coste, A., Jones, G.A., Streit, M., Wolfart, J.: Generalised Fermat hypermaps and Galois orbits. Glasg. Math. J. **51**, 289–299 (2009)

4. Du, S.-F., Jones, G.A., Kwak, J.H., Nedela, R., Škoviera, M.: Regular embeddings of $K_{n,n}$ where n is a power of 2. I. Metacyclic case. Eur. J. Combin. **28**, 1595–1609 (2007)
5. Du, S.-F., Jones, G.A., Kwak, J.H., Nedela, R., Škoviera, M.: Regular embeddings of $K_{n,n}$ where n is a power of 2. II: the non-metacyclic case. Eur. J. Combin. **31**, 1946–1956 (2010)
6. Erdős, P.: Some asymptotic formulas in number theory. J. Indian Math. Soc. **12**, 75–78 (1948)
7. Girondo, E., Torres-Teigell, D., Wolfart, J.: Shimura curves with many uniform dessins. Math. Z. **271**, 757–779 (2012)
8. Hall, P.: A note on soluble groups. J. Lond. Math. Soc. **3**, 98–105 (1928)
9. Hidalgo, R.: Homology closed Riemann surfaces. Q. J. Math. **63**, 931–952 (2012)
10. Hidalgo, R.: Edmonds maps on the Fricke-Macbeath curve. Ars Mat. Contemp. **8**, 275–289 (2015)
11. Huppert, B.: Über das Produkt von paarweise vertauschbaren zyklischen Gruppen. Math. Z. **58**, 243–264 (1953)
12. Huppert, B.: Endliche Gruppen I. Springer, Berlin/Heidelberg/New York (1967)
13. Itô, N.: Über das Produkt von zwei abelschen Gruppen. Math. Z. **62**, 400–401 (1955)
14. Jones, G.A.: Regular embeddings of complete bipartite graphs: classification and enumeration. Proc. Lond. Math. Soc. (3) **101**, 427–453 (2010)
15. Jones, G.A.: Characterisations and Galois conjugacy of generalised Paley maps. J. Combin. Theory Ser. B **103**, 209–219 (2013)
16. Jones, G.A., Streit, M., Wolfart, J.: Galois action on families of generalised Fermat curves. J. Algebra **307**, 829–840 (2007)
17. Jones, G.A., Nedela, R., Škoviera, M.: Regular embeddings of $K_{n,n}$ where n is an odd prime power. Eur. J. Combin. **28**, 1863–1875 (2007)
18. Jones, G.A., Nedela, R., Škoviera, M.: Complete bipartite graphs with a unique regular embedding. J. Combin. Theory Ser. B **98**, 241–248 (2008)
19. Lang, S.: Algebra. Addison-Wesley, Amsterdam (1965)
20. Lim, T.K., Praeger, C.E.: On generalised Paley graphs and their automorphism groups. Michigan Math. J. **58**, 293–308 (2009)
21. Margulis, G.: Discrete Subgroups of Semisimple Lie Groups. Springer, Berlin/Heidelberg/New York (1991)
22. Nedela, R., Škoviera, M.: Regular embeddings of canonical double coverings of graphs. J. Combin. Theory Ser. B **67**, 249–277 (1996)
23. Nedela, R., Škoviera, M., Zlatoš, A.: Regular embeddings of complete bipartite graphs. Discrete Math. **258**, 379–381 (2002)
24. Paley, R.E.A.C.: On orthogonal matrices. J. Math. Phys. Mass. Inst. Technol. **12**, 311–320 (1933)
25. Robinson, D.J.S.: A Course in the Theory of Groups. Springer, Berlin/Heidelberg/New York (1996)
26. Rose, J.S.: A Course on Group Theory. Cambridge University Press, Cambridge (1978)
27. Schneps, L.: Dessins d'enfants on the Riemann sphere. In: Schneps, L. (ed.) The Grothendieck Theory of Dessins d'Enfants. London Mathematical Society Lecture Note Series, vol. 200, pp. 47–77. Cambridge University Press, Cambridge (1994)
28. Singerman, D.: Finitely maximal Fuchsian groups. J. Lond. Math. Soc. (2) **6**, 29–38 (1972)
29. Streit, M.: Homology, Belyi functions and canonical curves. Manuscripta Math. **90**, 489–509 (1996)
30. Streit, M., Wolfart, J.: Characters and Galois invariants of regular dessins. Rev. Mat. Complut. **13**, 49–81 (2000)
31. Thompson, J.G.: Finite groups with fixed-point-free automorphisms of prime order. Proc. Natl. Acad. Sci. USA **45**, 578–581 (1959)
32. White, A.T.: Graphs of Groups on Surfaces. Mathematics Studies, vol. 188. North-Holland, Amsterdam (2001)
33. Wielandt, H.: Über das Produkt von paarweise vertauschbaren nilpotenten Gruppen. Math. Z. **55**, 1–7 (1951)

Part III
Applications

In this part of the book we describe some links between dessins and two other areas of mathematics, namely number theory and algebraic geometry.

The abc conjecture concerns the equation $a + b + c = 0$, proposing specific upper bounds on the size of the integers a, b and c in terms of the primes dividing them. Despite this apparently simple context, it is one of the most challenging open problems in number theory: if proved, it would have a number of deep recent results, such as the proofs of Fermat's Last Theorem and the Catalan and Mordell conjectures, together with many currently open problems, as simple corollaries. As evidence for the truth of the conjecture, we will show that similar statements about the degrees of polynomials and of meromorphic functions on compact Riemann surfaces are true. In the latter case, we will prove Zannier's result that the functions attaining the relevant upper bound are Belyĭ functions.

Elliptic curves are characterised among compact Riemann surfaces by having unramified self-covers, called endomorphisms. In the simplest case, these are obtained by multiplying elements of the universal covering space $\mathbb{C}$ by some integer. If an elliptic curve has endomorphisms obtained from multiplication by other complex numbers, it is said to have complex multiplication. We will show that an elliptic curve defined over a number field has complex multiplication if and only if each dessin on it has infinitely many self-covers of prime degree.

Complex surfaces (that is, 2-dimensional algebraic varieties over $\mathbb{C}$) are much harder to study than their comparatively well-understood 1-dimensional analogues, the algebraic curves. In 1978 Beauville showed how to construct complex surfaces with interesting properties from pairs of regular dessins which have the same automorphism group, acting freely on their product. In recent years these Beauville surfaces have been intensively studied by both geometers and group theorists, bringing new algebraic techniques into this area of geometry. In the final chapter of this book we will consider which groups, called Beauville groups, can be used in this construction, what topological properties the resulting Beauville surfaces have, and how the absolute Galois group acts on them.

Chapter 10
Arithmetic Aspects

Abstract In this chapter we discuss two links between dessins and arithmetic. The abc problem for integers concerns conjectured bounds on the size of integers a, b, c satisfying $a + b + c = 0$, in terms of the primes dividing them. Closely related to some of the deepest theorems and most difficult conjectures in number theory, it is a major open problem. However, its analogue for meromorphic functions on compact Riemann surfaces is comparatively easy to solve. In this case, Belyĭ functions provide extremal examples which show that the inequality in the main result is sharp.

In the second section, we prove a dessin-theoretic criterion for elliptic curves (those of genus 1) to have complex multiplication, that is, to have non-trivial endomorphisms. This criterion relies on a very special property of curves and dessins of genus 1, that they have infinitely many unramified self-coverings.

Keywords abc conjecture • Belyĭ function • Catalan conjecture • Complex multiplication • Elliptic curve • Endomorphism ring • Fermat's last theorem • Meromorphic function • Polynomial • Unramified self-cover

10.1 abc for Function Fields

10.1.1 Digression: Motivation from Number Theory

One of the most important contemporary Diophantine problems is the *abc conjecture*, first formulated by Masser [8] and Oesterlé [10] in the 1980s. In its simplest form it can be stated as follows.

Suppose that you are given a finite set of primes, and using only these as prime factors (repeated as often as you like), you are asked to find mutually coprime triples of integers a, b and c satisfying

$$a + b + c = 0 . \qquad (10.1)$$

Gareth A. Jones, J. Wolfart, *Dessins d'Enfants on Riemann Surfaces*, Springer Monographs in Mathematics, DOI 10.1007/978-3-319-24711-3_10

How large can these integers be? More specifically, can we find an upper bound for $\max\{|a|,|b|,|c|\}$ in terms of the *kernel*

$$K := \prod_{p|abc} p\,,$$

that is, the product of the prime divisors of abc, not counting multiplicities? (This is a useful measure of the size of the prime factors we are using.)

Example 10.1 The following examples, one trivial and two far more elaborate (due to de Weger and Reyssat—see Nitaj's website [9]) show that the largest integer can be very large, compared with K:

$$\begin{aligned} 1+2^3-3^2 &= 0 \text{ with } K = 2\cdot 3 = 6\,, \\ 1+2\cdot 3^7-5^4\cdot 7 &= 0 \text{ with } K = 2\cdot 3\cdot 5\cdot 7 = 210 \quad < \quad 5^4\cdot 7 = 4375\,, \\ 2+109\cdot 3^{10}-23^5 &= 0 \text{ with } K = 2\cdot 3\cdot 23\cdot 109 = 15042 < 23^5 = 6436343. \end{aligned}$$

The only known upper bounds are exponential in K, whereas experimental evidence strongly suggests that a polynomial (even quadratic) bound should be valid. The following example shows that an upper bound which is linear in K, that is, of the form CK for some constant C, will not work:

Exercise 10.1 Show that $2^{p(p-1)} \equiv 1 \bmod p^2$ for each odd prime p. Hence show that if $C > 0$ then there are infinitely many triples of the form

$$a = 1, \quad b = 2^{p(p-1)} - 1, \quad c = -2^{p(p-1)}$$

satisfying (10.1) with $|c| > CK$.

The abc conjecture asserts that if we replace K with $K^{1+\varepsilon}$ for any $\varepsilon > 0$, no matter how small, there are only finitely many triples satisfying (10.1) such that $\max\{|a|,|b|,|c|\} > K^{1+\varepsilon}$. Equivalently:

Conjecture For each $\varepsilon > 0$ there is a constant C_ε such that

$$\max\{|a|,|b|,|c|\} \;<\; C_\varepsilon\, K^{1+\varepsilon}$$

for each mutually coprime triple a, b, c satisfying (10.1).

The hope is that, although the quotient $q := \log(\max\{|a|,|b|,|c|\})/\log K$ exceeds 1 infinitely often (see Exercise 10.1), it exceeds each $1+\varepsilon > 1$ only finitely many times; for the triples given in Example 10.1, q is approximately 1.22629, 1.56789 and 1.62991. According to Nitaj's website [9], the last example is the current (January 2015) and already long-standing world record. It seems plausible that $q < 2$ for all such triples, so that

$$\max\{|a|, |b|, |c|\} < K^2.$$

Exercise 10.2 Assuming this inequality, prove Fermat's Last Theorem. (You may assume the classical proofs for exponents 3, 4 and 5.)

In the meantime many generalisations and more sophisticated versions of this conjecture have arisen. For these, for the history of the problem, and for many consequences and examples, including Mochizuki's recent claimed proof of the abc conjecture, see [9]. One reason for believing in some form of this conjecture is the fact that an analogous statement for function fields had been proved some years earlier by Stothers [12], and then in much more general form by Mason [7]. This can be formulated for arbitrary genera and arbitrary fields of constants, see Chap. 7 of [11]. For the proof on compact Riemann surfaces we will follow these general ideas, but first we illustrate the meaning of this function field statement in the simplest case; for this, Lang gave a very elementary proof [6], using only unique prime decomposition in $\mathbb{C}[x]$. Lang's proof is often reproduced in the literature, so we will not repeat it here. It is however worth studying since it shows clearly why the function field case is easier than the number field case: derivatives are a powerful tool!

10.1.2 The Polynomial Case

Proposition 10.1 *Let A, B, C be coprime polynomials in $\mathbb{C}[x]$, not all constant, satisfying $A + B + C = 0$. Let*

$$n_0 := |\{ w \in \mathbb{C} \mid ABC(w) = 0 \}|$$

be the number of distinct zeros of ABC, that is, the degree of the kernel

$$K := \prod (x - w)$$

where this product is over all irreducible factors $x - w$ of ABC, not counting multiplicities. Then we have

$$\max\{\deg A, \deg B, \deg C\} < n_0 .$$

The consequences of this result are far-reaching, as can be seen from Fermat's theorem for polynomials:

Proposition 10.2 *For any integer $n > 2$ the equation*

$$p^n + q^n + r^n = 0$$

has no solution p, q, r in coprime non-constant polynomials in $\mathbb{C}[x]$.

Proof To see how this statement follows from Proposition 10.1, let

$$m := \max\{\deg p, \deg q, \deg r\} ,$$

so that $m > 0$. Set

$$A := p^n , \quad B := q^n , \quad C := r^n ,$$

so that

$$n_0 = \deg K \leq 3m$$

because the kernel K of ABC is also the kernel of pqr. On the other hand, $A + B + C = 0$ and

$$\max\{\deg A, \deg B, \deg C\} = nm$$

would contradict Proposition 10.1. (Of course, the proof of Fermat's theorem for integers is rather more difficult than this!) □

Exercise 10.3 By analogy with the *Catalan conjecture* for positive integers (proved by Mihailescu in 2002), use Proposition 10.1 to show that there are no non-constant polynomials $f, g \in \mathbb{C}[x]$ satisfying $f^2 - g^3 = 1$.

10.1.3 abc on Riemann Surfaces

Theorem 10.1 *Let X be a compact Riemann surface of genus g. For non-constant meromorphic functions u, v on X with the property that $u + v = 1$, one has the inequality*

$$\deg u = \deg v \leq 2g - 2 + |\{x \in X \mid uv(x) = 0 \text{ or } \infty\}| .$$

(Note the contrast with the number field case:

- this is a theorem, not a conjecture;
- no ε and no constant C_ε are involved;
- the inequality is sharp, as we will see.)

Proof Note first that the degrees of u and v coincide, as do their poles, even with multiplicities. We have the same coincidence for the zeros of u and $1-v$ and for the zeros of v and $1-u$. By the Riemann-Hurwitz formula for $u : X \to \hat{\mathbb{C}}$ we have

$$2g-2 = -2\deg u + \sum_{x \in X}(\text{mult}_x u - 1) .$$

Restricting the summation to the zeros $z \in X$ of u gives

$$\deg u = \sum_{z}(\text{mult}_z u - 1) + |\{z \in X \mid u(z) = 0\}| ,$$

and we get similar formulæ for the points where $v(z) = 0$ (equivalently $u(z) = 1$) and for the poles of both functions. Replacing these three sums in the Riemann-Hurwitz formula we obtain

$$2g-2 = \deg u - |\{x \in X \mid uv(x) = 0 \text{ or } \infty\}| + \sum_{\substack{x \in X \\ u(x) \neq 0,1,\infty}} (\text{mult}_x u - 1) . \qquad (10.2)$$

The inequality in the theorem follows if we omit the last sum. □

This theorem implies Proposition 10.1, by the following argument. Put $X = \hat{\mathbb{C}}$, $g = 0$ and $u := -A/C$, $v := -B/C$. Then the degree of u and v is the maximum degree of A, B and C, and (not counting the point $x = \infty$) $|\{x \in \mathbb{C} \mid uv(x) = 0 \text{ or } \infty\}|$ is the number of zeros n_0 of the kernel K. Then the theorem gives

$$\max\{\deg A, \deg B, \deg C\} \leq -2 + n_0 + 1 ,$$

where the last term can be omitted if ∞ is not a zero or a pole of uv. In any case, the estimate of Proposition 10.1 is correct. □

Exercise 10.4 Show that the claim of Proposition 10.1 can be improved to

$$\max\{\deg A, \deg B, \deg C\} \leq n_0 - 2$$

if the three polynomials A, B and C have the same degree.

10.1.4 Belyĭ Functions: The Worst Case

In the number field case (Sect. 10.1.1) it is not at all clear how the constant C_ε in the abc conjecture could depend on ε. Therefore there is considerable research activity on 'bad cases', in which K is large in comparison with $\max\{|a|, |b|, |c|\}$: see [9] and the few examples given above after the conjecture. In contrast to that, Theorem 10.1 is sharp, as remarked first by Zannier [15].

Theorem 10.2 *Let X be a compact Riemann surface of genus g. For non-constant meromorphic functions u and v on X with the property that* $u + v = 1$, *the equality*

$$\deg u = \deg v = 2g - 2 + |\{x \in X \mid uv(x) = 0 \text{ or } \infty\}|$$

is valid if and only if u and $v = 1 - u$ *are Belyĭ functions on X.*

Proof Recall the proof of Theorem 10.1: Eq. (10.2) is made into an inequality by omitting the term

$$\sum_{\substack{x \in X \\ u(x) \neq 0,1,\infty}} (\text{mult}_x u - 1) ,$$

and this vanishes if and only if u, and hence also v, are Belyĭ functions. □

As the reader may imagine, by taking particular rational or integer values of Belyĭ functions, one might use this theorem as a possible tool in searching for those 'bad cases' in the original abc conjecture: compare the long list of references in [9], or Sect. 2.5.4 in [4].

10.2 Genus 1 Dessins and Complex Multiplication

The aim of this section is two-fold. Firstly, dessins on Belyĭ surfaces are not at all uniquely determined, as we have already seen in Exercise 1.14. Here we present another procedure, different from the composition of given Belyĭ functions with Shabat polynomials. Instead we use an easy application of Exercise 1.16 in the case of genus 1 curves to construct infinitely many new dessins from a given one by considering unramified covers of degree n^2, where n is an arbitrary positive integer.

Secondly, since all the relevant properties of a Belyĭ curve should be encoded in its dessins, one may ask how the existence of complex multiplications for an elliptic curve (automatically implying that it can be defined over a number field) is visible in its dessins. For most elliptic curves this is a difficult question as long as one considers only one dessin. It turns out that complex multiplication may be characterised by the existence of even more unramified *self-covers* than usual, as shown in the next section.

Remark 10.1 We note in passing that also in higher genera, deeper arithmetic properties of Belyĭ curves (for instance being a congruence subgroup Shimura curve, such as Klein's quartic, Bring's curve or the Fricke-Macbeath curve) can be characterised by the existence of several dessins of a specific type, see [4].

10.2.1 Unramified Self-covers of Elliptic Curves

We return to the elliptic curves already introduced in Chap. 1 as tori $T := \mathbb{C}/\Lambda$. Multiplying by a non-zero complex number, we may assume that Λ is the lattice $\mathbb{Z} + \mathbb{Z}\tau$ with *period quotient* $\tau \in \mathbb{H}$. For any positive integer n, the holomorphic function $\mathbb{C} \to \mathbb{C}$, $z \mapsto nz$ induces an unramified covering $\alpha_n : T \to T$ which has degree $|\Lambda : n\Lambda| = n^2$. If there is a Belyĭ function $\beta : T \to \hat{\mathbb{C}}$ of degree d, then Exercise 1.16 shows that $\beta \circ \alpha_n$ is also a Belyĭ function, now of degree $n^2 d$. We denote their respective dessins by $\mathscr{D}$ and $n^{-1}\mathscr{D}$, and call $n^{-1}\mathscr{D}$ an unramified *self-cover* of $\mathscr{D}$ of degree n^2. The following easy exercise shows that it is in fact reasonable to speak here about 'coverings of dessins':

Exercise 10.5 Show that the dessin $n^{-1}\mathscr{D}$ has a group A of automorphisms isomorphic to $C_n \times C_n$, acting freely on vertices and faces, such that

$$\mathscr{D} \cong A \backslash n^{-1}\mathscr{D} \,.$$

A comparison with the coverings introduced in Sect. 3.3.1 shows that self-covers are always unramified, and that the underlying base surface is isomorphic to the covering surface. One should note that such coverings exist only in genus 1.

Exercise 10.6 Show that for genus $g \neq 1$ there are no Riemann surfaces X with unramified coverings $X \to X$ of degree greater than 1.

The essential point of our consideration is contained in the following:

Lemma 10.1 *For any dessin on a Belyĭ curve of genus* 1 *and for any positive integer* n *there is a self-cover of degree* n^2.

Example 10.2 Let the torus T be defined by the affine equation

$$y^2 = (x+1)(x-1)\left(x - \cos\frac{3\pi}{10}\right).$$

On T we have a Belyĭ function $\beta : (x, y) \mapsto T_5^2(x)$, with the corresponding dessin $\mathscr{D}$ shown in the lower part of Fig. 10.1. (This idea is due to Friedrich Berg; see the definition of the Chebychev polynomials in Example 1.9 and Fig. 1.1.) As usual for dessins of genus 1, opposite sides of the boundary have to be identified to give a torus. Since the zeros of $\beta - 1$ are all of order 2, we are in the case of a map, and we may leave out all the black vertices, which can be thought of—up to homeomorphism—as the mid-points of edges. The upper part of Fig. 10.1 represents the self-cover $3^{-1}\mathscr{D}$ of degree 9.

Exercise 10.7 Let Γ be a subgroup of genus 1 of a cocompact Fuchsian triangle group Δ. Show that for each positive integer n there is a normal subgroup $\Gamma_n \lhd \Gamma$ with quotient $\Gamma/\Gamma_n \cong C_n \times C_n$ such that the quotient surfaces $\Gamma\backslash\mathbb{H}$ and $\Gamma_n\backslash\mathbb{H}$ are isomorphic.

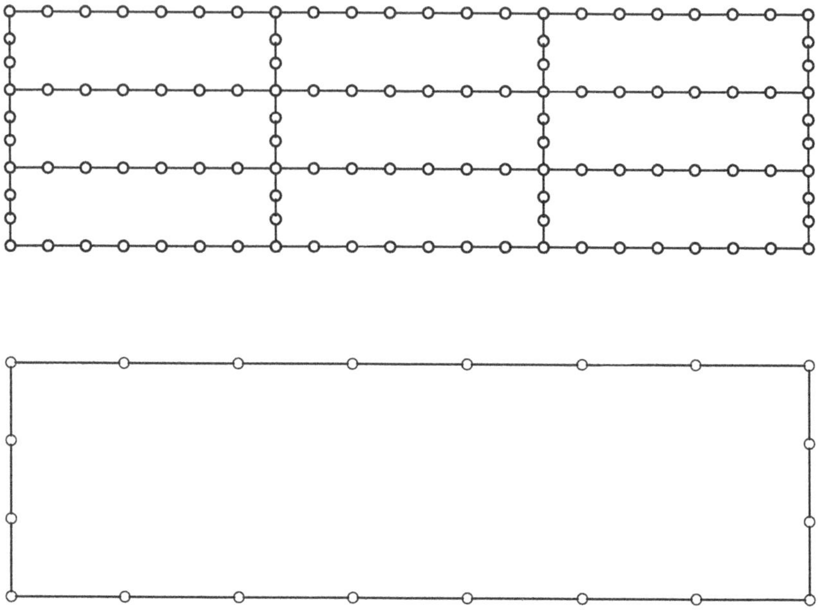

Fig. 10.1 A dessin $\mathscr{D}$ on the curve $y^2 = (x+1)(x-1)(x - \cos\frac{3\pi}{10})$ and its self-cover $3^{-1}\mathscr{D}$

10.2.2 Complex Multiplication by Roots of Unity

The algebraic background behind Lemma 10.1 is the fact that elliptic curves always have $\mathbb{Z}$ as a ring of endomorphisms, where each $n \in \mathbb{Z}$ acts as the endomorphism induced by $z \mapsto nz$; here we again write the elliptic curve as a torus $T = \mathbb{C}/\Lambda$ and consider $n = 0$ as the trivial endomorphism sending the curve to the point $0 \bmod \Lambda$. (Caution: the word *endomorphism* has to be understood here as indicating not only an endomorphism of the Riemann surface, but also an endomorphism of the group, fixing the point $0 \bmod \Lambda$. Thus the covering transformation $z \bmod \Lambda \mapsto nz \bmod \Lambda$ inducing the quotient map $n^{-1}\mathscr{D} \to \mathscr{D}$ counts as an endomorphism, whereas the (Riemann surface automorphism) $z \bmod \Lambda \mapsto (z+a) \bmod \Lambda$ does not if $a \notin \Lambda$! This also helps to explain our slightly inconsistent notation, writing $\Gamma\backslash\mathbb{H}$ for quotients of the *space* $\mathbb{H}$, but $\mathbb{C}/\Lambda$ for quotients of the *group* $\mathbb{C}$, as is conventional in Group Theory.)

If the endomorphism ring End T of T is strictly larger than $\mathbb{Z}$, we call T an elliptic curve with *complex multiplication*. This is the case if and only if the period quotient τ generates an imaginary quadratic extension K of $\mathbb{Q}$ (easy exercise!). If we denote the ring of integers of K by $\mathscr{O}_K$, and lift the endomorphisms to $\mathbb{C}$, then it is easy to see that End T is the subring of all $\alpha \in \mathscr{O}_K$ such that $\alpha\Lambda \subseteq \Lambda$. An *automorphism* of T occurs if α is a root of unity; since roots of unity generate imaginary quadratic fields if and only if they are of order 3, 4 or 6, we have the following:

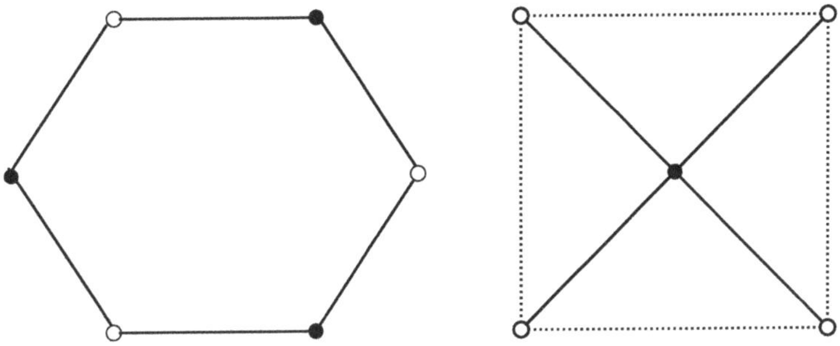

Fig. 10.2 Dessins on the elliptic curves $y^2 = x^3 - 1$ and $y^2 = x^3 - x$

Theorem 10.3 *A dessin $\mathcal{D}$ of genus 1 belongs to an elliptic curve with complex multiplication by $\mathbb{Z}[e^{2\pi i/3}]$ (or $\mathbb{Z}[i]$) if $\mathcal{D}$ has an automorphism of order 3 (or 4, respectively) fixing at least one vertex or face.*

We have already seen examples of such dessins in Figs. 7.2 and 7.1 respectively (Sect. 7.1). There are even simpler examples, such as those in Fig. 10.2: opposite sides of the boundaries have to be identified; the dotted boundary lines for the second example are not edges of the dessin.

Exercise 10.8 Explain why the dessins in Fig. 10.2 lead to the indicated equations.

Remark 10.2 Roughly speaking, Theorem 10.3 says that the existence of enough automorphisms with fixed points is a criterion for genus 1 dessins to have complex multiplication. This is no longer true for higher genera: there one has a natural generalisation considering the Jacobians of the curve (see [3] for their definition). They are called *of CM type* if they are isogenous to a product of simple abelian varieties with complex multiplication, that is, with a commutative endomorphism ring of maximal possible rank $2d$, where d denotes the dimension of the simple abelian variety. One might imagine that quasiplatonic curves should have Jacobians of CM type, but this is not true in general; for most quasiplatonic curves in genera 2 to 4 this is true, but it fails for the curves with Eqs. (5.7), (5.21), and (5.22) in Sects. 5.2.2 and 5.2.3, and for the Fricke-Macbeath curve in Sect. 5.1.2, see [13] or [14].

For a quite different link between (vertices of) dessins and complex multiplication of Jacobians, see [1].

10.2.3 Complex Multiplication, General Case

In the case of complex multiplication of T, the full endomorphism ring $\operatorname{End} T$ has the form

$$\operatorname{End} T = \mathbb{Z} + f\mathcal{O}_K \qquad \text{where} \qquad f \in \mathbb{N}$$

and $\mathscr{O}_K$ denotes the ring of integers of the imaginary quadratic field $K = \mathbb{Q}(\tau)$. By analogy with the positive integers considered above, we can now consider the action of an arbitrary non-zero $\alpha \in \operatorname{End} T$ on the dessins $\mathscr{D}$ for T via composition with the corresponding Belyĭ functions, $\beta \mapsto \beta \circ \alpha$. The target Belyĭ function defines a dessin $\alpha^{-1}\mathscr{D}$; this is a self-cover of $\mathscr{D}$ of degree $|\alpha|^2$ since the lift of α to $\mathbb{C}$ is just multiplication $z \mapsto \alpha z$, mapping the lattice Λ to the sublattice $\alpha\Lambda$ and multiplying the area of its fundamental parallelogram by the factor $|\alpha|^2$ ($\in \mathbb{N}$, of course). In the case of complex multiplication, $\operatorname{End} T$ itself can be considered as a lattice in $\mathbb{C}$. By counting its lattice points within distance at most x from 0, we obtain the second claim in the following criterion for the distinction of cases with and without complex multiplication.

Theorem 10.4 *If an elliptic curve T has no complex multiplication, any dessin $\mathscr{D}$ on T has at most $2\sqrt{x}$ self-covers of degrees at most x.*
If T has complex multiplication, the number of self-covers of $\mathscr{D}$ of degree at most x grows linearly with x.

Proof Any unramified covering $T \to T$ lifts to its universal cover $\mathbb{C} \to \mathbb{C}$, and by composing with a translation we may assume that this lift has the form

$$z \mapsto \alpha z \qquad \text{with} \qquad \alpha\Lambda \subseteq \Lambda\,.$$

If T has no complex multiplication, we have $\alpha = n \in \mathbb{Z}$. Then the covering has degree n^2, which proves the first claim. The factor 2 comes from the fact that $\mathscr{D}$ and $-\mathscr{D}$ might be different, but are always isomorphic. □

The number of self-covers shows that in the complex multiplication case there are many self-covers of non-square degree.

Example 10.3 Let T be given by the affine Legendre equation

$$y^2 = x(x-1)(x-\lambda) \qquad \text{where} \qquad \lambda = \frac{1}{2}(1+\sqrt{2})\,.$$

Using Eq. (1.2) at the end of Sect. 1.2.4, we obtain $J(\tau) = 3^{-3} \cdot 5^3$ as the corresponding value for the elliptic modular function J. According to Cremona's tables [2] the value $j = 12^3 J = 20^3$ determines an elliptic curve with complex multiplication by the ring $\mathscr{O}_K = \mathbb{Z}[\sqrt{-2}]$, that is, with $f = 1$, so we may assume that $\tau = \sqrt{-2}$. Any dessin $\mathscr{D}$ on T therefore has a self-cover $(\sqrt{-2})^{-1}\mathscr{D}$ of degree 2.

Exercise 10.9 Verify that $\beta(x, y) := \frac{1}{4}v(1-v)$ for $v := 2x(x-1) + \frac{1}{2}$ defines a Belyĭ function on T giving the right-hand dessin in Fig. 10.3.

In the case of Example 10.3 we have a self-cover of prime degree 2. This degree 2 is very special, of course. It is possible here because there is an endomorphism $\alpha \in \operatorname{End} T = \mathbb{Z}[\sqrt{-2}]$ with the property that $|\alpha|^2 = 2$. On the other hand, prime degree self-covers are typical for elliptic curves with complex multiplication:

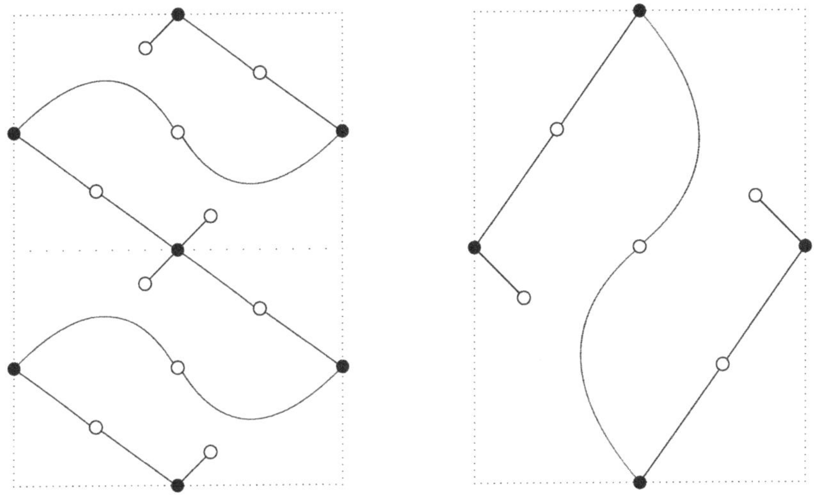

Fig. 10.3 $\mathscr{D}$ and its degree 2 self-cover $(\sqrt{-2})^{-1}\,\mathscr{D}$ on the elliptic curve with Legendre model $y^2 = x(x-1)(x - \frac{1}{2}(1+\sqrt{2}))$

Theorem 10.5 *Let T be an elliptic curve defined over a number field. It has complex multiplication if and only if every dessin on T has infinitely many self-covers of prime degree.*

Proof We already know that self-covers of elliptic curves without complex multiplication must have square degree. So, let $\mathscr{D}$ be a dessin on an elliptic curve T with complex multiplication. Recall that $\operatorname{End} T = \mathbb{Z} + f\mathcal{O}_K$ for some ideal $f\mathcal{O}_K$ in the ring of integers $\mathcal{O}_K$ of an imaginary quadratic field. We now use a strong version of Dirichlet's Prime Number Theorem in algebraic number fields (see Chap. VIII, §4 of [5]), concerning prime ideals in certain generalised ideal class groups: this guarantees the existence of infinitely many principal prime ideals, that is, prime elements $\alpha \in \mathcal{O}_K$, with $\alpha \equiv 1 \bmod f$ and with prime norms $|\alpha|^2$ in $\mathbb{Z}$; these primes $|\alpha|^2$ (which in fact have positive Dirichlet density in $\mathbb{Z}$) serve as the degrees for self-covers $\alpha^{-1}\mathscr{D}$ of prime degree. □

Exercise 10.10 By Theorem 3.10, the two dessins in Fig. 10.3 correspond to Fuchsian groups $\Gamma_2 \lhd \Gamma < \Delta(2,4,8)$. Determine the signatures of Γ and of Γ_2.

Remark 10.3 As Fig. 10.3 shows, it is relatively easy to draw dessins which are good candidates for being self-covers of prime degree. However, whether Theorem 10.5 is a useful criterion to detect complex multiplication depends on a computational problem: can one decide whether this prime degree covering defines a dessin on the same curve?

Exercise 10.11 Use Sects. 2.1.2 and 3.3.1 to describe the self-covers of the dessins in Exercise 10.5 in terms of algebraic bipartite maps.

References

1. Beazley Cohen, P., Wolfart, J.: Dessins de Grothendieck et variétés de Shimura. C. R. Acad. Sci. Paris Sér. I Math. **315**, 1025–1028 (1992)
2. Cremona, J.E.: Algorithms for Modular Elliptic Curves. Cambridge University Press, Cambridge (1992)
3. Farkas, H.M., Kra, I.: Riemann Surfaces. Springer, Berlin (1991)
4. Girondo, E., Torres-Teigell, D., Wolfart, J.: Shimura curves with many uniform dessins. Math. Z **271**, 757–779 (2012)
5. Lang, S.: Algebraic Number Theory. Addison-Wesley, Reading (1970)
6. Lang, S.: Die abc–Vermutung. Elemente der Math. **48**, 89–99 (1993)
7. Mason, R.: Diophantine Equations over Function Fields. London Mathematical Society Lecture Note Series, vol. 96. Cambridge University Press, Cambridge (1984)
8. Masser, D.W.: Open problems. In: Chen, W.W.L. (ed.) Proceedings of the Symposium on Analytic Number Theory. Imperial College, London (1985)
9. Nitaj, A.: The abc conjecture home page. http://www.math.unicaen.fr/$\sim$nitaj/abc.html (2013). Accessed 20 Jan 2015
10. Oesterlé, J.: Nouvelles approches du "théorème" de Fermat. Astérisque (Sém. Bourbaki Exp. 694) **161** (1988)
11. Rosen, M.: Number Theory in Function Fields. Graduate Texts in Mathematics, vol. 210. Springer, Berlin (2002)
12. Stothers, W.W.: Polynomial identities and hauptmoduln. Q. J. Math. (2) **32**, 349–370 (1981)
13. Wolfart, J.: Regular dessins, endomorphisms of Jacobians, and transcendence. In: Wüstholz, G. (ed.) A Panorama of Number Theory or the View from Baker's Garden, pp. 107–120. Cambridge University Press, Cambridge (2002)
14. Wolfart, J.: Triangle groups and Jacobians of CM type (2011). http://www.uni-frankfurt.de/50936228/Publikationen. Accessed 12 Dec 2014
15. Zannier, U.: On Davenport's bound for the degree of $f^3 - g^2$ and Riemann's existence theorem. Acta Arith. **71**, 107–137 (1995)

Chapter 11
Beauville Surfaces

Abstract Beauville surfaces are examples of complex surfaces, that is, they are 2-dimensional analogues of the complex algebraic curves (1-dimensional over $\mathbb{C}$) which we have considered so far. They are formed from certain pairs of regular dessins with the same automorphism group G (called a Beauville group) acting freely on their product. In the first section of the chapter we describe the basic theory of Beauville surfaces, in particular their construction by means of quasiplatonic (triangle) curves. We study Beauville's original example, and we classify and enumerate its generalisations, based on Fermat curves, in which G is an abelian group.

In the second section we consider which members of various other families of finite groups can be Beauville groups: these include symmetric groups, alternating groups and 2-dimensional projective linear groups over finite fields.

In the third and final section we discuss topological invariants of a Beauville surface, such as its Euler characteristic, fundamental group and automorphism group, and their behaviour under Galois actions. We consider conditions under which a Beauville surface has a real model, and we give examples of Serre's observation that Galois conjugation can change the topology of a complex variety.

Keywords Absolute Galois group • Algebraic fundamental group • Alternating group • Automorphism group • Beauville group • Beauville structure • Beauville surface • Burnside problem • Complex surface • Euler characteristic • Fermat curve • Fundamental group • Galois orbit • p-Group • Quasiplatonic curve • Real Beauville surface • Regular dessin • Simple group • Symmetric group

11.1 Basic Properties and First Examples

11.1.1 Basic Definitions

A Beauville surface S is formed from the cartesian product of two quasiplatonic curves of genus greater than 1, by factoring out the action of some finite group which acts freely on this product. Before making the definition more precise, we need to explain this use of the word 'surface'. It is used here in the sense of algebraic geometry, meaning that S is an example of a complex surface, that is, it is

Gareth A. Jones, J. Wolfart, *Dessins d'Enfants on Riemann Surfaces*, Springer Monographs in Mathematics, DOI 10.1007/978-3-319-24711-3_11

a 2-dimensional variety over the field $\mathbb{C}$. Thus each point in S has a neighbourhood homeomorphic to an open subset of $\mathbb{C}^2$, and hence of $\mathbb{R}^4$, so that S is in fact 4-dimensional over $\mathbb{R}$. By contrast, when we refer to a Riemann surface, we are using the word 'surface' in its topological sense, to denote an object which is 2-dimensional over $\mathbb{R}$, rather than $\mathbb{C}$: in algebraic geometry such an object is called a curve, since it is 1-dimensional over $\mathbb{C}$. This apparent confusion arises because we are working in the overlapping areas of algebraic geometry and topology, where $\mathbb{C}$ and $\mathbb{R}$ are respectively regarded as the fundamental fields of definition. To make our meaning clear, rather than referring simply to a surface, we will generally refer either to a Beauville surface or a complex surface, meaning an object which is 2-dimensional over $\mathbb{C}$, or to a Riemann surface, which is 2-dimensional over $\mathbb{R}$.

A *Beauville surface of unmixed type* is a compact complex surface S such that

(a) S is isogenous to a higher product, that is, $S \cong G\backslash(X_1 \times X_2)$ where X_1 and X_2 are algebraic curves of genus at least 2 and G is a finite group acting freely on $X_1 \times X_2$;
(b) G acts faithfully as a group of automorphisms of each X_i so that $G\backslash X_i$ is isomorphic to the projective line $\mathbb{P}^1(\mathbb{C})$ and the covering $X_i \to G\backslash X_i$ is ramified over three points.

(We will not consider the more difficult concept of a Beauville surface of mixed type, where G contains elements transposing the curves X_i. Note that here, as in earlier parts of this book, when denoting quotient spaces we write groups on the left of the spaces they act on; other authors in this field often use notations such as X_i/G and $(X_1 \times X_2)/G$.)

Condition (b) is equivalent to each curve X_i being a quasiplatonic Riemann surface, admitting a regular dessin $\mathscr{D}_i$ with automorphism group G. (In the literature about Beauville surfaces, quasiplatonic surfaces are often called *triangle curves*.) It follows from Belyĭ's Theorem that each X_i, and hence also S, is defined over $\overline{\mathbb{Q}}$, so we have a natural action of the absolute Galois group $\mathbb{G} = \mathrm{Gal}\,\overline{\mathbb{Q}}/\mathbb{Q}$ on Beauville surfaces, induced by its action on dessins.

A finite group G arises in the above way if and only if it has generating triples (x_i, y_i, z_i) for $i = 1, 2$, of orders l_i, m_i and n_i, such that

(1) $x_i y_i z_i = 1$ for each $i = 1, 2$,
(2) $l_i^{-1} + m_i^{-1} + n_i^{-1} < 1$ for each $i = 1, 2$, and
(3) $\Sigma_1 \cap \Sigma_2 = \{1\}$, where Σ_i consists of the conjugates in G of the powers of x_i, y_i and z_i.

We will call such a pair of triples (x_i, y_i, z_i) an *unmixed Beauville structure* for G, or simply a *Beauville structure*, of type $(l_1, m_1, n_1; l_2, m_2, n_2)$, and we will call G a *Beauville group*. Property (1) is equivalent to condition (b), with x_i, y_i and z_i representing the local monodromy permutations for the covering $X_i \to \mathbb{P}^1(\mathbb{C})$ over the three critical values (which, without loss of generality, can be assumed to be $0, 1$ and ∞). Property (2) is equivalent to each X_i having genus at least 2, so that $X_i \cong K_i\backslash\mathbb{H}$ for some surface group K_i; then K_i is a normal subgroup of a triangle group $\Delta_i = \Delta(l_i, m_i, n_i)$ with $\Delta_i/K_i \cong G$. Property (3) states that no non-identity power of x_1, y_1 or z_1 is conjugate in G to a power of x_2, y_2 or z_2; since Σ_i is the set of elements of G with fixed points in X_i (see Exercise 3.8), this is equivalent to G acting freely (that is, without fixed points) on $X_1 \times X_2$.

The groups G satisfying (1) and (2) for some i are simply the automorphism groups of regular dessins $\mathcal{D}_i$ of genus greater than 1, and we have seen many examples of these: indeed, most 2-generator finite groups arise in this way. It is less straightforward to find examples of groups G acting on pairs of dessins $\mathcal{D}_1$ and $\mathcal{D}_2$ in a 'disjoint' way, so that (3) is satisfied. Note that (3) is always satisfied if $l_1m_1n_1$ and $l_2m_2n_2$ are mutually coprime, since this implies that the cyclic subgroups generated by x_1, y_1 and z_1 (and hence also by their conjugates) have orders coprime to those generated by x_2, y_2 and z_2.

Bauer, Catanese and Grunewald have shown in [4] that properties (1) and (3) imply (2). Indeed, if $l_i^{-1} + m_i^{-1} + n_i^{-1} \geq 1$ for some i then $\mathcal{D}_i$ has genus 0 or 1. The automorphism groups G of such dessins are well known (see Sect. 6.2), and by inspection each of them is too 'small' to have two generating triples satisfying condition (3). Thus if G satisfies conditions (1) and (3) then there is no need to verify that it also satisfies condition (2).

11.1.2 Beauville's Original Example

Let $X_1 = X_2 = F_n$, the nth degree Fermat curve, given as a projective algebraic curve by

$$F_n = \{[x, y, z] \in \mathbb{P}^2(\mathbb{C}) \mid x^n + y^n + z^n = 0\}.$$

(In Example 1.5 we defined F_n by the equation $x^n + y^n = z^n$, but here it is more convenient to multiply the coordinate z by an nth root of -1, and to use this equivalent but more symmetric defining equation.) This curve has genus $(n-1)(n-2)/2$, so in order to obtain a Beauville surface we take $n > 3$.

Let $G = \mathbb{Z}_n \oplus \mathbb{Z}_n$, with a faithful action $\rho_1 : G \to \mathrm{Aut}\, X_1 = \mathrm{Aut}\, F_n$ given by

$$(j, k) : [x, y, z] \mapsto [\zeta_n^j x, \zeta_n^k y, z].$$

(Recall our notation $\mathbb{Z}_n := \mathbb{Z}/n\mathbb{Z}$ and $\zeta_n := e^{2\pi i/n}$.) This action of G is induced by taking Δ_1 to be the triangle group $\Delta(n, n, n)$, K_1 to be its commutator subgroup Δ_1' and $X_1 = K_1\backslash\mathbb{H}$ (see Example 5.5). With respect to the basis for $G = \Delta_1/\Delta_1'$ formed by the images of the generators X and Y of Δ_1, the corresponding generating triple for G consists of

$$x_1 = (1, 0), \quad y_1 = (0, 1), \quad z_1 = (-1, -1),$$

generating cyclic subgroups of order n which each fix n points

$$[0, \lambda z, z], \quad [\lambda z, 0, z], \quad [\lambda y, y, 0]$$

where $\lambda^n = -1$. Thus the set of elements with fixed points in this action of G is

$$\Sigma_1 = \langle x_1 \rangle \cup \langle y_1 \rangle \cup \langle z_1 \rangle = \{(j, k) \in G \mid j = 0 \text{ or } k = 0 \text{ or } j = k\}.$$

We now need to define a second action ρ_2 of G on F_n, such that the set Σ_2 of elements with fixed points satisfies $\Sigma_1 \cap \Sigma_2 = \{(0, 0)\}$. Now the matrix

$$A = \begin{pmatrix} 4 & 2 \\ 1 & 1 \end{pmatrix},$$

acting on the left of column vectors, sends x_1, y_1 and z_1 to the triple

$$x_2 = (4, 1), \quad y_2 = (2, 1), \quad z_2 = (-6, -2).$$

This is a generating triple provided $A \in \mathrm{GL}_2(\mathbb{Z}_n)$, that is provided n is odd, so that $\det A = 2$ is a unit in $\mathbb{Z}_n$. In this case, if α denotes the automorphism of G induced by A then we obtain a faithful action

$$\rho_2 = \alpha^{-1} \circ \rho_1 : G \to \mathrm{Aut}\, X_2 = \mathrm{Aut}\, F_n$$

of G on F_n with

$$\Sigma_2 = \langle x_2 \rangle \cup \langle y_2 \rangle \cup \langle z_2 \rangle = \{(j, k) \in G \mid j = 4k \text{ or } j = 2k \text{ or } j = 3k\}.$$

It is now straightforward to check that $\Sigma_1 \cap \Sigma_2 = \{(0, 0)\}$ if and only if n is coprime to 6, so for all such n we obtain a Beauville structure on the group $G = \mathbb{Z}_n \oplus \mathbb{Z}_n$.

In his original example [7, p. 159, Exercise 4], Beauville took $n = 5$, challenging the reader to find a suitable automorphism α in this case, and then to generalise his construction: 'Donner d'autres exemples analogues'. In 2000, Catanese [8, §3] showed that $\mathbb{Z}_n \oplus \mathbb{Z}_n$ is a Beauville group if and only if n is coprime to 6, and in 2005 he, together with Bauer and Grunewald [4, Theorem 3.4], showed that these are the only abelian groups with Beauville structures. The following proof is adapted from theirs:

Theorem 11.1 *An abelian group G is a Beauville group if and only if $G \cong \mathbb{Z}_n \oplus \mathbb{Z}_n$ where $n > 1$ and n is coprime to 6, in which case its Beauville structure has type $(n, n, n; n, n, n)$.*

Proof We have already shown that $\mathbb{Z}_n \oplus \mathbb{Z}_n$ is a Beauville group whenever $n > 1$ and n is coprime to 6. For the converse, suppose that G is an abelian group with a Beauville structure. Being finite and abelian, G is the direct sum $\oplus_p G_p$ of its Sylow

p-subgroups G_p, where p ranges over the primes dividing $|G|$. Each $g \in G$ has the form $g = (g_p)$, where the projection g_p of g into each G_p is a multiple of g. Any generating triple (x, y, z) for G projects onto a generating triple (x_p, y_p, z_p) for G_p; its elements are multiples of x, y and z, so the set Σ_p of non-identity multiples of x_p, y_p and z_p in G_p is contained in the set Σ of all multiples of x, y and z in G.

As a 2-generator finite abelian p-group, G_p has the form $\langle a \rangle \oplus \langle b \rangle$ where a and b have orders p^e and p^f with $e \geq f \geq 0$. The generating triple (x_p, y_p, z_p) must contain at least one element g of order p^e, for otherwise G_p could not have exponent p^e, so $p^{e-1}g \in \Sigma_p$. If $e > f$ then $p^{e-1}G_p = \langle p^{e-1}a \rangle \cong \mathbb{Z}_p$, so $p^{e-1}G_p \subseteq \Sigma_p \subseteq \Sigma$. This applies to all generating triples (x, y, z) for G, so there cannot be a pair of them satisfying condition (3). This contradicts the existence of a Beauville structure on G, so we must have $e = f$. This holds for all primes p dividing $|G|$, so $G_p \cong \mathbb{Z}_{p^e} \oplus \mathbb{Z}_{p^e}$ for each p and hence $G \cong \mathbb{Z}_n \oplus \mathbb{Z}_n$ for some $n > 1$.

For each prime p dividing n, any generating triple (x_p, y_p, z_p) for G_p must have type (p^e, p^e, p^e), so

$$\Sigma_p \cap p^{e-1}G_p = \langle p^{e-1}x_p \rangle \cup \langle p^{e-1}y_p \rangle \cup \langle p^{e-1}z_p \rangle \setminus \{0\},$$

the union of three distinct subgroups of order p in the group $p^{e-1}G_p \cong \mathbb{Z}_p \oplus \mathbb{Z}_p$, excluding 0. Since $\Sigma_p \subseteq \Sigma$, and since G has a Beauville structure, it follows that $p^{e-1}G_p$ must have at least six subgroups of order p. However $\mathbb{Z}_p \oplus \mathbb{Z}_p$ has only $p + 1$ such subgroups, so $p \geq 5$ and hence n is coprime to 6. Since (x_p, y_p, z_p) has type (p^e, p^e, p^e) for each p, it follows that any Beauville structure on G has type $(n, n, n; n, n, n)$. □

We can use Beauville's construction to find examples of non-abelian Beauville groups.

Theorem 11.2 *Let G be a finite group of exponent $n = p^e > 1$ for some prime $p \geq 5$, such that the abelianisation G/G' of G is isomorphic to $\mathbb{Z}_n \oplus \mathbb{Z}_n$. Then G is a Beauville group.*

Proof By Theorem 11.1 G/G' has a Beauville structure $(x_iG', y_iG', z_iG')_{i=1,2}$ of type $(n, n, n; n, n, n)$. Since G is a finite p-group, the Frattini subgroup $\Phi(G)$ of G, which is generated by the commutators and p-th powers in G [25, III.3.14(a)], is the inverse image of $p\mathbb{Z}_n \oplus p\mathbb{Z}_n$ under the natural epimorphism $G \to G/G' \cong \mathbb{Z}_n \oplus \mathbb{Z}_n$. Each triple (x_iG', y_iG', z_iG') maps onto a generating triple for $G/\Phi(G)$, and since elements of $\Phi(G)$ can be deleted from any generating sets for G [25, III.3.2(a)], it follows that each triple (x_i, y_i, z_i) generates G. Since each triple (x_iG', y_iG', z_iG') has type (n, n, n), each of x_i, y_i and z_i has order divisible by n; but this is the exponent of G, so they all have order n. It follows from this that the two generating triples (x_i, y_i, z_i) for G inherit the Beauville property from their images (x_iG', y_iG', z_iG') in G/G'. □

Corollary 11.1 *Let G be a 2-generator finite group of exponent p for some prime $p \geq 5$. Then G is a Beauville group.*

Proof Since G is a 2-generator p-group, $G/\Phi(G) \cong \mathbb{Z}_p \oplus \mathbb{Z}_p$. Since G has exponent p, all p-th powers are the identity, so $\Phi(G) = G'$. Hence $G/G' \cong \mathbb{Z}_p \oplus \mathbb{Z}_p$, so Theorem 11.2 applies. □

The condition that G should be finite cannot be removed from these two results: Adyan's work [1] on the unrestricted Burnside problem shows that for all odd $n \geq 665$ there are infinite 2-generator groups of exponent n. On the other hand, by Kostrikin's solution [29] of the restricted Burnside problem for prime n, and Zelmanov's extension [39, 40] to all n, for each r and n there is a largest finite r-generator group $R(r, n)$ of exponent n, and all others are quotients of it. Even when $r = 2$ (the case we are interested in here), this group tends to be very large: for instance, Havas, Wall, and Wamsley [24] have shown that $|R(2, 5)| = 5^{34}$, Hall and Higman [23] have shown that $|R(2, 6)| = 2^{28} \cdot 3^{25}$, and O'Brien and Vaughan-Lee [33] have shown that $|R(2, 7)| = 7^{20416}$. For a detailed survey of the restricted Burnside problem, see [37].

Since p-groups tend to have many quotients, these results show that there is no shortage of groups satisfying the hypotheses of Theorem 11.2 or Corollary 11.1. The following exercise gives some specific examples.

Exercise 11.1 Let W be the semidirect product of a normal subgroup $B = \mathbb{Z}_n^n$ by a cyclic group A of order n which acts by conjugation on B by cyclically permuting coordinates. (In group-theoretic terminology and notation, this is the *wreath product* $C_n \wr C_n$ of two cyclic groups of order n.) Let $G = W/D$ where D is the diagonal subgroup of B, consisting of the vectors $(b_1, \ldots, b_n)$ with $b_1 = \ldots = b_n$. Show that G is a 2-generator group of exponent n, with abelianisation $G/G' \cong \mathbb{Z}_n \oplus \mathbb{Z}_n$.

These results provide examples of p-groups with Beauville structures for all primes $p \geq 5$. In [14], Fuertes, González-Diez and Jaikin-Zapirain have constructed rather more complicated examples, both of order p^{12}, for $p = 2$ and 3. (Subsequently Barker, Boston and Fairbairn [3] have shown that the smallest Beauville p-groups for these primes have orders 2^7 and 3^5.) Thus there are Beauville p-groups for all primes p.

Exercise 11.2 Do dihedral groups have Beauville structures?

11.1.3 Enumeration of Beauville Surfaces

In this section we will compute the number of non-isomorphic Beauville surfaces obtained from the group $G = \mathbb{Z}_n \oplus \mathbb{Z}_n$ (see [20]). Although the two curves X_1 and X_2 are isomorphic (to the Fermat curve F_n), the actions of G on them differ by an automorphism of G, and different choices for this automorphism may give rise to non-isomorphic Beauville surfaces.

If any abelian group is generated by a pair of elements of orders l and m, then it has order at most lm. It follows that any generating triple for G must have type (n, n, n). The automorphism group $\mathrm{GL}_2(\mathbb{Z}_n)$ of G, consisting of the 2×2 matrices over $\mathbb{Z}_n$ whose determinants are units in $\mathbb{Z}_n$, acts transitively on such triples, so by applying an automorphism we may assume that the first triple in any Beauville structure on G is the *standard triple*

$$x_1 = (1, 0), \quad y_1 = (0, 1), \quad z_1 = (-1, -1),$$

as in Beauville's original example (Sect. 11.1.2.) The second triple x_2, y_2, z_2 then differs from the first by an automorphism of G, induced by a matrix

$$A = \begin{pmatrix} a & b \\ c & d \end{pmatrix} \in \mathrm{GL}_2(\mathbb{Z}_n)$$

(acting on column vectors), so that

$$x_2 = (a, c), \quad y_2 = (b, d), \quad z_2 = (-a - b, -c - d).$$

Now Σ_1 is as defined in the preceding section, so $\langle x_2 \rangle$ has trivial intersection with Σ_1 if and only if a, c and $a - c$ are units in $\mathbb{Z}_n$. The corresponding conditions for $\langle y_2 \rangle$ and $\langle z_2 \rangle$ are that b, d and $b - d$ are units, and that $a + b, c + d$ and $a + b - c - d$ are units. Let $\mathfrak{F}_n$ denote the set of matrices $A \in \mathrm{GL}_2(\mathbb{Z}_n)$ satisfying all these conditions.

Lemma 11.1 *For each $n \geq 1$ we have*

$$|\mathfrak{F}_n| = n^4 \prod_{p|n} \left(1 - \frac{1}{p}\right)\left(1 - \frac{2}{p}\right)\left(1 - \frac{3}{p}\right)\left(1 - \frac{4}{p}\right), \tag{11.1}$$

where p ranges over the distinct primes dividing n.

Proof First suppose that n is a prime p. Then G can be regarded as a 2-dimensional vector space over $\mathbb{F}_p$, and its non-identity cyclic subgroups $\langle (u, v) \rangle$ correspond bijectively to the points u/v in the projective line $\mathbb{P}^1(\mathbb{F}_p) = \mathbb{F}_p \cup \{\infty\}$ over $\mathbb{F}_p$. For instance, the standard triple corresponds to the points $\infty, 0, 1$. Those second triples which form a Beauville structure with this first triple correspond to ordered choices of three points from $\mathbb{P}^1(\mathbb{F}_p) \setminus \{\infty, 0, 1\}$, giving $(p-2)(p-3)(p-4)$ possibilities. Each choice represents $p - 1$ such generating triples, differing from each other by scalar multiplication, so we obtain $(p-1)(p-2)(p-3)(p-4)$ second triples and hence the same number of matrices A in $\mathfrak{F}_p$, as required.

Now suppose that $n = p^e$ for some prime p and integer $e > 1$. The units in $\mathbb{Z}_n$ are the elements mapping onto the units in $\mathbb{Z}_p$ under the natural epimorphism $\mathbb{Z}_n \to \mathbb{Z}_p$. Since this mapping is p^{e-1}-to-1, and each A has four entries, it follows that $|\mathfrak{F}_n| = p^{4(e-1)}|\mathfrak{F}_p|$. This proves the result for prime powers. Since both sides of the equation are multiplicative functions of n (the left-hand side by the Chinese Remainder Theorem), the result follows for all n. □

Note that this formula confirms the result that G yields no Beauville surfaces if n is divisible by 2 or 3. It also shows that $0 \le |\mathfrak{F}_n|/n^4 \le 1$ for all $n \ge 1$.

Exercise 11.3 Find an increasing sequence $(n_k)_{k\in\mathbb{N}}$ of integers n_k coprime to 6 such that $|\mathfrak{F}_{n_k}|/n_k^4 \to 1$ as $k \to \infty$. Do the same again, but now with $|\mathfrak{F}_{n_k}|/n_k^4 \to 0$ as $k \to \infty$.

One can show (see [20]) that two matrices $A, B \in \mathrm{GL}_2(\mathbb{Z}_n)$ yield isomorphic Beauville surfaces if and only if

$$B = PA^{\pm 1}Q,$$

where P and Q are elements of the subgroup

$$\left\{\begin{pmatrix}1&0\\0&1\end{pmatrix}, \begin{pmatrix}0&1\\1&0\end{pmatrix}, \begin{pmatrix}-1&0\\-1&1\end{pmatrix}, \begin{pmatrix}1&-1\\0&-1\end{pmatrix}, \begin{pmatrix}0&-1\\1&-1\end{pmatrix}, \begin{pmatrix}-1&1\\-1&0\end{pmatrix}\right\} \cong S_3$$

of $\mathrm{GL}_2(\mathbb{Z}_n)$ permuting the standard triple. (Think of P and Q as permuting the ramification points on the two curves X_i, and inversion as transposing the curves.) These transformations $A \mapsto B$ of $\mathfrak{F}_n$ form a group W of order 72, a semidirect product of $S_3 \times S_3$ by S_2, with the latter transposing the direct factors by conjugation. (This is the wreath product $S_3 \wr S_2$ of S_3 by S_2.) The isomorphism classes of Beauville surfaces obtained from G therefore correspond to the orbits of W on $\mathfrak{F}_n$. The number $\Theta(n)$ of such orbits (and hence of non-isomorphic Beauville surfaces) is given by the Cauchy-Frobenius counting formula (sometimes known as Burnside's Lemma):

$$\Theta(n) = \frac{1}{|W|}\sum_{w\in W} |\mathrm{Fix}(w)|,$$

where $\mathrm{Fix}(w)$ is the set of matrices $A \in \mathfrak{F}_n$ fixed by $w \in W$.

When $w = 1$ we have $\mathrm{Fix}(w) = \mathfrak{F}_n$, so $|\mathrm{Fix}(w)|$ is given by Lemma 11.1. Of the eight conjugacy classes of non-identity elements of W, only three consist of elements with fixed points in $\mathfrak{F}_n$: there are four elements $g = (g_1, g_2) \in S_3 \times S_3$ with each g_i of order 3, there are six involutions $h \in W \setminus (S_3 \times S_3)$, and there are twelve elements gh of order 6 in $W \setminus (S_3 \times S_3)$, where g and h are commuting elements of the two preceding classes. We therefore have

$$\Theta(n) = \frac{1}{72}\left(|\mathfrak{F}_n| + 4|\mathrm{Fix}(g)| + 6|\mathrm{Fix}(h)| + 12|\mathrm{Fix}(gh)|\right). \tag{11.2}$$

In computing $|\mathrm{Fix}(w)|$ for $w = g, h$ and gh, it is sufficient to assume that n is a prime power p^e, and then use the fact that $|\mathrm{Fix}(w)|$ is a multiplicative function of n. In the prime power case, by choosing explicit elements $g, h \in W$ and then counting matrices A fixed by g, h and gh, we find that

$$|\mathrm{Fix}(g)| = \begin{cases} p^{2e}\left(1-\frac{1}{p}\right)\left(1-\frac{4}{p}\right) \text{ if } p \equiv 1 \bmod 3, \\ p^{2e}\left(1-\frac{1}{p}\right)\left(1-\frac{2}{p}\right) \text{ if } p \equiv 2 \bmod 3, \end{cases}$$

$$|\mathrm{Fix}(h)| = p^{2e}\left(1-\frac{3}{p}\right)\left(1-\frac{5}{p}\right),$$

and

$$|\mathrm{Fix}(gh)| = \begin{cases} 2 \text{ if } p \equiv 1 \bmod 3, \\ 0 \text{ if } p \equiv 2 \bmod 3. \end{cases}$$

The formulae for general n are then found by multiplying these functions as p^e ranges over the prime powers appearing in the factorisation of n. For large n, the formula for $\Theta(n)$ is dominated by the first summand in Eq. (11.2), corresponding to $w = 1$, so we have

$$\Theta(n) \sim \frac{1}{72}|\mathfrak{F}_n| = \frac{n^4}{72}\prod_{p|n}\left(1-\frac{1}{p}\right)\left(1-\frac{2}{p}\right)\left(1-\frac{3}{p}\right)\left(1-\frac{4}{p}\right)$$

as $n \to \infty$.

For instance, taking $n = 5$ we have $|\mathfrak{F}_5| = 24$, $|\mathrm{Fix}(g)| = 12$ and $|\mathrm{Fix}(h)| = |\mathrm{Fix}(gh)| = 0$, so $\Theta(5) = 1$. Thus Beauville's example is the unique Beauville surface obtained from the group $G = \mathbb{Z}_5 \oplus \mathbb{Z}_5$.

Exercise 11.4 Show that for primes $p \equiv 1 \bmod 3$ we have

$$\Theta(p) = \frac{1}{72}\left(p^4 - 10p^3 + 54p^2 - 118p + 154\right).$$

Find the corresponding formula for odd primes $p \equiv 2 \bmod 3$, and evaluate $\Theta(p)$ for $p = 7, 11$ and 13.

11.2 Beauville Structures for Specific Families of Groups

11.2.1 Beauville Surfaces and Simple Groups

In studying which non-abelian finite groups are Beauville groups, it is natural to start with the non-abelian finite simple groups (see Sect. 2.2 for a brief description of them). This is partly because a great deal is already known about how these groups can arise as quotients of triangle groups, and hence which quasiplatonic curves they can act on. In the rest of this chapter we will use *ATLAS* notation [10] for finite simple groups, so in particular $\mathrm{PSL}_2(\mathbb{F}_q)$ is denoted by $L_2(q)$, $\mathrm{SL}_2(\mathbb{F}_q)$ by $SL_2(q)$, and so on. We start with the smallest of the non-abelian finite simple groups, namely A_5.

Theorem 11.3 *The alternating group A_5 is not a Beauville group.*

Proof Each non-identity element of A_5 has order $2, 3$ or 5. If $\{l_i, m_i, n_i\} \subseteq \{2, 3\}$ then the triangle group $\Delta(l_i, m_i, n_i)$ is solvable, so it cannot have the nonsolvable group A_5 as an epimorphic image. It follows that any generating triple (x_i, y_i, z_i) for A_5 must contain an element of order 5. Now the Sylow 5-subgroups of A_5 are cyclic, and are all conjugate to each other, so given any two elements of order 5, each is conjugate to a power of the other. Any two generating triples (x_i, y_i, z_i) therefore fail condition (3), so they cannot form a Beauville structure for A_5. □

Exercise 11.5 Prove that $SL_2(5)$ is not a Beauville group. (Note that $SL_2(5)$ is a double covering of $L_2(5) \cong A_5$; it is sometimes called the *binary icosahedral group*.)

Bauer, Catanese, and Grunewald [4] conjectured that every non-abelian finite simple group except A_5 has a Beauville structure. As evidence, they verified that the following simple groups are Beauville groups:

1. the alternating groups A_n for all sufficiently large n;
2. the projective special linear groups $L_2(p)$ for primes $p > 5$;
3. the Suzuki groups $Sz(2^e)$ for primes $e > 3$;
4. all non-abelian simple groups of order at most 50,000 except A_5.

They also showed that the almost simple group S_n is a Beauville group for all $n \geq 7$, as is the quasisimple group $SL_2(p)$ for all primes $p > 5$. (A group G is *almost simple* if $S \leq G \leq \mathrm{Aut}\, S$ for some non-abelian simple group S; a *quasisimple* group is a perfect central extension of a simple group.)

Subsequently Fuertes and González-Diez [12] have extended these results, showing that A_n is a Beauville group for all $n \geq 6$, as is S_n for all $n \geq 5$. Garion and Penegini [16] have shown that $L_2(q)$ is a Beauville group for every prime power $q > 5$, as are the following simple groups provided $q = p^e$ is sufficiently large: $L_3(q), U_3(q), Sz(2^e), R(3^e)$, and also $G_2(q)$ and ${}^3D_4(q)$ if $p > 3$. Fuertes and Jones [13] have shown that $L_2(q)$ and $SL_2(q)$ are Beauville groups for all prime powers $q > 5$, as are the Suzuki groups $Sz(2^e)$ and the 'small' Ree groups $R(3^e)$ for all odd $e \geq 3$.

Recently Garion, Larsen, and Lubotzky [15] have shown that all but finitely many non-abelian finite simple groups are Beauville groups. Guralnick and Malle [22] have extended this to all such groups except A_5, thus proving the conjecture, while Fairbairn, Magaard, and Parker [11] have extended this further to all finite quasisimple groups except $A_5 \cong L_2(5)$ and its double cover $SL_2(5)$.

In subsequent sections we will construct Beauville structures for the groups A_n and $L_2(q)$, but first it is useful to consider a rather easier case, namely the symmetric groups.

11.2.2 Beauville Structures for Symmetric Groups

We will need to prove that various triples of permutations (x_i, y_i, z_i) generate the symmetric group S_n (or, in the next section, the alternating group A_n). A widely-used method of proving that various permutations generate either the alternating or the symmetric group is to use a theorem of Jordan (see [38, Theorem 13.9]), which states that if a permutation group G of degree n is primitive (preserves no non-trivial equivalence relation) and contains a cycle of prime length $l \leq n-3$, then it contains the alternating group (so $G = A_n$ or S_n). A recent extension of this [27], using the classification of finite simple groups, removes the often restrictive requirement that l should be prime. To apply this method, one must first verify that the given group G is primitive: this is immediate if G is transitive and n is prime, or if G is doubly transitive (easy exercises!). However, in other cases this can be less straightforward, since one has to prove a negative claim, that no non-trivial equivalence relation is invariant under G (or equivalently, under a generating set for G). On the other hand, it is usually much easier to verify that permutations generate a *transitive* group, simply by inspecting their cycle decompositions. One can then apply the following lemma [28, Lemma 6.4(1)]:

Lemma 11.2 *Let H be a transitive subgroup of S_n. If H has an element h with cycle structure c, d for coprime integers $c, d \geq 2$, then $H \geq A_n$.*

Proof First we show that H is primitive, so suppose that it is imprimitive, preserving an equivalence relation with n/b equivalence classes (called blocks) of size b, a proper divisor of n. Let h have cycles C and D, of coprime lengths $c < d$. If C contains a block it is a union of blocks (exercise!), as therefore is D, so b divides both c and d, contradicting their coprimality. The same applies to D, so each block meets both C and D. Thus $\langle h \rangle$ permutes the blocks transitively, so they all meet C in the same number r of points, giving $c = rn/b$. Similarly, $d = sn/b$ for some s, so n/b divides both c and d, again a contradiction. Hence H is primitive.

Since c and d are coprime, h^d is a cycle of length c. Since H is primitive and $1 < c < n/2$, a theorem of Margraff (see [38, Theorem 13.5]) implies that $H \geq A_n$. □

Theorem 11.4 *The symmetric group S_n is a Beauville group for each $n \geq 8$.*

Proof We need two generating triples for S_n. For the first generating triple we follow Example 6.1 in Sect. 6.1, and define

$$x_1 = (1, 2, \ldots, n), \quad y_1 = (1, 2) \quad \text{and} \quad z_1 = (n, n-1, \ldots, 2).$$

As noted there this triple, of type $(n, 2, n-1)$, generates S_n. The corresponding regular dessin is the minimal regular cover of the genus 0 dessin shown in Fig. 6.1. The subset Σ_1 of S_n corresponding to this triple consists of the following permutations:

- the conjugates of powers of x_1, namely those permutations with n/l cycles of length l for some l dividing n;
- the conjugates of y_1, namely the transpositions;
- the conjugates of powers of z_1, with one fixed point and $(n-1)/l$ cycles of length l for some l dividing $n-1$.

We form the second triple by taking

$$x_2 = (1, 2, \ldots, c)(c+1, c+2, \ldots, n) \quad \text{and} \quad y_2 = (1, n)(c+1, c+2)(n-2, n-1),$$

where c and $d := n - c \geq 5$ are chosen so that x_2 satisfies the hypotheses of Lemma 11.2: if $n = 2m + 1$ is odd we take $c = m$, so that x_2 has coprime cycle-lengths c and $d = m + 1$; similarly, if $n = 2m \equiv 0 \bmod 4$ we take $c = m - 1$ and $d = m + 1$, and if $n = 2m \equiv 2 \bmod 4$ we take $c = m - 2$ and $d = m + 2$. (We will use these two triples again in Example 11.3, in Sect. 11.3.2; the corresponding dessins are the minimal regular covers of the genus 0 dessins shown in Fig. 11.1.)

By inspection $\langle x_2, y_2 \rangle$ is transitive, and it contains x_2 and an odd permutation y_2, so by Lemma 11.2 it is S_n. The permutation $z_2 := (x_2 y_2)^{-1}$ has two fixed points and a cycle of length $n-2$, so this triple has type $(c(n-c), 2, n-2)$. The corresponding set Σ_2 consists of the following elements:

- the conjugates of powers of x_2, that is, the permutations with c/k cycles of length k and d/l of length l, where k divides c and l divides d;
- the conjugates of y_2, with three transpositions and $n-6$ fixed points;
- the conjugates of powers of z_2, with two fixed points and $(n-2)/l$ cycles of length l for some l dividing $n-2$.

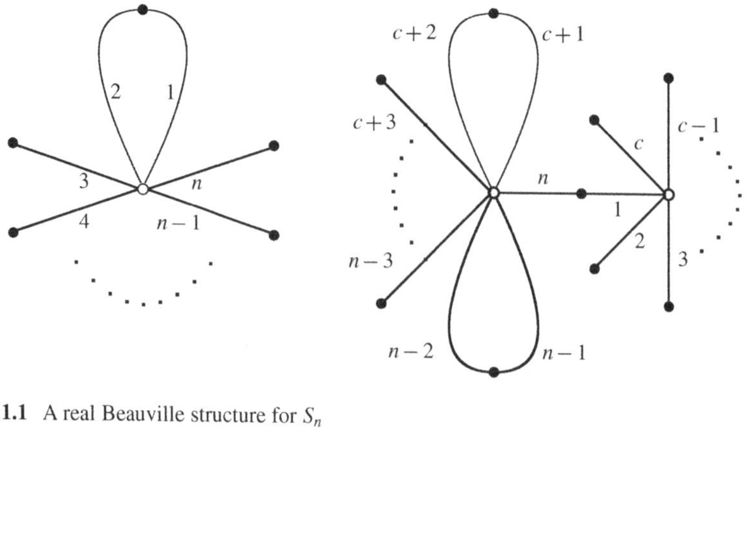

Fig. 11.1 A real Beauville structure for S_n

Since $d \geq c > 2$ it follows by comparing cycle structures that $\Sigma_1 \cap \Sigma_2 = \{1\}$, so this is a Beauville structure for S_n. □

This construction fails for $n = 7$ since then z_1^3 is conjugate to y_2. In fact, S_n is a Beauville group if and only if $n \geq 5$. Bauer, Catanese and Grunewald have given Beauville structures for all sufficiently large n in [5], and Fuertes and González-Diez have dealt with the remaining small values in [12]. As is often the case with infinite families of finite groups, small examples require separate arguments.

Exercise 11.6 Find Beauville structures for S_n in the cases $n = 5, 6$ and 7, and prove that there are none when $n \leq 4$.

11.2.3 Beauville Structures for Alternating Groups

We saw in Theorem 11.3 that A_5 is not a Beauville group. In fact, A_n is a Beauville group if and only if $n \geq 6$. Here we will prove this for $n \geq 8$ as a corollary to a more general theorem based on ideas of Fuertes and González-Diez in [12].

Theorem 11.5 *Suppose that G_0 is a group which has a Beauville structure of type $(2q_1, 2, r_1; 2q_2, 2, r_2)$ with each r_i odd, and which has a subgroup G of index 2. Then G has a Beauville structure of type $(q_1, r_1, r_1; q_2, r_2, r_2)$.*

Proof The given Beauville structure corresponds to a free action of G_0 on the product of two quasiplatonic curves X_i of genera $g_i \geq 2$. Clearly, the subgroup G also acts freely on $X_1 \times X_2$, giving a complex surface $G\backslash(X_1 \times X_2)$ which is a double cover of the Beauville surface $G_0\backslash(X_1 \times X_2)$. It is therefore sufficient to verify that the actions of G on the curves X_i induce coverings $X_i \to G\backslash X_i \cong \mathbb{P}^1(\mathbb{C})$ ramified over three points.

We use the fact that if r is odd then the triangle group $\Delta_0 = \Delta(2q, 2, r)$ has a unique subgroup of index 2, namely the triangle group $\Delta = \Delta(q, r, r)$ (see [35] or Theorem 3.11). Given any epimorphism $\theta : \Delta_0 \to G_0$, the kernel is contained in a subgroup $\theta^{-1}(G)$ of index 2, which must be Δ, so θ restricts to an epimorphism $\Delta \to G$. Applying this argument to the triangle groups $\Delta(2q_i, 2, r_i)$ associated with the actions of G_0 on X_i for $i = 1, 2$, we see that the coverings induced by G have the form

$$X_i \to G\backslash X_i \cong \Delta(q_i, r_i, r_i)\backslash\mathbb{H} \cong \mathbb{P}^1(\mathbb{C}),$$

ramified over three points. The actions of G on the curves X_i therefore correspond to a Beauville structure on G of the required type. □

Corollary 11.2 *The alternating group A_n is a Beauville group for each $n \geq 8$.*

Proof The two triangle groups corresponding to the Beauville structure for S_n constructed in the proof of Theorem 11.4 have types $(n, 2, n-1)$ and $(c(n-c), 2, n-2)$, with c odd if n is even, so (after possibly transposing their first and last periods) we see that their types each have the form $(2q, 2, r)$ with r odd. Theorem 11.5 therefore implies that A_n is a Beauville group. □

Exercise 11.7 Find Beauville structures for A_6 and A_7.

11.2.4 Beauville Structures for $L_2(q)$

The basic facts we will need about the groups $L_2(q) = \mathrm{PSL}_2(\mathbb{F}_q)$ are summarised in Sect. 5.5.

Theorem 11.6 *For each odd prime power $q = p^e \geq 13$ the group $L_2(q)$ is a Beauville group.*

Proof Let $G = L_2(q) = SL_2(q)/\{\pm I\}$. We will choose $X_i, Y_i \in SL_2(q)$ for $i = 1, 2$ so that their images $x_i, y_i \in G$ generate G, and then define z_i to be the image of $Z_i := (X_iY_i)^{-1}$, so that $x_iy_iz_i = 1$. Any matrix $A \in SL_2(q)$ has eigenvalues λ and μ with $\lambda\mu = \det A = 1$, so the value of the trace $\operatorname{tr} A = \lambda + \mu$ of A uniquely determines the pair $\{\lambda, \mu\}$ and hence almost uniquely determines the orders of A and of its image in G. (The only exceptions arise when $\lambda = \mu = \pm 1$, corresponding to elements of G of order 1 or p.) This shows that the orders l_i, m_i and n_i of x_i, y_i and z_i can be controlled by choosing X_i, Y_i and X_iY_i to have appropriate traces.

Let

$$X_1 = \begin{pmatrix} 0 & 1 \\ -1 & a \end{pmatrix} \quad \text{and} \quad Y_1 = \begin{pmatrix} b & -1 \\ 1 & 0 \end{pmatrix}, \quad \text{so} \quad Z_1 = \begin{pmatrix} 1 & 0 \\ b-a & 1 \end{pmatrix}. \tag{11.3}$$

We can choose $a \in \mathbb{F}_q$ so that $\pm a$ is the trace of an element of order $(q+1)/2$ in G, and then put $b = -a$, so that x_1 and y_1 have orders $l_1 = m_1 = (q+1)/2$, while z_1 has order $n_1 = p$.

By inspecting the maximal subgroups of G, listed in Proposition 5.3, we see that since $(q+1)/2 > 5$, so that subgroups isomorphic to A_4, S_4 and A_5 are excluded, no maximal subgroup contains elements of orders $(q+1)/2$ and p. Thus the triple (x_1, y_1, z_1) generates G.

We will choose $X_2, Y_2 \in SL_2(q)$ so that x_2, y_2 and z_2 have orders $l_2 = m_2 = n_2 = (q-1)/2$. It then follows from Proposition 5.3 that x_2 and y_2 generate G provided they have no common fixed point in $\mathbb{P}^1(\mathbb{F}_q)$. Since $l_1m_1n_1$ is coprime to $l_2m_2n_2$ we therefore have a Beauville structure on G. To achieve this, let

$$X_2 = \begin{pmatrix} c & 0 \\ 0 & c^{-1} \end{pmatrix} \quad \text{and} \quad Y_2 = \begin{pmatrix} x & y \\ z & w \end{pmatrix}, \quad \text{so} \quad Z_2 = \begin{pmatrix} c^{-1}w & -cy \\ -c^{-1}z & cx \end{pmatrix}, \tag{11.4}$$

where $xw - yz = 1$. We choose c to be a primitive root for $\mathbb{F}_q$, so $l_2 = (q-1)/2$. If we choose x and w so that $x + w = c + c^{-1}$, then $\operatorname{tr} Y_2 = \operatorname{tr} X_2$ and so $m_2 = l_2$. Now

$$\operatorname{tr} Z_2 = (c - c^{-1})x + (c + c^{-1})c^{-1},$$

with $c - c^{-1} \neq 0$ since $c \neq \pm 1$, so for a fixed X_2 there is a bijection between values of x in $\mathbb{F}_q$ and of $\operatorname{tr} Z_2$. The fixed points of x_2 are 0 and ∞, and y_2 fixes these as $y = 0$ or $z = 0$ respectively, so we need to choose Y_2 so that $yz \neq 0$, that is, $xw \neq 1$. Since $x + w = \operatorname{tr} X_2$ we have $xw = 1$ if and only if $\{x, w\} = \{c, c^{-1}\}$, so by taking $x \neq c^{\pm 1}$ we can obtain any value for $\operatorname{tr} Z_2$ except $c^2 + c^{-2}$ and 2. In particular, we can choose x so that $\operatorname{tr} Z_2 = \operatorname{tr} X_2$, so $n_2 = l_2$ as required. □

One can extend this result to all prime powers $q > 5$. If $q = 11$ then the above construction of the generating triple (x_1, y_1, z_1) is still valid, but a triple (x_2, y_2, z_2) of elements of order $(q-1)/2 = 5$ could generate a subgroup $H \cong A_5$. However, a simple calculation within A_5 shows that x_2, y_2 and z_2 would be conjugate in H and hence in G, so a triple such as

$$x_2 = \pm\begin{pmatrix} 2 & 0 \\ 0 & 6 \end{pmatrix}, \quad y_2 = \pm\begin{pmatrix} 0 & 1 \\ -1 & -3 \end{pmatrix}, \quad z_2 = \pm\begin{pmatrix} 4 & -2 \\ -5 & 0 \end{pmatrix}, \tag{11.5}$$

all of order 5 but with different traces ± 3, ± 3 and ± 4, must generate G. Thus $L_2(11)$ is a Beauville group.

The case $q = 7$ is covered by the proof by Bauer, Catanese, and Grunewald in [4] that $L_2(p)$ is a Beauville group for each prime $p > 5$. Since $L_2(9)$ is isomorphic to A_6 (through its action on the six cosets of a subgroup isomorphic to A_5), the case $q = 9$ is covered by the result of Fuertes and González-Diez [12] that A_n is a Beauville group for each $n \geq 6$ (also see Exercise 11.7).

Exercise 11.8 Deal with the cases $q = 7$ and $q = 9$ yourself.

Exercise 11.9 Adapt the proof of Theorem 11.6 to deal with the case where $q = 2^e \geq 8$, so that $G = SL_2(q)$.

11.3 Further Properties of Beauville Surfaces

11.3.1 Invariants of a Beauville Surface

Two general theorems of topology state that if X and Y are topological spaces with Euler characteristics $\chi(X)$ and $\chi(Y)$, then $X \times Y$ has Euler characteristic $\chi(X)\chi(Y)$, and that if X is an n-sheeted unbranched covering of Y then $\chi(X) = n\chi(Y)$. It follows immediately that a Beauville surface $S = G\backslash(X_1 \times X_2)$ has Euler characteristic

$$\chi(S) = \frac{\chi(X_1)\chi(X_2)}{|G|} = \frac{4(g_1 - 1)(g_2 - 1)}{|G|},$$

where each X_i has genus g_i.

Example 11.1 The Fermat curves F_n have genus $(n-1)(n-2)/2$, so the Beauville surfaces S based on them, with $G = \mathbb{Z}_n \oplus \mathbb{Z}_n$ (see Sect. 11.1.2), have Euler characteristic

$$\chi(S) = (n-3)^2.$$

Another general theorem [2, Theorem 5.13] states that if X is a simply connected and path-connected topological space, and H is a group of self-homeomorphisms acting properly discontinuously and without fixed-points on X (that is, so that each $x \in X$ has a neighbourhood U with $g(U) \cap U = \emptyset$ for all $g \neq 1$ in G), then the fundamental group $\pi_1(H\backslash X)$ of the quotient space $H\backslash X$ is isomorphic to H. We can use this to describe the fundamental group of a Beauville surface $S = G\backslash(X_1 \times X_2)$.

Each curve X_i has the form $X_i \cong K_i\backslash\mathbb{H}$ with $G \cong \Delta_i/K_i$ for some hyperbolic triangle group Δ_i and normal surface subgroup K_i of Δ_i, so that $X_1 \times X_2 \cong (K_1 \times K_2)\backslash\mathbb{H}^2$. Now $\mathbb{H}^2$ is simply connected (since $\mathbb{H}$ is), and $K_1 \times K_2$ acts properly discontinuously, without fixed points, on $\mathbb{H}^2$ (since each K_i acts in this way on $\mathbb{H}$), so

$$\pi_1(X_1 \times X_2) \cong K_1 \times K_2 \cong \pi_1 X_1 \times \pi_1 X_2.$$

The Beauville structure on G gives us epimorphisms $\theta_i : \Delta_i \to G$ for $i = 1, 2$, and by the construction of S we have $S \cong K\backslash\mathbb{H}^2$ where

$$K := \{(g_1, g_2) \in \Delta_1 \times \Delta_2 \mid \theta_1(g_1) = \theta_2(g_2)\}$$

acts freely on $\mathbb{H}^2$ (by the Beauville property of G), so

$$\pi_1 S \cong K.$$

There is an epimorphism $K \to G$, $(g_1, g_2) \mapsto \theta_1(g_1) = \theta_2(g_2)$, with kernel $K_1 \times K_2$, so $\pi_1 S$ has a normal subgroup isomorphic to $\pi_1 X_1 \times \pi_1 X_2$, with quotient isomorphic to G. This normal subgroup corresponds to the regular covering of S by $X_1 \times X_2$, with G the group of covering transformations.

We have standard presentations for the triangle groups Δ_i, and hence an obvious finite presentation for $\Delta_1 \times \Delta_2$. This group contains K with finite index, so in theory one could use the Reidemeister-Schreier method [31, §II.4] to find a finite presentation for K, and thus for $\pi_1 S$. However, the index is equal to $|G|$, which tends to be large, so in practice this is not straightforward. In general, if this method is applied to a subgroup of index m in a group defined by r generators and s relations, one obtains a presentation with $1 + m(r-1)$ generators and ms relations. Here each Δ_i has two generators and three relations, so their product has four generators and ten relations (three from each factor, and four making the two pairs of generators commute), giving a presentation for $\pi_1 S$ with $1+3|G|$ generators and $10|G|$ relations. Even in the smallest case, where $G = \mathbb{Z}_5 \oplus \mathbb{Z}_5$, this gives 76 generators and 250 relations, though some may be redundant.

Catanese [9, Theorem 3.3] has proved that Beauville surfaces are rigid, that is, if S is a Beauville surface and S' is any complex surface with $\chi(S') = \chi(S)$ and $\pi_1 S' \cong \pi_1 S$, then S and S' are diffeomorphic (isomorphic as differentiable real manifolds). By [4, 18, 19] it follows that S' is biholomorphic to S or its complex conjugate surface $\overline{S}$, obtained by applying complex conjugation to the coefficients of the polynomials defining S. (A *biholomorphic map* is a holomorphic bijection; the inverse of such a map is also holomorphic.)

This is in complete contrast with the situation for complex algebraic *curves*: for each $g > 0$ there are uncountably many isomorphism classes of Riemann surfaces of genus g, all mutually homeomorphic, and hence with the same Euler characteristic $2 - 2g$, and with isomorphic fundamental groups

$$\Pi_g = \langle a_1, b_1, \ldots, a_g, b_g \mid \prod_{i=1}^{g} [a_i, b_i] = 1 \rangle.$$

Another invariant of a Beauville surface is its automorphism group $\mathrm{Aut}\, S$; here we outline some results explained in more detail in [26]. It follows from the rigidity of Beauville surfaces that if $S = G \backslash (X_1 \times X_2)$ as before, then any automorphism α of S lifts to an automorphism $\overline{\alpha}$ of $X_1 \times X_2$, which either preserves the curves X_i or transposes them via suitable isomorphisms between them. It follows that $\mathrm{Aut}\, S$ has a subgroup of index at most 2 consisting of *direct automorphisms*, those for which $\overline{\alpha}$ preserves the curves X_i.

The action of G on each X_i induces an action of $G \times G$ on $X_1 \times X_2$. Only the elements of $Z \times Z$ (where $Z := Z(G)$, the centre of G) are compatible with taking the quotient by G (exercise!). By the Beauville construction, the diagonal subgroup of $Z \times Z$ acts trivially on S, so there is a faithful action of Z on S (for instance, fixing one X_i and acting naturally on the other). This is the group $\mathrm{Inn}\, S \cong Z$ of *inner automorphisms* of S; it is finite and abelian, and the following result shows that these are the only restrictions on its structure:

Theorem 11.7 *Every finite abelian group A is isomorphic to* $\mathrm{Inn}\, S$ *for some Beauville surface S.*

Outline Proof We need to find a Beauville group G with $Z(G) \cong A$. If $|A| = 1$ we can take $G = S_n$ for any $n \geq 5$, so let us assume that $|A| \geq 2$ and write

$$A \cong C_{m_1} \times \cdots \times C_{m_k}$$

where each $m_i \geq 2$. The special linear group $SL_n(q)$ has centre

$$Z(SL_n(q)) = \{\lambda I_n \mid \lambda \in \mathbb{F}_q,\ \lambda^n = 1\} \cong C_m$$

where $m = \gcd(n, q-1)$. By a result of Lucchini [30], given any integer $t \geq 7$, there exists $d_t \in \mathbb{N}$ such that $SL_n(q)$ is a quotient of $\Delta(2, 3, t)$ for all $n \geq d_t$ and all prime powers q. Moreover, if t is prime then this quotient is smooth, that is, a quotient by a torsion free normal subgroup, so it follows from Exercise 11.10 that $SL_n(q)$, which has no subgroups of index 2, is also a smooth quotient of $\Delta(t,t,t)$.

Given any prime $t \geq 7$, one can choose distinct pairs (n_j, q_j) for $j = 1, \ldots, k$, with each q_j a prime power, such that

1. $\gcd(n_j, q_j - 1) = m_j$ for $j = 1, \ldots, k$;
2. $n_j \geq d_t$ for $j = 1, \ldots, k$;
3. the quotient groups $L_{n_j}(q_j)$ are mutually non-isomorphic non-abelian simple groups.

(For (3), one simply has to avoid the solvable groups $L_2(2)$ and $L_2(3)$, and the isomorphisms $L_2(4) \cong L_2(5)$ and $L_2(7) \cong L_3(2)$.) By (2) the groups $SL_{n_j}(q_j)$ are all quotients of $\Delta(t,t,t)$. Now (3) implies that no two of them have a non-trivial epimorphic image in common, so a simple argument by induction on k (exercise!) shows that their cartesian product

$$G := SL_{n_1}(q_1) \times \cdots \times SL_{n_k}(q_k)$$

is also a quotient of $\Delta(t,t,t)$. If we apply this argument to two distinct primes $t_1, t_2 \geq 7$, then we see that G has a Beauville structure of type $(t_1, t_1, t_1; t_2, t_2, t_2)$, so it is a Beauville group. By (1) it has centre $Z(G) \cong A$, as required. □

Exercise 11.10 Show that if a group G is a smooth quotient of $\Delta(2,3,t)$, where t is odd, and G has no subgroups of index 2, then G is also a smooth quotient of $\Delta(t,t,t)$.

One can show that $\operatorname{Inn} S$ is a normal subgroup of $\operatorname{Aut} S$ with

$$\operatorname{Out} S := \operatorname{Aut} S / \operatorname{Inn} S$$

isomorphic to a subgroup of the wreath product

$$W = S_3 \wr S_2 \cong (S_3 \times S_3) \rtimes S_2 \,.$$

(We first met this group W in a similar context in Sect. 11.1.3.) Here the complement S_2 acts by transposing the curves X_i if there are suitable isomorphisms between them, and each copy of S_3 permutes the critical values of a curve X_i, if this is compatible with the Belyĭ function $\beta_i : X_i \to \mathbb{P}^1(\mathbb{C})$. The first condition requires the period triples (l_i, m_i, n_i) of the Beauville structure to be equal (up to a permutation of the periods), and the second requires (l_i, m_i, n_i) to have repeated periods. In most cases $\operatorname{Out} S$ is trivial, but there are examples, for instance with $G = L_2(5^2) \times L_2(3^3)$ and type $(13, 13, 13; 13, 13, 13)$, where $\operatorname{Out} S \cong W$: see [26] for details.

Since W is finite and solvable, of derived length 3, and $\operatorname{Inn} S$ is finite and abelian, we have:

Corollary 11.3 *If S is a Beauville surface then* $\operatorname{Aut} S$ *is a finite solvable group, of derived length at most* 4.

Exercise 11.11 Describe the automorphism groups of the Beauville surfaces constructed from Fermat curves in Sect. 11.1.3.

11.3.2 Real Beauville Surfaces

The consideration of rigidity in the preceding section raises two questions concerning a Beauville surface S and its complex conjugate $\overline{S}$:

1. Is there a biholomorphic map $\sigma : S \to \overline{S}$?
2. Is S real, that is, does S have a model defined over $\mathbb{R}$? This is equivalent to the existence of a biholomorphic map $\sigma : S \to \overline{S}$ such that $\overline{\sigma} \circ \sigma$ is the identity.

(See Exercise 4.11 for the corresponding property for curves.)

In [4, Proposition 3.11] Bauer, Catanese and Grunewald give necessary and sufficient conditions, in terms of the corresponding Beauville structure on G, for a Beauville surface $S = G\backslash(X_1 \times X_2)$ to have these properties. If a Beauville structure consists of triples (x_i, y_i, z_i) for $i = 1, 2$, they define its *inverse* to be the Beauville structure with triples $(x_i^{-1}, x_i z_i, z_i^{-1})$ for $i = 1, 2$. They then define a certain group $A_{\mathbb{U}}(G)$ of transformations of the Beauville structures on G, and they show that S has property (1) or (2) if and only if the corresponding Beauville structure is respectively sent to or interchanged with its inverse by some element of $A_{\mathbb{U}}(G)$.

Example 11.2 The group $A_{\mathbb{U}}(G)$ contains Aut G, with its natural action on Beauville structures. Every abelian group G has an automorphism inverting all its elements, so the Beauville surfaces with abelian Beauville groups, namely those based on Fermat curves considered in Sect. 11.1.2, are real.

The full definition of the group $A_{\mathbb{U}}(G)$ in [4, §3.1] is rather complicated, so here we will merely state and use a simpler necessary and sufficient condition for S to be real, applicable to Beauville structures satisfying some fairly weak restrictions (see [4, Corollary 3.13]).

We say that a triple (x, y, z) is *inverted* by an automorphism α of G if α sends two of x, y and z to their inverses. (No automorphism can invert all three of x, y and z, unless they commute with each other.) In this case, any two of x, y and z can be inverted by an automorphism: for instance, if α inverts x and y, then by following α with conjugation by x^{-1} or y we obtain automorphisms inverting x and z, or y and z. This condition is equivalent to the regular dessin corresponding to the triple being symmetric, that is, having an orientation-reversing automorphism. Then we have the following:

Theorem 11.8 *Let* $S = G\backslash(X_1 \times X_2)$ *be a Beauville surface corresponding to a Beauville structure* $(x_i, y_i, z_i)_{i=1,2}$ *of type* $(l_1, m_1, n_1; l_2, m_2, n_2)$ *on a group* G. *Suppose that for each* i *the integers* l_i, m_i *and* n_i *are mutually distinct, and that* *the sets $\{l_i, m_i, n_i\}$ for $i = 1, 2$ are distinct. Then the following are equivalent:*

1. *the triples (x_i, y_i, z_i) are inverted by automorphisms α_i of G for $i = 1, 2$ such that $\alpha_1 \circ \alpha_2^{-1}$ is an inner automorphism;*
2. *there is a biholomorphic map $S \to \overline{S}$;*
3. *S is real.*

The restrictions on the periods l_i, m_i and n_i prevent $A_{\mathbb{U}}(G)$ from containing certain transformations which either permute the members of a triple or transpose the two triples; this ensures that in proving the theorem one has to deal with a smaller group $A_{\mathbb{U}}(G)$ than in the general case, leading to a simpler statement of the theorem.

The Beauville structures for the groups A_n and $L_2(q)$ which were constructed in Corollary 11.2 and Theorem 11.6 do not satisfy these restrictions since they each involve at least one triple with repeated periods. However, the Beauville structure for S_n constructed in Theorem 11.4 does satisfy them, so Theorem 11.8 can be applied in this case, as shown in the following example:

Example 11.3 Let $G = S_n$ with $n \geq 8$, and let (x_i, y_i, z_i) for $i = 1, 2$ be the generating triples of type $(n, 2, n-1)$ and $(c(n-c), 2, n-2)$ used in the proof of Theorem 11.4. Specifically, for the first triple we take

$$x_1 = (1, 2, \ldots, n) \quad \text{and} \quad y_1 = (1, 2),$$

and for the second we take

$$x_2 = (1, 2, \ldots, c)(c+1, c+2, \ldots, n) \quad \text{and} \quad y_2 = (1, n)(c+1, c+2)(n-2, n-1),$$

where $c = m$ if $n = 2m+1$ is odd, $c = m-1$ if $n = 2m \equiv 0 \bmod 4$, and $c = m-2$ if $n = 2m \equiv 2 \bmod 4$. In each case we take $z_i = (x_i y_i)^{-1}$.

The resulting Beauville structure has type $(n, 2, n-2; c(n-c), 2, n-1)$, which satisfies the hypotheses of Theorem 11.8, so we look for a pair of automorphisms α_i of S_n inverting these two triples. The corresponding dessins are the minimal regular covers $\hat{\mathcal{D}}_i$ of the genus 0 dessins $\mathcal{D}_i$ shown in Fig. 11.1; these dessins $\mathcal{D}_i$ are visibly invariant under reflections, inverting the generators x_i and y_i, and hence so are the dessins $\hat{\mathcal{D}}_i$, so we can take α_1 and α_2 to be the inner automorphisms of S_n induced by the involutions

$$(1, 2)(3, n)(4, n-1)\ldots \quad \text{and} \quad (2, c)(3, c-1)\ldots(c+1, n-1)(c+2, n-2)\ldots$$

corresponding to these reflections. Then $\alpha_1 \circ \alpha_2^{-1}$ is also an inner automorphism, so by Theorem 11.8 the corresponding Beauville surface is real.

Example 11.4 Let us now make a small change to the preceding example, using the same permutations x_1, y_1 and x_2 as before, but redefining

$$y_2 = (1, n)(c+1, c+2)(n-3, n-2)$$

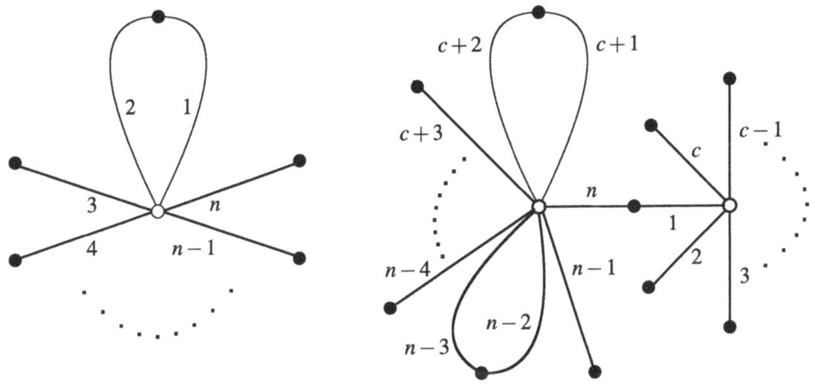

Fig. 11.2 A non-real Beauville structure for S_n

(so we now need $n \geq 10$). As before we obtain a Beauville structure on S_n, of type $(n, 2, n-1; c(n-c), 2, n-2)$ which satisfies the hypotheses of Theorem 11.8. The corresponding dessins are now the minimal regular covers $\hat{\mathscr{D}}_i$ of the genus 0 dessins $\mathscr{D}_i$ shown in Fig. 11.2.

In this case, however, it is easy to see that the symmetry of the second dessin is lost, and no inner automorphism of S_n inverts both x_2 and y_2 (exercise: give an algebraic proof of this!). Since the only automorphisms of S_n, for $n \neq 6$, are inner (see [25, Satz II.5.5(a)]), it follows from Theorem 11.8 that the corresponding Beauville surface S is not biholomorphic to $\overline{S}$; in particular, it is not real.

11.3.3 The Action of the Absolute Galois Group on Beauville Surfaces

If two dessins are conjugate under the absolute Galois group $\mathbb{G} = \mathrm{Gal}\,\overline{\mathbb{Q}}/\mathbb{Q}$, then by Theorem 4.11 they have the same genus, so their underlying surfaces are homeomorphic to each other. This may seem rather surprising, since the action of an element of $\mathbb{G}$ (other than complex conjugation or the identity) is very far from continuous.

This property of complex varieties is restricted to those of dimension 1, that is, to algebraic curves. In 1964 Serre [34] gave an example of a pair of complex surfaces, defined over $\overline{\mathbb{Q}}$, which are conjugate under $\mathbb{G}$ but have non-isomorphic fundamental groups, so they are not homeomorphic to each other. Since then, Bauer, Catanese, and Grunewald [4] have produced examples of Beauville surfaces with this property. We will give a simple example here.

If S is a Beauville surface $G\backslash(X_1 \times X_2)$ then each X_i is a quasiplatonic curve, so it is defined over $\overline{\mathbb{Q}}$, as is the corresponding Belyĭ function $\beta_i : X_i \to \mathbb{P}^1(\mathbb{C})$ and hence also the group G inducing this regular covering. It follows that there is a natural action of $\mathbb{G}$ on Beauville surfaces, each $\sigma \in \mathbb{G}$ sending S to $S^\sigma = G^\sigma\backslash(X_1^\sigma \times X_2^\sigma)$ (it is straightforward to check that this is again a Beauville surface).

To explain our example, we need a sufficient condition for two Beauville surfaces to be non-homeomorphic. The following theorem of Catanese [8] essentially states that the topology of a Beauville surface determines the quasiplatonic curves X_i (up to complex conjugation) and the group G. (See also [18, Theorem 6] for a proof by González-Diez and Torres-Teigell using uniformisation theory.)

Theorem 11.9 *If* $S = G\backslash(X_1 \times X_2)$ *and* $S' = G'\backslash(X_1' \times X_2')$ *are Beauville surfaces with* $\pi_1 S \cong \pi_1 S'$ *then* $G \cong G'$ *and, possibly after transposing factors, each* X_i *is isomorphic to* X_i' *or* $\overline{X_i'}$.

In particular, these conclusions hold if S and S' are homeomorphic to each other.

The example we use is based on work of González-Diez and Torres-Teigell [18]. Extending a result of Macbeath [32] for the case $n = 7$, Streit [36] showed that for each $n \geq 7$ and each prime $p \equiv \pm 1 \bmod n$ the group $G := L_2(p)$ has (up to automorphisms) $\phi(n)/2$ generating triples (x_1, y_1, z_1) of type $(2, 3, n)$, one for each conjugacy class of elements of order n in G. He showed that the corresponding $\phi(n)/2$ quasiplatonic curves are defined over the field $\mathbb{Q}(\zeta_n) \cap \mathbb{R} = \mathbb{Q}(\zeta_n + \zeta_n^{-1})$, and that they form a single orbit under the action of $\mathbb{G}$. We will take X_1 to be one of these curves. Since the moduli field is real, we have $\overline{X}_1 \cong X_1$.

We will now construct a second generating triple (x_2, y_2, z_2) of type (p, p, p) for G. The elements of order p in G are the non-identity elements of trace ± 2, forming a single orbit under $\mathrm{Aut}\, G = PGL_2(p)$. Applying a suitable automorphism, we may therefore assume that

$$x_2 = \pm\begin{pmatrix} 1 & 1 \\ 0 & 1 \end{pmatrix} \quad \text{and} \quad y_2 = \pm\begin{pmatrix} a & b \\ c & 2-a \end{pmatrix}$$

for some $a, b, c \in \mathbb{F}_p$ with $a(2-a) - bc = 1$. Then

$$z_2 := (x_2 y_2)^{-1} = \pm\begin{pmatrix} 2-a & a-b-2 \\ -c & a+c \end{pmatrix},$$

so we require $c = 0$ or $c = -4$. If $c = 0$ then x_2 and y_2 both fix ∞ in the action of G on $\mathbb{P}^1(\mathbb{F}_p)$, so they cannot generate G. We therefore take $c = -4$, so $b = (a-1)^2/4$. This gives us p triples, one for each $a \in \mathbb{F}_p$. Now x_2 generates its own centraliser in $\mathrm{Aut}\, G$, and it is straightforward to check that this centraliser acts transitively on the p possible choices for y_2, so the corresponding triples are equivalent under $\mathrm{Aut}\, G$. Since x_2 and y_2 do not commute, it follows from Dickson's description of the maximal subgroups of G (see Sect. 5.5) that each triple generates G provided $p \geq 7$. Thus G has, up to automorphisms, a unique generating triple (x_2, y_2, z_2) of type (p, p, p). By this uniqueness, the corresponding quasiplatonic curve X_2 is defined over $\mathbb{Q}$. Since p is coprime to $2, 3$ and n, the triples (x_i, y_i, z_i) for $i = 1, 2$ form a Beauville structure for G, so let $S = G\backslash(X_1 \times X_2)$ be the corresponding Beauville surface.

By its construction, X_1 is a member of an orbit of $\mathbb{G}$ consisting of $\phi(n)/2$ mutually non-isomorphic curves $X_1^\sigma = \overline{X_1^\sigma}$, $\sigma \in \mathbb{G}$. None of these is isomorphic to the curve $X_2 = \overline{X}_2$: for instance, one can easily check that they have different genera. It follows that S is contained in an orbit of at least $\phi(n)/2$ mutually non-homeomorphic Beauville surfaces. By suitable choices of n one can make this orbit arbitrarily large. A modification of this construction in [21], using groups $G = PGL_2(p)$ instead of $L_2(p)$, provides exact values, rather than lower bounds, for the size of the orbits, and also gives m-parameter families of Galois conjugate but non-homeomorphic complex surfaces for each $m \geq 0$.

These examples illustrate a curious phenomenon, first observed by Serre in [34]. Although these complex surfaces have mutually non-isomorphic fundamental groups $\pi_1 S$, their algebraic fundamental groups $\pi_1^{\mathrm{alg}} S$ are all isomorphic. This group $\pi_1^{\mathrm{alg}} S$ is defined to be the profinite completion

$$\widehat{\pi_1 S} = \lim_{\leftarrow} (\pi_1 S)/N$$

of $\pi_1 S$, that is, the projective limit of all the finite quotients $(\pi_1 S)/N$ of $\pi_1 S$. The explanation of this Galois invariance is that the normal subgroups N of finite index in $\pi_1 S$, together with their quotient groups, correspond to the finite regular coverings of S and their covering groups; these are algebraically defined, so they and their projective limit are preserved by the action of $\mathbb{G}$, whereas the topologically defined group $\pi_1 S$ is not.

We close by referring back to Remark 4.2 in Sect. 4.2.4, where we noted that González-Diez and Jaikin-Zapirain [17] have proved that $\mathbb{G}$ acts faithfully on regular dessins. Now Beauville surfaces are constructed from pairs of regular dessins, and it follows from this, together with the rigidity results mentioned earlier, that $\mathbb{G}$ also acts faithfully on the set of all Beauville surfaces; for another proof and further consequences, see [6]. In principle, therefore, one can 'see' the whole of the Galois theory of algebraic number fields by looking at the actions of $\mathbb{G}$ on quasiplatonic curves and on Beauville surfaces.

References

1. Adyan, S.I.: The Burnside problem and identities in groups (Russian). Izdat. Nauka, Moscow (1975)
2. Armstrong, M.A.: Basic Topology. Springer, Berlin (1983)
3. Barker, N., Boston, N., Fairbairn, N.: A note on Beauville p-groups. Exp. Math. **21**(3), 298–306 (2012)
4. Bauer, I., Catanese, F., Grunewald, F.: Beauville surfaces without real structures I. In: Geometric Methods in Algebra and Number Theory. Progress in Mathematics, vol. 235, pp. 1–42. Birkhäuser, Boston (2005)
5. Bauer, I., Catanese, F., Grunewald, F.: Chebycheff and Belyi polynomials, dessins d'enfants, Beauville surfaces and group theory. Mediterr. J. Math. **3**, 121–146 (2006)

6. Bauer, I., Catanese, F., Grunewald, F.: Faithful actions of the absolute Galois group on connected components of moduli spaces. Invent. Math. **199**, 859–888 (2015)
7. Beauville, A.: Surfaces algébriques complexes, Astérisque 54, Soc. Math. France, Paris (1978). English translation: Complex algebraic surfaces. London Mathematical Society Student Texts Series, vol. 34 (2nd edn.). Cambridge University Press, Cambridge (1996)
8. Catanese, F.: Fibred surfaces, varieties isogenous to a product and related moduli spaces. Am. J. Math. **122**, 1–44 (2000)
9. Catanese, F.: Moduli spaces of surfaces and real structures. Ann. Math. **158**, 577–592 (2003)
10. Conway, J.H., Curtis, R.T. Norton, S.P., Parker, R.A., Wilson, R.A.: ATLAS of Finite Groups. Clarendon Press, Oxford (1985)
11. Fairbairn, B,T., Magaard, K., Parker, C.W.: Generation of finite quasisimple groups with an application to groups acting on Beauville surfaces. Proc. Lond. Math. Soc. (3) **107**, 744–798 (2013)
12. Fuertes, Y., González-Diez, G.: On Beauville structures on the groups S_n and A_n. Math. Z. **264**, 959–968 (2010)
13. Fuertes, Y., Jones, G.A.: Beauville surfaces and finite groups. J. Algebra **340**, 13–27 (2011)
14. Fuertes, Y., González-Diez, G., Jaikin-Zapirain, A.: On Beauville surfaces. Groups Geom. Dyn. **5**, 107–119 (2011)
15. Garion, S., Larsen, M., Lubotzky, A.: Beauville surfaces and finite simple groups. J. Reine Angew. Math. **666**, 225–243 (2012)
16. Garion, S., Penegini, M.: New Beauville surfaces and finite simple groups. Manuscripta Math. **142**, 391–408 (2013)
17. González-Diez, G., Jaikin-Zapirain, A.: The absolute Galois group acts faithfully on regular dessins and on Beauville surfaces. Proc. Lond. Math. Soc. (3) **111**(4), 775–796 (2015)
18. González-Diez, G., Torres-Teigell, D.: Non-homeomorphic Galois conjugate Beauville structures on $PSL(2, p)$. Adv. Math. **229**, 3096–3122 (2012)
19. González-Diez, G., Torres-Teigell, D.: An introduction to Beauville surfaces via uniformization. Contemp. Math. **575**, 123–151 (2012)
20. González-Diez, G., Jones, G.A., Torres-Teigell, D.: Beauville surfaces with abelian Beauville group. Math. Scand. **114**, 191–204 (2014)
21. González-Diez, G., Jones, G.A., Torres-Teigell, D.: Arbitrarily large Galois orbits of non-homeomorphic surfaces. arXiv:1110:4930 [math.AG] (2011). Accessed 20 Jan 2015
22. Guralnick, R., Malle, G.: Simple groups admit Beauville structures. J. Lond. Math. Soc. (2) **85**, 694–721 (2012)
23. Hall, P., Higman, G.: On the p-length of p-soluble groups and reduction theorems for Burnside's problem. Proc. Lond. Math. Soc. (3) **6**, 1–42 (1956)
24. Havas, G., Wall, G.E., Wamsley, J.W.: The two generator restricted Burnside group of exponent five. Bull. Aust. Math. Soc. **10**, 459–470 (1974)
25. Huppert, B.: Endliche Gruppen I. Springer, Berlin (1967)
26. Jones, G.A.: Automorphism groups of Beauville surfaces. J. Group Theory **16**, 353–381 (2013)
27. Jones, G.A.: Primitive permutation groups containing a cycle. Bull. Aust. Math. Soc. **89**, 159–165 (2014)
28. Jones, G.A.: Characteristically simple Beauville groups, I: cartesian powers of alternating groups. In: Aravinda, C.S., Goldman, W.M., et al. (eds.) Geometry, Groups and Dynamics. Contemporary Mathematics, vol. 629, pp. 289–306. Amer. Math. Soc., Providence (2015)
29. Kostrikin, A.I.: The Burnside problem (Russian). Izv. Akad. Nauk SSSR Ser. Mat. **23**, 3–34 (1959)
30. Lucchini, A.: $(2, 3, k)$-Generated groups of large rank. Arch. Math. **73**, 241–248 (1999)
31. Lyndon, R.C., Schupp, P.E.: Combinatorial Group Theory. Springer, Berlin (1977)
32. Macbeath, A.M.: Generators of the linear fractional groups. In: Number Theory (Proceedings of Symposia in Pure Mathematics, vol. XII Houston, TX, 1967), pp. 14–32. American Mathematical Society, Providence, RI (1969)
33. O'Brien, E.A., Vaughan-Lee, M.: The 2-generator restricted Burnside group of exponent 7. Int. J. Algebra Comput. **12**, 575–592 (2002)

34. Serre, J.-P.: Variétées projectives conjuguées non homéomorphes. C. R. Acad. Sci. Paris **258**, 4194–4196 (1964)
35. Singerman, D.: Finitely maximal Fuchsian groups. J. Lond. Math. Soc. (2) **6**, 29–38 (1972)
36. Streit, M.: Field of definition and Galois orbits for the Macbeath-Hurwitz curves. Arch. Math. **74**, 342–349 (2000)
37. Vaughan-Lee, M.: The Restricted Burnside Problem. Clarendon Press, Oxford (1993)
38. Wielandt, H.: Finite Permutation Groups. Academic, New York (1964)
39. Zelmanov, E.I.: Solution of the restricted Burnside problem for groups of odd exponent (Russian). Izv. Akad. Nauk SSSR Ser. Mat. **54**, 42–59, 221 (1990). English translation in Math. USSR-Izv. **36**, 41–60 (1991)
40. Zelmanov, E.I.: Solution of the restricted Burnside problem for 2-groups (Russian). Mat. Sb. **182**, 568–592 (1991). English translation in Math. USSR-Sb. **72**, 543–565 (1992)

Hints for Selected Exercises

In the following we give additional hints for some of the more difficult exercises.

1.1 The image of such a function f would be a compact subset of $\mathbb{C}$, hence $|f|$ would have a maximal value. Is that possible?

1.3 Given a meromorphic function f, find a rational function r with the same zeros and poles in $\mathbb{C}$, of the same orders. Now consider the behaviour of the quotient f/r at ∞ and prove that f/r is a constant.

1.4 Apply the Riemann-Hurwitz formula (Proposition 1.2) to the projection $(x, y) \mapsto x$.

1.5 Use an affine transformation $x \mapsto ax + b$ to put the equation into Weierstrass normal form.

1.6 For all $w \in \Lambda$, we have $\wp'(z + w) - \wp'(z) \equiv 0$ by rearrangement of the series. What does this imply about $\wp(z + w) - \wp(z)$? To determine the constant, put $z = -w/2$ and prove that $\wp$ is an even function.

1.7 The biholomorphic function a cannot be transcendental by the consequence of Picard's theorem that transcendental functions take almost all values infinitely often. Which polynomials a are possible?

1.8 Suppose that $a, b, c, d \in \mathbb{Z}$ satisfy $ad - bc \equiv 1 \bmod n$. Recall the Chinese Remainder Theorem, and prove that you can replace $a, b, c, d \in \mathbb{Z}$ with other representatives of their residue classes mod n such that the greatest common divisor $(c, d) = 1$ and moreover $ad - bc = 1$.

Gareth A. Jones, J. Wolfart, *Dessins d'Enfants on Riemann Surfaces*, Springer Monographs in Mathematics, DOI 10.1007/978-3-319-24711-3

1.9 Show that any coprime pair of integers a, c can be completed to some element $\tau \mapsto (a\tau + b)/(c\tau + d)$ of Γ. To show that $\Gamma(2)$ has three orbits, represented by ∞, 0 and 1, use parity considerations. How does Γ act on the set consisting of these three orbits?

1.10 Prove that $g_2(\omega) = 0$ using $\omega\Lambda = \Lambda$ and the definition of G_2, and a similar symmetry property for g_3 when the lattice has $\tau = i$.

1.12 Use Proposition 1.4 for the pole orders n and $n + 1$ at P.

1.13 Choose one further point P_{k+1} and apply Proposition 1.5 to construct two differentials with suitable residues at $P_1, \dots, P_k$.

1.21 For the lazy reader: look at Fig. 2.3, and identify opposite sides of the outer hexagon to form a torus.

1.23 See Sect. 4.2.2.

1.24 For Lemma 1.2 recall that non-constant holomorphic functions are locally biholomorphic at all points except ramification points. For Lemma 1.3 observe that $|V| < \deg p$, and that the set of zeros of $p' \in \mathbb{Q}[x]$ (= ramification points of p) is invariant under algebraic conjugations, so the same is true for V. The next polynomial $q \in \mathbb{Q}[x]$ we are looking for—sending V into $\mathbb{Q}$—can therefore be chosen for example as $\prod_{z \in V}(x - z)$. Prove that in this case $q \in \mathbb{Q}[x]$ and $\deg q < \deg p$.

2.2 For (a), look at Fig. 6.1. To determine G in (c), consider instead xy, yx and y as generators.

2.4 If e is an ordered k-tuple then $G_e \cong S_{n-k}$, fixing the k points and permuting their complement. Now show that $N_G(G_e) \cong S_k \times S_{n-k}$ and $C \cong S_k$.

2.7 When do the type and the genus determine the number of edges?

2.9 See Sect. 3.3.1.

2.10 Start with Examples 1.11 and 1.13.

3.7 Show first that via a hyperbolic motion we can assume that $p = -x + iy$ and $p' = +x + iy$ for some $x > 0$. Then prove that the imaginary axis is the line in question, for example by a symmetry argument.

3.8 For the equivalence (1) $\Leftrightarrow$ (2) recall the classification of the elements $\gamma \neq 1$ in $\mathrm{PSL}_2(\mathbb{R})$ into elliptic, hyperbolic and parabolic elements. The implication (3) $\Rightarrow$ (1) is obvious; to prove (1) $\Rightarrow$ (3), consider the fixed point p of γ;

the tessellation of $\mathbb{H}$ by Δ-images of F shows that p must be a vertex of some $\delta(F), \delta \in \Delta$. What does this mean for γ ?

3.9 Recall the presentation of Δ given in Example 3.4, and try to construct an epimorphism onto some nontrivial abelian group.

3.11 Recall that $\mathrm{PSL}_2(\mathbb{Z}/n\mathbb{Z})$ is an epimorphic image of the modular group $\mathrm{PSL}_2(\mathbb{Z}) \cong \Delta(2,3,\infty)$, and observe that this epimorphism maps γ_∞ to an element of order n.

3.13 Imitate Example 3.8, mapping $\Delta(2,3,4n)$ onto $\Delta(2,3,4)$ in (H), for example, so the monodromy group is $\Delta(2,3,4)$ acting on the cosets of a subgroup $\Delta(1,4,4)$; what are these groups?

3.16 The composition $\Delta(7,7,7) <_3 \Delta(3,3,7) <_8 \Delta(2,3,7)$ is inclusion (A), namely $\Delta(7,7,7) <_{24} \Delta(2,3,7)$. The monodromy group is $\mathrm{PSL}_2(\mathbb{F}_7)$ acting on the cosets of a subgroup of index 24, that is, a Sylow 7-subgroup.

3.19 Consider an epimorphism $\Delta(2,3,2n) \rightarrow S_3$ and recall that $\Delta(2,3,2n)$ is maximal; prove that the normaliser of Δ also normalises its commutator subgroup Δ'.

3.22 First find the reflections in the sides of the triangle T, then consider their products. See the book by W. Magnus (Noneuclidean Tesselations and Their Groups. Academic Press, New York (1974)) for this and many other beautiful hyperbolic tessellations.

4.2 First check that K_∞ is closed under field operations, and that each element is an algebraic number. Now look at Example 4.4 and consider $\varprojlim G_K$, where K ranges over the cyclotomic fields $\mathbb{Q}(\zeta_n)$ for $n \in \mathbb{N}$. Example 4.4 gives a lot of information about the projective limit.

4.3 For $\overline{\mathbb{Q}}$, count polynomials. For $\mathbb{G}$, if p is any prime then $\mathbb{G}$ has $\varprojlim G_K$ as an epimorphic image, where K ranges over the cyclotomic fields $\mathbb{Q}(\zeta_{p^e})$ for $e \in \mathbb{N}$; this group is uncountable.

4.4 Use Tychonoff's Theorem on products of compact spaces, and the fact that each equation $\rho_{K,L}(g_K) = g_L$ defines a closed subset of Π.

4.5 Multiplication is continuous, so the cosets of an open or closed subgroup are also open or closed.

4.6 Use Exercise 4.2 and the Kronecker-Weber Theorem, that all abelian extensions of $\mathbb{Q}$ are contained in cyclotomic fields.

4.7 Show that $|f^{-1}(x)|$ is an integer-valued continuous function of x.

4.8 Consider the universal covering space of Y.

4.10 Take $\tau = \mathrm{id}$ with an arbitrary σ in Weil's condition.

4.12 Show that neither of the two possible isomorphisms $f : X \to \overline{X}$ satisfies the condition found in Exercise 4.11.

4.14 Use the facts that complex conjugation inverts x and y, and that $x^i y^j$ is conjugate to $(x^{-i}y^{-j})^{-1}$. A similar argument applies to $[x, y]$.

5.1 Refer to Exercise 3.2, and use the fact that if two elements commute then each permutes the set of fixed points of the other.

5.2 Consider the three types of euclidean triangle groups; have a look at Figs. 7.1 and 7.2 or Sect. 6.2. Which tori can have automorphisms with fixed points of order 3 or 4?

5.4 Consider $\Delta(2, 3, 8)$. For examples, see Sects. 5.2.1 and 5.2.2.

5.5 Consider the canonical epimorphism $h : \Delta \to \Delta/K$. What can the h-images of the torsion elements of Δ look like? Remember Exercise 3.8 !

5.6 See Example 6.2.

5.7 Take $p = 13$ in Example 5.11.

5.8 Suppose that (l, m, n) is the type of the dessin, let r be the least common multiple of l, m, n, and let $\Delta := \Delta(r, r, r)$. Prove that the automorphism group of the dessin is isomorphic to a quotient Δ/K for a normal subgroup $K \rhd \Delta'$, and remember Example 3.11 or Example 5.1.

5.18 Show that α^2 generates a normal subgroup of G with quotient $\Delta(2, 2, 4)$.

5.19 Use a method similar to that for Exercise 5.18.

6.1 Try identifying opposite sides of an octagon or decagon.

7.2 The map for A_4 is a truncated tetrahedron, with small triangles replacing the four vertices of the tetrahedron. For S_4 it is a truncated cube.

8.1 Prove that $x \mapsto x^j$ extends to an isomorphism of algebraic (bipartite) maps.

8.2 Translation of the ideas explained in the previous example is easy. However, to give the generator of the quadratic field of definition K in the case $q = 16, \zeta := \zeta_{15}$, as $\zeta + \zeta^2 + \zeta^4 + \zeta^8$ is not very satisfactory since K should be generated by a square root of an integer. Remember that the prime 2 splits in K, that $K < \mathbb{Q}(\zeta)$, and prove that 2 remains prime in the subfields $\mathbb{Q}(\sqrt{-3})$ and $\mathbb{Q}(\sqrt{5})$ of $\mathbb{Q}(\zeta)$. Is there another quadratic subfield of $\mathbb{Q}(\zeta)$ serving as splitting subfield for 2? Remember that 2 splits in $\mathbb{Q}(\sqrt{n})$ if and only if $n \equiv 1 \bmod 8$.

9.2 Prove that these conditions are necessary by taking X and Y to be stabilisers of a white and a black vertex; prove that they are sufficient by defining the edges and vertices of $K_{n,n}$ to be the elements of G and the cosets of X and Y in G, respectively.

9.4 Since x_S and x_T commute and have coprime orders s and t, x has order $st = n$, as has y. Since $G = ST$, each element $g \in G$ has the form $x_S^i y_S^j x_T^k y_T^l = x_S^i x_T^{k'} y_S^j y_T^l$ for some i, j, k, l and k', so $g = x^u y^v$ for some u, v and thus $G = XY$ where $X = \langle x \rangle$ and $Y = \langle y \rangle$. Since $|G| = s^2t^2 = |X||Y|$ we must have $X \cap Y = 1$.

9.8 The xy-model is singular only at the points with $x = 0$ and $x = 1$. Deduce the original equation from the new equations in x, y, z and w, and try to use the new variables as chart maps at points with $x = 0$ or $x = 1$.

9.9 Consider the edges of $\mathscr{D}$ in terms of the coordinates z and w of the previous exercise!

9.10 Use Exercises 9.9 and 2.1.

9.11 Show first that G cannot be commutative. Its structure follows from an application of Sylow's theorems. The numbers of white and black vertices and the topological shape of the dessin result from simple counting, and the fields of definition follow from Theorem 9.5 by passing to a dual dessin.

10.7 Use Lemma 10.1 and Theorem 3.10 to prove the existence of $\Gamma_n < \Delta$. To see that $\Gamma_n \lhd \Gamma$, one may either lift the endomorphism α_n to $\mathbb{H} \to \mathbb{H}$, or use Theorem 3.11 with a suitable epimorphism $\Gamma \to C_n \times C_n$.

10.9 Determine the ramifications of $(x, y) \mapsto x$, then follow Belyĭ's algorithm.

11.1 In any group, $(ab)^n = a^n b^{a^{n-1}} b^{a^{n-2}} \dots . b^a b$.

11.3 For the second part, if $p_1, p_2, \dots, p_k$ are the first k primes, then $\prod_{i=1}^k (1 - \frac{1}{p_i}) \to 0$ as $k \to \infty$.

11.6 For S_3 and S_4 imitate the proof of Theorem 11.3, respectively using 3- and 4-cycles. For S_5 you could take $x_1 = (1, 2, 5)$, $y_1 = (1, 4, 5)(2, 3)$, $x_2 = (2, 3, 4, 5)$ and $y_2 = (1, 3, 4, 5)$; use Jordan's Theorem with cycles y_1^3 and $y_2^{-1} z_2^2$ to show that

$\langle x_i, y_i \rangle = S_5$ for $i = 1, 2$. Similarly S_6 and S_7 have Beauville structures of types $(2, 5, 6; 4, 4, 4)$ and $(2, 12, 12; 2, 6, 7)$. (Here, and in the next exercise, we follow the paper by Y. Fuertes and G. González-Diez, On Beauville structures on the groups S_n and A_n, Math. Z. **264**, 959–968 (2010).)

11.7 You could find Beauville structures of type $(3, 5, 5; 4, 4, 4)$ for A_6 and of type $(5, 5, 5; 3, 7, 7)$ for A_7; see the hint for the preceding exercise.

11.10 Pull the epimorphism $\Delta(2, 3, t) \to G$ back to $\Delta(2, 3, 2t)$, and then restrict it to the normal subgroup $\Delta(t, t, t)$—see Theorem 3.12(c).

11.11 Consider which matrices $A \in \mathfrak{F}_n$ are fixed by various elements $w \in W$.

Index

Abc conjecture, 213
Absolute Galois group, 91
Algebraic closure, 89
Arithmetically defined, 119

Beauville structure, 226
Beauville surface, 225
Belyĭ function, 21, 26
Belyĭ pair, 26
Biholomorphic map, 241
Bring's curve, 124
Burnside's Lemma, 232

Cauchy–Frobenius formula, 232
Character, 126
Character table, 126
Chebyshev polynomial, 27
Chiral pair, 164
Clean Belyĭ function, 32
Cocompact, 68
Cocycle condition, 100
Commensurable, 119
Complex multiplication, 220
Conformal structure, 80
Congruence subgroup, 119
Cori representation, 49
Covering
  of algebraic maps, 80
  canonical regular, 51
  cyclic, 96
  Galois, 95
  group, 95
  map, 60
  normal, 80, 95
  ramified, 94
  regular, 80
  2–sheeted, 11
  universal, 60
Critical value, 11
Curve
  Accola–Maclachlan, 140
  affine, 7
  algebraic, 13
  Belyĭ, 26
  Bring's, 139
  elliptic, 14, 219
  hyperelliptic, 7
  Lefschetz, 121
  projective, 9
  Shimura, 119
  triangle, 116, 226
  Wiman, 142
Cusp, 65

Dart, 52
Degree, 8
Dessin, 30, 42, 45
  regular, 47
  uniform, 72
Dirichlet problem, 25
Discontinuous, 61
Discriminant, 14
Dynkin diagram, 56

Elliptic function, 16
Euler characteristic, 12

Gareth A. Jones, J. Wolfart, *Dessins d'Enfants on Riemann Surfaces*, Springer Monographs in Mathematics, DOI 10.1007/978-3-319-24711-3

Fermat curve, 6, 42, 136, 197, 227
Field of definition, 26, 96, 142
Frattini subgroup, 229
Free product, 19, 79
Fricke–Macbeath surface, 120
Frobenius kernel, 169
Fuchsian group, 61
Fundamental region, 15, 18, 64, 68

Galois correspondence, 90
Galois extension, 90
Genus, 12
Graph
- bipartite, 30, 42
- Cayley, 166
- complete, 164
- complete bipartite, 42, 197
- generalised Paley, 194
- Paley, 194
- ribbon, 52

Great dodecahedron, 124
Group
- acting faithfully, 104
- acting primitively, 195
- acting regularly, 44
- affine, 145, 168
- algebraic fundamental, 247
- almost simple, 234
- alternating, 55, 237
- arithmetic, 64
- Beauville, 226
- binary icosahedral, 234
- cartographic, 45
- covering, 60
- dihedral, 78, 230
- discrete, 63
- finitely generated, 78
- finitely presented, 78
- Fischer–Griess monster, 56, 120
- free, 79
- free abelian, 79
- Frobenius, 169
- Fuchsian, 61
- general linear, 145
- generator, 78
- isobicyclic, 200
- Mathieu, 53, 56
- modular, 18
- monodromy, 44, 74
- normaliser, 116
- presentation, 19, 68
- profinite, 92
- projective linear, 145
- quasisimple, 234
- Ree, 56, 120
- relation, 78
- representation, 125
- simple, 54, 233
- simple of Lie type, 56
- special linear, 145
- sporadic, 56
- surface, 62
- Suzuki, 56
- twisted of Lie type, 56
- unitary, 56

Holomorphic mappings, 7
Hurwitz bound, 117
Hurwitz group, 118, 158
Hurwitz surface, 118
Hyperbolic
- motions, 63
- plane, 18

Hyperelliptic involution, 99
Hypermap, 48

Icosahedron, 124
Inverse triple, 243
Irreducible representation, 126
Isomorphism of dessins, 51

Jacobian, 221
James representation, 50

Klein's quartic, 77, 118, 136
Krull topology, 93

Lattice, 15
Lattice basis, 15
Legendre normal form, 15
Lie algebra, 55

Macbeath–Hurwitz surface, 120
Map, 41
- algebraic, 52
- algebraic bipartite, 45, 80
- automorphism, 45
- Biggs, 170
- bipartite, 30, 42
- Cayley, 165

modular, 159
Paley, 195
regular, 153
type, 44
universal bipartite, 82
Möbius function, 129
Modular function, 20, 57
Modular group, 18, 65, 85
Moduli field, 97, 142
Monodromy group, 44, 74
Monsieur Mathieu, 53
Multiplicity, 8, 29

Normal extension, 90

Orientation, 11

Period quotient, 219
Permutation group, 46
Petrie length, 120
Petrie polygon, 119

Quasiplatonic, 116
Quotient space, 15

Ramification order, 8
Ramified, 8
Regular dessin, 47
Relator, 79
Renormalisation, 28, 49
Representation, 125
Riemann mapping theorem, 20, 60
Riemann–Roch theorem, 24, 25, 95
Riemann sphere, 6

Schwarz's reflection principle, 66
Schwarz triangle function, 20
Self–cover, 219
Semidirect product, 138, 201
Shabat polynomials, 28
Shimura curve, 119
Smooth quotient, 241
Splitting field, 184
Surface, 225
Beauville, 225
Belyĭ, 26
Fricke–Macbeath, 120
Hurwitz, 118
Macbeath–Hurwitz, 120
Surface group, 62

Teichmüller space, 71
Torus, 16, 17
Triangle group
euclidean, 67
spherical, 67
Triangulation, 12, 29
Type, 44, 48

Uniformisation, 60
Unmixed type, 226
Unramified, 8
Upper half plane, 18

Walsh representation, 49
Weierstrass function, 16
Weierstrass normal form, 14
Weierstrass points, 99
Wilson's operations, 179
Word, 78, 103
reduced, 79
Wreath product, 138, 230

Lightning Source UK Ltd.
Milton Keynes UK
UKOW06n1330090516

273869UK00001B/25/P